D0893706

Biometrical Genetics
The study of continuous variation

Biometrical Genetics

The study of continuous variation

Sir Kenneth Mather

C.B.E., D.Sc, F.R.S.

John L. Jinks

D.Sc, F.R.S.

PROFESSORS OF GENETICS IN THE UNIVERSITY OF BIRMINGHAM

Third edition

London   New York

CHAPMAN AND HALL

First published 1949 by
Methuen and Co Ltd
Second edition published 1971 by
Chapman and Hall Ltd
11 New Fetter Lane, London EC4P 4EE
Third edition 1982

© *1982 K. Mather and J. L. Jinks*

Printed in Great Britain at the
University Press, Cambridge

Typeset by
Macmillan India Ltd Bangalore

ISBN 0 412 22890 4

British Library Cataloguing in Publication Data

Mather, *Sir* Kenneth

Biometrical genetics.—3rd ed
1. Genetics—Statistical methods
2. Biometry
575.1′028   QH441

ISBN 0–412–22890–4

Contents

To R. A. FISHER
Statistician and Geneticist

Preface to the first edition

The properties of continuous variation are basic to the theory of evolution and to the practice of plant and animal improvement. Yet the genetical study of continuous variation has lagged far behind that of discontinuous variation.

The reason for this situation is basically methodological. Mendel gave us not merely his principles of heredity, but also a method of experiment by which these principles could be tested over a wider range of living species, and extended into the elaborate genetical theory of today. The power of this tool is well attested by the speed with which genetics has grown. In less than fifty years, it has not only developed a theoretical structure which is unique in the biological sciences, but has established a union with nuclear cytology so close that the two have become virtually a single science offering us a new approach to problems so diverse as those of evolution, development, disease, cellular chemistry and human welfare. Much of this progress would have been impossible and all would have been slower without the Mendelian method of recognizing and using unit differences in the genetic materials.

These great achievements should not, however, blind us to the limitations inherent in the method itself. It depends for its success on the ability to assign the individuals to classes whose clear phenotypic distinctions reveal the underlying genetic differences. A certain amount of overlap of the phenotypic classes can be accòmmodated by the use of genetical devices; but where the variation in phenotype is fully continuous in its frequency distribution, so that no such classes can be defined, the method cannot be used. A different approach is required, one based on the use of measurement rather than frequency.

The first steps were taken nearly 40 years ago, when the theory of cumulative factors or multifactorial inheritance, as it was variously called, was formulated. The full implications of this theory have, however, only gradually become realized. In the same way the special types of experiment and statistical analysis necessary for the study of continuous variation have only gradually become available. Nevertheless, though slow, progress has been real and we are now in a position to see not merely how continuous variation can be explained genetically, but also how experiments can be conducted enabling us to understand and to measure the special genetical quantities in terms of which continuous variation can be analysed and its behaviour in some measure predicted.

The present book does not aim at covering the whole literature of the subject. I have concentrated attention rather on trying to show the kind of evidence upon which the genetical theory of continuous variation is based, to bring out the special problems which it raises, to see how the familiar genetical concepts must be adapted to their new use, and to outline an analytical approach which can help us to understand our experimental results, particularly those which can be obtained from plant material. In doing so I have assumed some knowledge of genetics and statistics. To have done otherwise would have made the text unnecessarily long, for this information can be gained from a variety of other sources.

The data with which I have had to work have been limited by the paucity of experiments adequate in both scope and description and I have therefore been unable to try out the methods, which are described, in as wide a variety of circumstances as could have been wished. These methods are in no sense exclusive or final; indeed their limitations require no stressing. But improvements can be brought about only as more and better experiments are undertaken; and such experiments cannot be planned until we have explored the scope and limitations of those we already have. Improvement of experiment, refinement of analysis and development of theory must be simultaneous and progressive.

Among the experiments upon which I have been able to draw none has been more instructive than that on ear conformation in barley, hitherto unpublished. This experiment was made in collaboration with Dr Ursula Philip, now of the Department of Zoology, King's College, Newcastle-on-Tyne, and I wish to express my indebtedness to her for allowing its results to be published in this way.

April 1947   K. M.

Preface to the second edition

Much has happened in the genetical study of continuous variation since the Preface to the first edition of this book was written 23 years ago. Theory has been developed and extended to include, notably, interaction between non-allelic genes and also interaction between genotype and environment. New methods of genetical analysis have been introduced, among which special mention must be made of diallel crosses. Statistical techniques have been refined and extensive experimental programmes have been carried out. There is, of course, still much to do, but we are encouraged by representations from many people to feel that a new edition of *Biometrical Genetics* revised and extended to cover these new developments would meet a growing need.

We have based our treatment very largely on the approach set out in the first edition as extended particularly by the Department of Genetics in the University of Birmingham with which we have both been associated for so many years. We have not aimed at giving the more general mathematical treatments in all their complexity, even where these are available, or to cover such matters as selection procedures: accounts of these can be found elsewhere (as, for example, in Kempthorne's *An Introduction to Genetic Statistics* and Falconer's *Introduction to Quantitative Genetics*) and they are in any case extraneous to our theme which is, to quote from the Preface to the first edition, 'to show the kind of evidence upon which the genetical theory of continuous variation is based, to bring out the special problems which it raises, to see how the familiar genetical concepts must be adapted to this new use, and to outline an analytical approach which can help us to understand our experimental results'. We shall be content if our readers feel we have gone some way to achieving this aim.

We are indebted to many colleagues, past and present, both for the encouragement they have given us in planning this new edition and their help in preparing it. We would mention especially Dr B. W. Barnes and Dr R. Killick for allowing us to use unpublished data, Dr J. M. Perkins for carrying out certain calculations for us, Dr Killick, Dr Perkins and Mr D. Hay for checking a number of the formulae, and Miss S. M. Evans for her help in the preparation of the typescript. We hope that they will feel the trouble they took on our behalf not to have been fruitless.

April 1970
K. M.
J. L. J.

Preface to the third edition

The 11 years that have elapsed since the preface to the second edition was written have seen a continuing development in the genetical theory and analysis of continuous (or, as it is frequently termed, quantitative) variation. This is especially so in relation to genotype × environment interaction, the prediction of the range of true-breeding genotypes that can be extracted from a cross between two parental lines, the use of the triple test cross as a means of analysing the genetical variation, and in both the genetical and statistical planning of experiments, or, where experiments are impossible as in man, of the necessary observations.

These developments were foreshadowed in the previous edition, but we have now sought to bring out more clearly the ways in which the methods of Biometrical Genetics can be used to gain a greater understanding of the genetical properties of natural and quasi-natural populations, including those of man himself, and of the potentialities open to the plant breeder in the manipulation of his material. There is still, of course, much to do; but we hope that we are at least beginning to dispel the notion, still too widely held, that Biometrical Genetics is no more than an esoteric form of genetical endeavour having little but theoretical interest.

Not infrequently in the past the question has been asked, by reviewers and others, as to why we make what to some seems excessive use of our own experiments to provide illustrative material, rather than drawing on a wider range of data from the literature. There are several reasons for this. In some cases no similar experiment has been carried out elsewhere, and in any case if we are undertaking a novel type of analysis we may well need access to the original observations – which are available only from our own experiments. Perhaps, however, the chief reason is that we have been carrying out this kind of investigation with *Nicotiana rustica* and certain lines of *Drosophila melanogaster* for nearly 40 years. This has given us not only a close and comprehensive knowledge of our living material, but also series of linked observations each of which can, over time, aid in the interpretation of the rest, as is well illustrated by the frequent use we make of the cross between our lines 1 and 5 of *N. rustica*.

We have taken the opportunity of this edition, in general, to change the presentation of metrical data from inches to centimetres. We have, however, retained the use of inches in certain long-standing experiments.

Again we tender our thanks to those of our colleagues who have assisted in the preparations of this edition, and particularly to Dr P. D. S. Caligari, for allowing us access to certain of his original observations before his own account of them was published.

May 1981   K. M.
                                                                          J. L. J.

1   *The genetical foundation*

1. BIOMETRY AND GENETICS

The growth of genetical science as we know it today began with the rediscovery of Mendel's work in 1900. Nevertheless, at the time of that event there were already genetical investigations in active progress; investigations which, although contributing relatively little to the development of genetical theory, still have an importance of their own. These were begun by Francis Galton, who published a general account of his methods and findings in *Natural Inheritance* (1889), and were continued by Karl Pearson and his pupils. From them the application of statistical mathematics to biological problems received a great impetus, and if only for this reason they mark a significant step in the growth of quantitative biology.

The relative failure of this work in its avowed purpose, that of elucidating the relations of parent to offspring in heredity, stems from a variety of causes. Mendel himself regarded the failure of his predecessors as due to their experiments not making it possible 'to determine the number of different forms under which the offspring of hybrids appear, or to arrange these forms with certainty according to their separate generations, or definitely to ascertain their statistical relations'. While Galton's work can hardly be regarded as failing in the third respect, the nature of the material he chose rendered it impossible for him to succeed in the other two. His extensive use of human data, with its small families and genetically uncertain ancestries, introduced difficulties enough; but it was the choice of metrical or quantitative characters, like stature in man, that foredoomed the work from the point of view of the laws of inheritance. These characters show continuous gradations of expression between wide extremes, the central expression being most common in any family or population, and the frequency of occurrence falling away as we proceed towards either extreme (see Fig. 2). The distributions of frequencies of the various grades of expression sometimes, as with stature in man, approximate closely to the Normal curve; but while retaining the same general shape they depart in other cases from this precise form, for example by being asymmetrical. The simple Mendelian ratios, with their clear implication of the particulate or discontinuous nature of hereditary constitution and transmission, depend on the use of characters by which individuals could be classed unambiguously into a few (usually two) distinct groups; they cannot come from continuous variation. Indeed, Mendel himself deliberately neglected such variation in his material, presumably with the clear recognition that it could only have a distracting influence in his analyses.

Yet this continuous variation could not be completely overlooked. Darwin himself had emphasized the importance of small cumulative steps in evolutionary change, and observation on any living species, especially the most familiar of all, man, showed how much of the variation between individuals was of this kind. The genetical problem of continuous variation remained, therefore, a challenge to geneticists; the more so as biometrically, Galton and Pearson had clearly shown such variation to be at least in part hereditary, even though they had failed to discover the mode of transmission. Neither the Galtonian nor the Mendelian method was of itself capable of supplying the solution. The understanding of continuous variation awaited a fusion of the two methods of approach, the genetical and the biometrical, for each supplied what the other lacked. The one gave us the principles on which the analysis must be based; the other showed the way in which to handle continuous variation, the way of representing it in a form which made fruitful analysis possible.

Fusion was, however, delayed by a rivalry which arose between Biometricians and Mendelians as soon as Mendel's work was rediscovered. This was aggravated by divergent opinions on the importance of continuous and discontinuous variation in evolutionary change, and exacerbated by the polemics of the protagonists. In time, attempts to reconcile the two views became welcome to neither party. The original discordance seems to have arisen because neither side understood the full implications of Mendel's fundamental separation of determinant and effect, of genotype and phenotype. The Biometricians seem to have regarded continuous somatic variation as implying continuous genetic variation, and the Mendelians seem to have considered discontinuous genetic variation as incompatible with anything but obviously discontinuous somatic variation. Indeed, de Vries took continuity of variation in the phenotype as a criterion of its non-heritability.

Two important steps had to be taken, therefore, before the biometrical and genetical methods could be brought together. In 1909 Johannsen published his *Elemente der exakten Erblichkeitslehre*. In it he described the experiments with beans which led him to formulate his pure line theory. In particular he showed that heritable and non-heritable agencies were jointly responsible for the variation in seed weight with which he was concerned; that their effects were of the same order of magnitude; and that there was no means, other than the breeding test, of distinguishing between their contributions to the variation. The relations between genotype and phenotype were thus becoming clearer. The effects of discontinuity of the genotype could be smoothed out and continuous variation realized in the phenotype by the action of the environment.

In the same year a second Scandinavian geneticist, Nilsson-Ehle, took the other step. He found that in wheat and oats there existed hereditary factors whose actions were very similar, if not exactly alike. There were, for example, three such factors for red versus white grain in wheat. Any one of them, when segregating alone, gave an F_2 ratio of 3 red:1 white. Two of them segregating together gave 15:1 for red:white, and all three together gave a 63:1 ratio. That the red-grained plants in these F_2's were of various genetical constitutions could be shown by growing F_3 families. Some of these gave 3 red:1 white, others 15:1, others 63:1 and still others all red. Yet there were no detectable differences in colour between plants owing their redness to the different factors. There were certainly some differences in redness, but these appeared to be associated more with the number of factors than with the particular factors present. The first degree of redness would be given equally by the three genotypes Aabbcc, aaBbcc and aabbCc; the second by the six genotypes AAbbcc, aaBBcc, aabbCC, AaBbcc, AabbCc and aaBbCc; and so on. It thus appeared that different factors could have similar actions, and actions which were, at least in some measure, cumulative.

These factors in wheat and oats had effects sufficiently large for Mendelian analysis to be possible; but it was realized by Nilsson-Ehle, and also independently by East, that similar factors of smaller individual action could account for continuous quantitative variation if enough of them were segregating . Each factor would be inherited in the Mendelian way, and its changes would be discontinuous or qualitative. Yet with a number of such factors, having similar and cumulative action, many different dosages would be possible, of which the intermediate ones would be the most common (Fig. 1). With phenotypic expression proportional to factor dosage, variation would be quantitative, would follow Galton's frequency curves and would be nearly continuous. Continuity would be completed by the blurring effect of non-heritable agencies, which would of course make the phenotypic ranges of the various genotypes overlap.

During the next 10 years this multiple factor hypothesis, as it was called, was applied to data from a variety of organisms, notably by East and his collaborators, and by Fisher. The former showed that the inheritance of a number of continuously variable characters in tobacco and maize could be fully accounted for on this view (e.g. East, 1915; Emerson and East, 1913). Fisher carried the integration of biometry and genetics still further. He demonstrated that the results of the Biometricians themselves, particularly the correlations which they had found between human relatives, must follow on the new view (Fisher, 1918). From the Biometricians' own data he was able to produce evidence of dominance of the multiple factors, and he attempted the first partition of continuous variation into the components which the multiple factor hypothesis led him to expect.

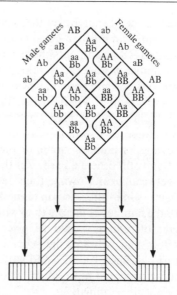

Figure 1 The polygenic or multiple factor theory. The distribution of phenotypes obtained in an F$_2$ with two genes of equal and additive effect but without dominance, neglecting non-heritable variation. The phenotypic expression is proportional to the number of capital letters in the genotype. There would be seven phenotypic classes with three such genes, nine classes with four genes and $2k + 1$ classes with k genes.

2. POLYGENIC SEGREGATION AND LINKAGE

The essential features of the multiple factor hypothesis are two: that the governing factors or genes are inherited in the Mendelian fashion; and that they have effects on the character under observation similar to one another, supplementing each other and sufficiently small in relation to the non-heritable variation, or at least in relation to the total variation, for discontinuities to become indiscernible in the phenotypic distribution. In this way smooth, continuous variation of the phenotype could arise from discontinuous, quantal variation of the genotype.

There is an obvious danger in postulating these multifactorial or polygenic systems. The constituent genes are so alike in their effects and so readily mimicked by non-heritable agencies, that they cannot easily be identified individually within the systems. Since such genes obviously cannot be followed by the simple Mendelian technique, how may we be sure that they are in truth borne on the chromosomes and so subject to Mendelian inheritance?

On the negative side there is the evidence of reciprocal crosses. Though these sometimes differ a little in respect of continuously variable characters, presumed to be under polygenic control, they do so no more often than is the case with discontinuously variable characters. The two parents therefore generally contribute equally to the genotype of the offspring in the way expected of nuclear heredity, and not unequally as might be expected if inheritance was of some other kind.

More positive evidence is, however, available. The properties characteristic of nuclear-borne genes are two, namely segregation and linkage. Although neither segregation nor linkage of the genes under discussion can be observed by the usual methods, the necessary tests can be made in other ways.

If we take two different inbred, and therefore very nearly true-breeding strains, both they and their F_1 will show variation virtually only in so far as non-heritable agencies are at work. But genetical segregation of the nuclear genes which differentiate the parents will occur in F_2, and the heritable variation to which it leads will be added to the non-heritable. The F_2 should therefore be more variable than the parents and the F_1: its frequency distribution will be broader and flatter. Furthermore, as Mendel showed, the genes at each locus are homozygous in half the F_2 individuals. Segregation will still occur in F_3 families, but it will be for only half the gene pairs on average. The average variation of F_3 families will therefore lie between that of F_2 on the one hand, and parents and F_1 on the other; but the F_3 families will differ among themselves, some having variances approaching one extreme, some the other and most being intermediate. At the same time the homozygous genes by which the F_2 individuals differed will give rise to differences between the mean phenotypes of the F_3 families, and these means will be correlated with the phenotypes of the F_2 parents. Even where the parental strains are not nearly true breeding the F_2 will generally (though not inevitably) show greater variation than either F_1 or parents.

Thus the necessary test of segregation is to be found in the relative variation of the different generations following crossing. It is sufficient to say that whenever a critical test has been made, and a very large number have now been made, the results have accorded with the expectation based on nuclear inheritance. A characteristic case is shown in Fig. 2.

Tests of linkage, the second property of nuclear genes, may be of two kinds. We may seek for linkage of the quantitative genes (or polygenes if we name them after the polygenic variation they determine) with genes of major effect, capable of being followed by Mendelian methods. Or we may seek for linkage between polygenes themselves.

The first case of apparent linkage between polygenes and a major gene was reported by Sax (1923). He crossed a strain of *Phaseolus vulgaris*, having large coloured seeds, with another whose seeds were small and white. Seed size showed itself to be a continuously variable character, but pigmentation proved to be due to a single gene difference, the F_2 giving a ratio of 3 coloured-: 1 white-seeded plant. By means of F_3 progenies the coloured F_2 plants were further classified into homozygotes and heterozygotes. On weighing the beans from the three classes of F_2 plant, PP, Pp and pp (P giving pigment and p no pigment), the average bean weights shown in Table 1 were obtained.

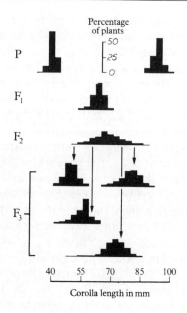

Figure 2 The inheritance of corolla length in *Nicotiana longiflora* (East, 1915). For ease of presentation, the results are shown as the percentage frequencies with which individuals fall into classes, each covering a range of 3 mm in corolla length and centred on 34, 37, 40, etc., mm. This grouping is quite artificial and the apparent discontinuities spurious: corolla length actually varies continuously. The means of F_1 and F_2 are intermediate between those of the parents. The means of the four F_3 families are correlated with the corolla length of the F_2 plants from which they came, as indicated by the arrows. Variation in parents and F_1 is all non-heritable, and hence is less than that in F_2 which shows additional variation arising from the segregation of the genes concerned in the cross. Variation in F_3 is on the average less than that of F_2 but greater than that of parents and F_1. Its magnitude varies among the different F_3's, according to the number of genes which are segregating.

TABLE 1
Bean weight (cg) in a *Phaseolus vulgaris* F_2 (Sax, 1923)

Number of plants	Colour constitution	Average seed weight
45	PP	$30{\cdot}7 \pm 0{\cdot}6$
80	Pp	$28{\cdot}3 \pm 0{\cdot}3$
41	pp	$26{\cdot}4 \pm 0{\cdot}5$

The standard errors show the differences in seed weight to be significant. As in the parents, P is associated with large seeds and p with small ones. Indeed, the average weight is nearly proportional to the number of P alleles present.

This is not, of course, final evidence of linkage of one or more polygenes governing seed weight with the major gene governing pigmentation. The effect could be due to a pleiotropic secondary effect of P itself. Such a criticism was, however, ruled out in other experiments. Rasmusson (1935) investigated the variation of flowering-time in crosses of the garden pea. Flowering-time was expressed as a deviation, in terms of days, from the average flowering-time of certain standard varieties grown each year for this purpose. A positive deviation indicated later flowering and a negative deviation earlier flowering than the standard.

To take one of his crosses as an example, the variety Gj with coloured flowers gave a mean flowering-time of 8·5, while Bism with white flowers gave − 9·3. Colour versus white in the flower depends on a single major gene, A–a. In the F_2 between these varieties the coloured plants had a mean flowering-time of $5·37 \pm 0·31$ and the whites $2·11 \pm 0·76$. The difference is significant and the coloureds are later than the whites, as would be expected from the parents. The difference is smaller than that between the parents, but this only shows that the association between pigmentation gene and the polygenic system governing flowering-time is incomplete.

So far the results are like those in the beans; but the cross had been made on an earlier occasion, and from it an early-flowering coloured strain (HRT-II) had been selected. Its mean flowering-time was nearly as early as that of Bism, namely − 6·1. In the F_2 of HRT-II × Bism, the coloureds gave a mean of $−7·97 \pm 0·36$ and the whites $− 8·30 \pm 0·81$. The flowering-time difference associated with colour had vanished. Hence the difference in the original cross must have been due to one or more flowering-time genes linked with the major gene governing pigmentation. HRT-II contained the recombinant chromosome carrying the colour gene from Gj and early-flowering gene or genes from Bism.

TABLE 2
Flowering-time in peas (Rasmusson, 1935)

Cross	Mean flowering-time of F_2 plants		Flowering-time difference (A − a)
	Coloured (A)	White (a)	
Gj × Bism	5·37	2·11	3·26
HRT-II × Bism	− 7·97	− 8·30	0·33
HRT-II × St	− 1·24	1·63	− 2·87

To round off the case, it may be observed that in a cross of HRT-II with a late-flowering white variety, St, the coloured plants of F_2 had a mean flowering-time of $− 1·24 \pm 0·20$ and the whites $1·63 \pm 0·23$. The relation is here reversed in the way to be expected from linkage.

Many cases of linkage between major genes and polygenes controlling continuous variation have been reported, although in the majority of them the possibility of pleiotropic action of the major gene has not been finally excluded by the demonstration of recombination. In *Drosophila melanogaster*, where all the chromosomes can be marked by major genes, it was soon shown that they all carry polygenes affecting a single continuously variable character, such as egg size (Warren, 1924). In many of these experiments it has also been shown, by comparing a number of unmarked chromosomes with the same marked tester chromosome, that the differences cannot be due to the major gene itself. There must be polygenes acting and indeed the contributions that the genes in each of the three larger chromosomes make to the differences between the lines under test can be assayed. An early example of such chromosome assays is given by Mather (1942), but the technique will be illustrated by the more extensive data of Mather and Harrison (1949).

The distribution among the chromosomes of the genes affecting the number of abdominal chaetae, a continuously variable character,* was followed in twelve lines. Two of these, Or, itself an inbred stock, and Sk, were parental to the other ten in the sense that the ten were derived by selection in various ways from the cross of Or and Sk. The ten selected lines were distinguished by numbers as shown in Table 3, which also gives for each of them the average number of chaetae borne by female flies. The chaeta numbers of males were not included in the analysis because, having only a single X chromosome, the comparison upon which the assays are based can be made only for their autosomes, whereas in females it can be made for sex-chromosomes as well.

Males from each of the lines were crossed with females from a common tester stock of constitution ClB/ + ; Cy L^4/Pm; H/Sb In$_3$ (R)Mo. Females of the type ClB/ + ; Pm/ + ; Sb In$_3$ (R) Mo/ + were taken from this F$_1$ and were crossed back to males of the parental line undergoing assay. The three large chromosomes from the tester stock were thus distinguished each by its own dominant marker gene. Furthermore, recombination between the chromosomes from the tester stock and their homologues from the line under assay would be so reduced by the inversions in the tester chromosomes, for each chromosome to be treated sufficiently well as a genetic unit. True, while the Pm inversion reduces recombination in the second chromosome to negligible proportions, a little recombination follows from double crossing-over inside the ClB inversion of the X chromosome and the 3(R) Mo inversion allows even more recombination in the third chromosome. Furthermore, the small fourth chromosome was not marked and followed in the experiment. Nevertheless, as we shall see, the loss of efficiency from the assays consequent on this recombination and the segregation of the fourth chromosome was not large.

---

* Or rather 'quasi-continuous', to use Grüneberg's (1952) term, since the number of hairs must be an integer, fractional hairs being impossible. Aside from the discontinuities imposed by this limitation, springing from the nature of the character, the distribution is continuous and of the Normal type.

TABLE 3
Chromosome assays in *Drosophila melanogaster* (Mather and Harrison, 1949)

Line	Chaeta number of line	Difference between + and B, Pm, Sb flies		Effects of chromosome			Total	Excess of line over Or (43·53)
		A	B	X	II	III		
Sk	37·63	−9·0	−5·6	−1·85	−1·83	−0·80	−4·48	−5·90
1	36·00	−6·3	−4·9	−0·54	−1·59	−0·98	−3·11	−7·53
2	37·04	−6·6	−6·0	−3·09	−1·67	−0·81	−5·57	−6·49
3	44·35	−2·1	−1·0	0·15	−0·21	0·23	0·17	0·82
4	50·65	0·6	3·5	2·15	−0·57	3·21	4·79	7·12
5	52·26	2·3	3·5	1·33	−0·14	3·01	4·20	8·73
6	52·60	2·0	3·7	2·17	0·01	2·80	4·98	9·07
7	59·53	5·0	6·0	1·96	2·47	3·10	7·53	16·00
8	63·35	8·8	7·0	1·83	2·69	4·16	8·68	19.82
9–1	59·19	6·6	6·2	1·15	2·65	3·62	7·42	15·66
9–2	70·25	13·4	11·6	2·17	2·22	8·09	12·48	26·72

A = difference from line and F_1
B = difference from backcross classes

The females from the backcross of F_1 to the experimental line fall into eight classes each recognizable by the marker genes it carries. If a chromosome originating from the tester line is denoted by T, and one from the line under assay by L, the constitutions of the eight classes are set out in Table 4. The chaeta numbers were counted on a maximum number of five females from each class and the excess of chaeta-producing power of the L chromosome over its T homologue can be found for each of the three chromosomes, X, II and III by the comparisons also shown in Table 4. They are, of course, the kind of orthogonal comparisons employed in the analysis of variance and were

TABLE 4
Significance of the marker genes in Mather and Harrison's chromosome assays

Class	Phenotype	Sources of chromosome			Comparisons measuring effects of		
		X	II	III	X	II	III
1	B Pm Sb	T, L	T, L	T, L	−	−	−
2	B Pm	T, L	T, L	2L	−	−	+
3	B Sb	T, L	2L	T, L	−	+	−
4	B	T, L	2L	2L	−	+	+
5	Pm Sb	2L	T, L	T, L	+	−	−
6	Pm	2L	T, L	2L	+	−	+
7	Sb	2L	2L	T, L	+	+	−
8	wild-type	2L	2L	2L	+	+	+

T, L = one chromosome each from the tester and the line
2L = both chromosomes from the line

indeed used by Mather and Harrison to derive analyses of variance. We will, however, confine our attention to the linear comparisons. Nor shall we derive estimates of the effects of interactions among the chromosomes though these can be obtained, and indeed were discussed by Mather and Harrison.

Before we proceed further with the analysis, however, we will pause to note that the classes, B, Pm, Sb and wild-type, are, apart from the effects of recombination and the fourth chromosome, of the same constitution as the F_1 females used as parents in the backcross and females from the line undergoing assay. So, within these limitations, comparisons of the difference in chaeta number between these two classes in the backcross with the difference between the F_1 females and the line females will give a measure of the proportion of the genetic effect on chaeta number, traceable to differences in genes borne by the chromosomes. These differences in chaeta number are set out in columns A and B of Table 3, A showing the difference between line and F_1, and B the difference between the backcross classes. With the genetic effects determined entirely by chromosomal genes the entries in columns A and B should be alike, apart from sampling variation, and the regression of the B on the A differences over the various lines should be 1. The A and B entries for each of the eleven lines for which Mather and Harrison give these differences (they do not give the differences for the Or line) are on the whole closely alike and, more particularly, the slope of the regression line of B on A passing through the origin is 0·82. Thus 82 % of the effect is clearly traceable to genes carried by the chromosomes and, bearing in mind that the consequences of residual recombination in the three major chromosomes and of any genes carried on the fourth chromosome will be to reduce the regression coefficient, we can have little doubt that 100 % of the genetic differences between the chaeta numbers of these eleven lines and the tester is in fact to be ascribed to chromosomal genes. The loss of efficiency in the assays due to residual recombination and the fourth chromosome would thus be no more than 18 %.

To return to the main analysis, it is obvious that as the A difference, and with it the B difference, varies from line to line there must be gene differences between the lines themselves undergoing assay. The comparisons set out in Table 4 allow us to estimate the excess in chaeta-producing power of each of the three large chromosomes (L) from any line over its homologue (T) from the tester. We can then remove the effects of the tester chromosomes, and with them the effects of the major genes they carry, by comparisons between these excesses of the lines over tester. Thus it was found that the excess chaeta-producing power of the Or X chromosome over the X from the tester was 0·79, while that of the Sk X over tester was − 1·06 and that of the line 4 X was 2·94. Thus using the Or X as the new standard of reference the line 4 X has an excess of 2·94 − 0·79 = 2·15 in chaeta-producing power over the Or X, while the Sk X has an excess of − 1·06 − 0·79 = − 1·85 over Or X.

The effects of the three major chromosomes from each of the other eleven lines, using the Or chromosomes as the standard of reference in this way, are set out in Table 3. The total effect of all three chromosomes from the line is shown in the penultimate column and this may be compared with the excess in chaeta number of the line females themselves (shown in the second column) over the Or females. It is clear that all these larger chromosomes carry genes affecting chaeta number which, by comparison with the genes of the Or chromosomes, they may be reducing, hardly affecting, or increasing to various degrees. The total effect measured by the assay for the three chromosomes from any line tallies well with the excess of the tester females over Or females (final column) as found from direct counts of the chaetae borne by females, except that in general the chromosome assays account for only about half the difference in chaeta number between the line and Or. The slope of the regression line of the assay totals on the excess of chaetae over Or (passing through the origin), which is the best measure of the proportion of the difference accounted for by the assay, is 0·48. It is not surprising that the assays fail to account for the whole of the differences, for they depend ultimately on comparisons between flies which are T, L and others which are 2L in respect of the chromosomes in question. They will thus measure the effects of genes in any line only to the extent that these are recessive to their alleles in the tester chromosomes. If a line gene is fully dominant its effects will not appear at all in the comparisons and in the absence of dominance only half the gene's effect will appear in the comparison. Given, therefore, that the genes whose effects we are assaying show no dominance, or that they are equally and equally often dominant and recessive to their alleles in the tester, we should expect the assays to account directly for half the differences in chaeta number between line and tester and hence, by derivation, between line and Or, as was in fact observed.

One last point must be made before we leave these assays. The difference in chaeta number between lines 1 and 7 is $59·53 - 36·00 = 23·53$ of which $10·64$ was accounted for directly by the assay. Yet both lines were derived from the same cross of Or and Sk and, although the assays were not carried out until many generations later, the difference in chaeta number between these two lines was present to the full no more than 30 generations after the cross was made. Other experiments (e.g. Mather, 1941; Mather and Wigan, 1942) show that such a difference is far too large to be accounted for by mutation. It must, therefore, have arisen by redistribution of the genes in each of the chromosomes consequent on recombination. Clearly polygenes recombine with one another as they do with major genes. They are in fact transmitted and reshuffled in inheritance in just the same way as the major genes familiar from genetical experimentation of the classical kind.

Stocks in which the chromosomes carry dominant marker genes and inversions, like the one used by Mather and Harrison, can also be used to build up sets of homozygous lines carrying the three major chromosomes from any two wild-type stocks in all the eight possible combinations (Cooke and Mather, 1962; Kearsey and Kojima, 1967). The wild-type stocks are crossed with that carrying the marked inversion chromosomes and the wild-type chromosomes are then carried heterozygous against their marked homologues until they have been brought together in each of the eight combinations. Similar heterozygotes are then mated together and their wild-type progeny, which will be true breeding for the relevant combination of wild-type chromosomes, are used to found the desired line. The marked chromosomes are thus eliminated at the last stage in the construction of each line. Such a set of eight substitution lines, as they are called, can be used for assaying the differences between the homologous chromosomes from the two parental wild-type stocks.

Unlike Mather and Harrison's assays, those made by using sets of substitution lines will measure in full the combined effects of all the genes carried by the chromosomes, irrespective of their dominance relations, though subject of course to such erosion as may have resulted from any residual recombinations between wild-type and marked inversion chromosomes. This erosion may be expected to be somewhat greater in substitution lines than in the type of assay that Mather and Harrison used, as the wild-type chromosomes are held heterozygous against the marked chromosomes for several generations, rather than just one.

An example of this type of assay is analysed in detail by Mather and Jinks (1977) using data from Caligari and Mather's (1975) analysis of the difference in sternopleural chaeta number between two inbred lines, Samarkand and Wellington. On the present occasion, however, we will take a different example for reasons which will become apparent in the next section. Mather and Hanks (1978) record several analyses of the numbers of coxal chaetae in *Drosophila* using substitution lines. One of these sets was between the Samarkand (S) and 6CL (L) inbreds, and the average numbers of chaetae borne on the front coxae of males of each of the eight lines in this set are shown in Table 5. Since all combinations of the X, II and III chromosomes are present equally in the set of eight lines, we can obtain estimates of the effects on the front coxal chaeta number of the gene differences in each of the chromosomes by using orthogonal functions such as are employed in the analysis of variance. Indeed the results lead directly to an analysis of variance.

The function relating to the effects of the X, II and III chromosomes are also set out in the table. Thus for the X chromosome we find $d_X = (\overline{LLL} + \overline{LLS} + \overline{LSL} + \overline{LSS} - \overline{SLL} - \overline{SLS} - \overline{SSL} - \overline{SSS})$ where d_X measures the amount by which the chaeta number of the L line exceeds the overall mean, $m(= 29{\cdot}163)$, and equally the amount by which the chaeta number of S falls short of m, by reason of the genes carried on chromosome X. d_{II} and d_{III} similarly measure the effects of chromosomes II and III. Four more comparisons can be made measuring the effects of interaction between, respectively, chromosomes X and II, X and III, II and III and X, II and III. It will be seen that d_X and d_{III} are both substantial while d_{II} is quite small, thus suggesting that both the X and III chromosomes are affecting the chaeta number while chromosome II is not.

TABLE 5
Chromosome assay using substitution lines (Number of front coxal chaetae–Mather and Hanks, 1978)

Line	LLL	LLS	LSL	LSS	SLL	SLS	SSL	SSS	
Observed Number (O)	28.80	32.00	27.90	33.75	25.90	29.15	25.65	30.10	$m = 29.163$
X	+	+	+	+	−	+	−	−	$d_X = 1.450$
II	+	+	−	−	+	+	−	−	$d_{II} = -0.188$
III	+	−	+	−	+	−	+	−	$d_{III} = -2.088$
Expected Number (E)	28·525	32·701	28·525	32·701	25·625	29·801	25·625	29·801	
	$(m + d_X + d_{III})$	$(m + d_X - d_{III})$	$(m + d_X + d_{III})$	$(m + d_X - d_{III})$	$(m - d_X + d_{III})$	$(m - d_X - d_{III})$	$(m - d_X + d_{III})$	$(m - d_X - d_{III})$	
$(O - E)$	0·275	−0·701	−0·625	1·049	0·325	−0·651	0·025	0·299	$S(O - E)^2 = 2·678$

Analysis of variance

Item	df	SS	MS	VR	P
X	1	16·820	16·820	25·96	<0·001
II	1	0·281	0·281	0·43	0·7–0·5
III	1	34·861	34·861	53·80	<0·001
Ints.	4	2·396	0·599	0·92	0·5–0·3
Total	7	54·359			
Error Var.	32		0·648		

The significance of the d value can be tested by an analysis of variance as set out in the lower part of the table. Since each of our comparisons corresponds to 1 degree of freedom (df), the sum of squares (SS) to which it leads is also the mean square (MS). The SS ($=$ MS) for the X chromosome is found as $\frac{1}{8}(\overline{LLL} + \overline{LLS} + \overline{LSL} + \overline{LSS} - \overline{SLL} - \overline{SLS} - \overline{SSL} - \overline{SSS})^2$ and so on.

The large experiment from which these data were taken was carried out in duplicate so enabling the error variance of single observations to be calculated from the differences between duplicates. The estimate of error variance so obtained for the front coxal chaeta numbers in the L/S assays was found to be 0·648 as shown in the analysis of variance. Testing the MS for the three individual chromosomes against this error variance shows the effects of X and III to be highly significant while that for II has a probability between 0·5 and 0·7, so confirming that the substitution lines provide no evidence of an effect of this chromosome on the chaetae number. The four interaction comparisons jointly account for an SS of 2·396 which corresponds to 4df one from each comparison. Thus the MS for the pooled interactions is 0·599 which, with a probability between 0·5 and 0·3, is not significant when compared with the error variance. There is thus no evidence for interaction among the chromosomes in exerting their effects on the front coxal chaeta number.

The value found for d_X is 1·450 while that for d_{III} is $-2·088$, the negative sign indicating that whereas the X chromosome from L acted to raise the chaeta number and the X from S to lower it relative to the overall mean, chromosome III from L acted to lower the chaeta number and the III from S to raise it. Thus the effects of the two chromosomes tend to balance one another in the L and S inbred lines. Since neither chromosome II nor the interactions had any significant effect we can omit them from further consideration and find an expected chaeta number (E) for each of the eight substitution lines, using $m = 29·165$, $d_X = 1·450$ and $d_{III} = -2·088$, as shown in the table. These expectations can be compared with the numbers observed (O) by finding $(O - E)^2$ for the eight substitution lines and summing them, which gives $S(O - E)^2 = 2·678$. Since in finding E we have assumed $d_{II} = 0$, as well as all four interactions to be 0, $S(O - E)^2$ will correspond to 5df so giving an MS $= 0·536$ for the deviation of O from E. As expected, this is not significant when tested against the error variance.

3. LOCATING THE GENES

Within the limitations imposed by residual recombination, chromosome assays assess the total effect of all the genes within a chromosome in respect of which lines differ; but they give no indication of the actual location of the genes along the chromosome. To obtain information about the locations of the genes, and hence about their distribution within the chromosome, a tester chromosome must be used which is marked in appropriate places along its length by a series of major mutants, and recombination must be allowed to proceed normally. Furthermore, since the marker genes in the tester chromosome must in general be recessive, the F_1 must be backcrossed to the tester stock. In consequence the effects of the genes in the chromosome under analysis will appear only to the extent that they are dominant to their alleles in the tester, except of course in special cases such as the X chromosome in *Drosophila* where observations can be made on the male progeny of F_1 females.

Consider a gene pair A–a affecting the expression of a continuously varying character such that the heterozygote adds an increment a to the expression of the character and the homozygote reduces the expression by the same increment, both measurements being made from the mean expression m. Let this gene be linked to a marker, G–g, with p_{ag} as the frequency of recombinations between them. Then if the line is AAGG and the tester aagg, their F_1 will be AaGg and the backcross to the tester will consist of the four genotypes:

Genotype	Frequency	Marker phenotype	Expression of metrical character
AaGg	$\frac{1}{2}(1-p_{ag})$	G	$m+a$
aaGg	$\frac{1}{2}p_{ag}$	G	$m-a$
Aagg	$\frac{1}{2}p_{ag}$	g	$m+a$
aagg	$\frac{1}{2}(1-p_{ag})$	g	$m-a$

On average the individuals of marker phenotype G will show an expression $(m+a)(1-p_{ag})+(m-a)p_{ag}=m+a(1-2p_{ag})$. Similarly the average expression of the g individuals is $m-a(1-2p_{ag})$, and the difference between the average of the two marker classes is $2a(1-2p_{ag})$.

Extending consideration to more than one gene affecting the metrical character, it is not difficult to show that the average expression of the character in G individuals is $m+S[a(1-2p_{ag})]$ and in g individuals is $m-S[a(1-2p_{ag})]$ the difference between the two marker classes thus being $2S[a(1-2p_{ag})]$, where S indicates summation over all the A–a gene pairs. This result is independent of the occurrence of interference in the crossing-over within the chromosome and of how the A–a genes are arranged along the length of the chromosome relative to G–g. It should be noted that a will itself be negative where the allele coming in from the line chromosome reduces the expression of the character relative to the allele from the tester. Hence $S[a(1-2p_{ag})]$ must take into account the signs of the effects of the genes from the line and it may even be negative where the effects of these genes are preponderantly towards reduction of the character.

Proceeding to the case of two marker genes, G–g and H–h, with H–h to the right of G–g and showing a recombination value of p with it, let A–a fall to the left of G–g. The difference of average expression in the two marker classes G and g will of course still be $2a(1-2p_{ag})$ irrespective of the phenotype in respect of H–h. Similarly the difference of average expression in the H and h marker classes will be $2a(1 - 2p_{ah})$, but this will necessarily be smaller than $2a(1 - 2p_{ag})$ since $p_{ah} > p_{ag}$. Furthermore, when we take into consideration all four marker phenotypes, GH, Gh, gH and gh, it can be shown that, provided that there is no double crossing-over, the average expressions in classes GH and Gh are alike, as are those of gH and gh. This reveals that A–a is to the left of G–g: if it had been to the right of H–h the average expressions in GH and gH, and in Gh and gh, would have been alike with GH and Gh, and also gH and gh, differing.

When A–a lies between G–g and H–h, the average expression in the marker classes and their relations with one another take on a different form. Apart from the effects of double crossing-over (which will be absent or at least rare unless G–g and H–h are genetically a long way apart) the GH chromosome will always carry A and the gh chromosome allele a (Fig. 3). Thus the mean expressions in marker classes GH and gh will be $m + a$ and $m - a$ respectively, and half this difference will afford a direct estimate of a. Where, as before, p_{ag} is the frequency of recombination between G–g and A–a and p_{ah} that between A–a and H–h, $p_{ag} + p_{ah} = p$ in the absence of disturbance from double crossing-over. Then in a proportion p_{ag}/p of cases the recombinant chromosomes will be Gah and gAH, with average expressions $m - a$ and $m + a$. In the remaining p_{ah}/p of cases the recombinant chromosomes will be GAh and gaH with expressions $m + a$ and $m - a$. Thus the average expression of marker class Gh will be $m + a(p_{ah} - p_{ag})/p$ and that of gH will be $m - a(p_{ah} - p_{ag})/p$. In consequence, $\frac{1}{2}(\overline{Gh} - \overline{gH})$ provides an estimate of $a(p_{ah} - p_{ag})/p$. The third comparison, which completes the analysis, is $\frac{1}{4}(\overline{GH} - \overline{Gh} - \overline{gH} + \overline{gh})$ which, however, has an expected value of 0 and so provides only an estimate of error variation and is uninformative about the effect and location of A–a.

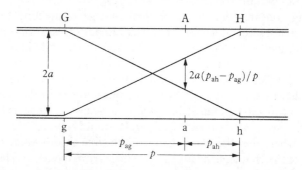

Figure 3 Diagram to show the effects of a gene-pair, A–a, on a metrical character in the four classes distinguished by the combinations of the marker genes G–g and H–h, between which loci A–a is situated.

The value of p can, of course, be found from the frequencies of the four marker classes. The experiment thus yields estimates of a, $a(p_{ah} - p_{ag})/p$ and p. So in addition to a, the effect of A–a on the expression of the continuously varying character, we can find p_{ag} (or $p_{ah} = p - p_{ag}$) which serves to locate the locus of A–a in relation to those of G–g and H–h. The calculation can be illustrated by reference to one of a number of experiments described by Wolstenholme and Thoday (1963) aimed at locating genes affecting the number of sternopleural chaetae and borne on chromosome III of *Drosophila melanogaster*. The marker genes clipped wing (cp, locus 45.3) and Stubble bristle (Sb, locus 58.2) were used in this particular experiment. The mean numbers of sternopleural chaetae observed for the four marker classes were $+$ $+$, 20·62; $+$ Sb, 19·19; cp $+$, 18·95; cp Sb, 18·00. Then $a = \frac{1}{2}(20·62 - 18·00)$ $= 1·31$, and $a(p_{ah} - p_{ag})/p = \frac{1}{2}(19·19 - 18·95) = 0·12$. The authors do not record the frequencies of the four marker classes and a direct extimate of p is thus not available. In the absence of such an estimate we may use the standard map distance between the two marker loci, which is $58·2 - 45·3 = 12·9$ so giving $p = 0·129$ which is small enough for double crossing-over to be negligible, if not completely absent. Then $(p_{ah} - p_{ag}) = (0·12 \times 0·129)/1·31$ $= 0·0118$. Since $(p_{ah} + p_{ag}) = p = 0·129$, we find $p_{ah} = 0·070$ and $p_{ag} = 0·059$, so placing the locus of A–a at 5·9 map units to the right of cp, that is at $45·3 + 5·9$ $= 51·2$ on the standard map. We shall return to this example later for, as Wolstenholme and Thoday showed, the situation was more complex than we have assumed it to be for our illustrative purposes.

Thus, in principle, we have means of finding the locus of a gene contributing to quantitative variation, if we choose two marker genes between which the gene in question falls. The difficulty lies, of course, in separating the gene in question from its fellows so that its individual effects can be observed. If it is not so isolated the effects observed on the continuously varying character, as expressed in the recombinant and non-recombinant marker classes, will be the resultant of the action of all the genes affecting the character. As we have seen, the resultant effects of genes outside the marked segment will be of the form $S[a(1 - 2p_{ag})]$ while that of genes inside the segment will be $S[a(p_{ah} - p_{ag})]$. Certain of the difficulties that ensue have been discussed at some length by McMillan and Robertson (1974).

An overall analysis of the general distribution of polygenic activity along a chromosome can be made using a multiple marked tester chromosome (Mather, 1942). This kind of analysis, which was used for the effects of the X chromosome on sternopleural chaeta number in *Drosophila* by Wigan (1949), may be illustrated by reference to the effects on front coxal chaeta number of the X chromosomes from the L and S inbred lines, already given as an example of chromosome assay using substitution-lines (Mather and Hanks, 1978). The two Xs, L and S, were each made heterozygous in females with a tester chromosome carrying the marker genes yellow body (y^2, 0·0), apricot eye (w^a, 1·5), crossveinless wings (cv, 13·7), cut wings (ct, 20·0), vermillion eye (v, 33·0) and forked bristles (f, 56·7), the II and III chromosomes being homozygous and so giving rise to no segregational variation in the progeny. The male offspring were, of course, scorable for all the marker genes, and the front coxal chaetae number was recorded for 10 males from each of the two non-recombinant classes and each of the single recombinant classes, of which there were eight since recombinants between y^2 and w^a were neglected. No double recombinants revealed by the marker genes were used, and the genetical distances between adjacent markers are sufficiently small for double cross-overs within a marked segment to be sufficiently rare as to be negligible, except perhaps in the segment delimited by v and f.

Now the $y^2 w^a cv + + +$ recombinant between the L and tester (T) chromosomes, for example, will have the constitution of T to the left of cv and that of the L chromosome to the right of the ct locus. The $+ + + ct v f$ recombinant will be the reverse having the constitution of L to the left of the cv locus and T to the right of ct. Thus if, in respect of any gene pair affecting chaeta number, we regard the allele in the L chromosome as increasing the mean number of chaetae by a and that in the T chromosome as reducing it by the same increment, the mean chaeta number of males with the complete chromosome from T will be $m - S(a)$, bearing in mind that a can take sign. So in respect of chaeta genes to the left of cv, the difference in chaeta numbers between the recombinants $+ + + ct v f$ and $y^2 w^a cv + +$ will include an item $2S_{PL}(a)$ and in respect of genes to the right of ct an item $S_{PR}(a)$, where S_{PL} is the partial sum of a over the relevant genes to the left and $S_{PR}(a)$ the partial sum over the relevant genes to the right.

The contribution that chaeta genes lying in the region between the loci of cv and ct will make to the difference in chaeta number follows from our previous discussion of chaeta gene pair A–a in relation to marker genes G–g and H–h. If now we write A for the chaeta allele in the L chromosome and a for that in the T chromosome, $+^{cv}$ (the wild-type allele of cv) for G with cv for g, and $+^{ct}$ for H with ct for h, the contribution of A–a to the difference $(+ + + ct v f)$ $- (y^2 w^a cv + + +)$ will be $2a(p_{act} - p_{acv})/p$ where p is the recombination value between cv and ct. Where A–a lies closer to the locus of ct than to that of cv, i.e. $p_{acv} > p_{act}$, this contribution will be negative like the contribution of any gene to the right of ct; but where A–a lies nearer to the locus of cv than to that of ct $p_{acv} < p_{act}$ and the contribution will be positive like that of any gene to the left of cv. Where A–a lies mid-way on the genetical map between the two marker loci $p_{acv} = p_{act}$ and the gene will contribute nothing to the difference.

So in these two recombinant chromosomes the chaeta genes to the left of the mid-point between the loci of cv and ct may be said to be associated with cv in the sense that their alleles in the L chromosome were marked by $+^{cv}$, the allele at the cv locus in this chromosome, while their alleles in the T chromosome are marked by cv. In the same way the chaeta genes to the right of the mid-point are associated with ct. Association is complete for chaeta genes outside the cv-ct segment, but, because of the recombination between cv and ct, it is incomplete for genes within the segment, becoming weaker as the chaeta gene approaches the mid-point, and vanishing at the mid-point itself.

Recombination in this region in effect divides the chromosome into two at the mid-point of the cv-ct segment.

Moving on to y^2 w^a cv ct + + and + + + + v f, the recombinants between ct and v, by the same argument we have divided the X into two parts at the mid-point of the ct-v segment: all chaeta genes to the left of ct will be completely associated and those in the left half of the ct-v segment incompletely associated, with the ct locus, while all genes to the right of v will be completely, and those in the right half of the ct-v segment incompletely, associated with the v locus. Thus, taking the two sets of recombinants together, if we denote the mean chaeta numbers of y^2 w^a cv + + +, y^2 w^a cv ct + +, + + + + v f and + + + ct v f by (a), (b), (c) and (d) respectively, (b) differs from (a) only by the effects of genes associated with ct and (d) differs from (c) in the same way. So both (a) − (b) and (c) − (d) provide estimates of the effects of chaeta genes associated with the ct locus. Apart from sampling variation (a) − (b) will differ from (c) − (d) only by effects of any interaction which there may be between the genes associated with it and any other chaeta genes in the chromosome, and which will inflate the difference between (a) − (b) and (c) − (d). Thus $\frac{1}{2}$[(a) − (b) + (c) − (d)] provides an estimate of the resultant effect of chaeta genes associated with ct, and $\frac{1}{2}$[(a) − (b) − (c) + (d)] provides a test for interaction of these genes with those in other parts of the chromosome. Mather and Hanks (1978) carried out the experiment we have been discussing, and other similar experiments, in triplicate and so obtained from the differences among replicate observations an estimate of error variation as $V = 0.1193$ for 36 df, with which $\frac{1}{2}$[(a) − (b) − (c) + (d)] can be compared to test for interaction. No evidence of interaction was found.

Now, where the wild-type chromosome was L, (a) − (b) and (c) − (d) provide estimates of the difference between the alleles of the genes associated with the cut locus in L and those in T. Similarly, we can find the S − T difference when the wild-type chromosome is S. Then taking $\frac{1}{2}$[(L − T) − (S − T)] = $\frac{1}{2}$(L − S) provides an estimate of the difference in effect of the relevant genes in the X chromosomes of L and S, while $\frac{1}{2}$[(L − T) + (S − T)] = $\frac{1}{2}$(L + S − 2T) is an estimate of the difference between the mean of L and S on the one hand and T on the other. These estimates will both have an error variance of $2V/4 = 0.05966$ and hence a standard error of $\sqrt{0.05966} = 0.244$.

The ct locus has been taken as an example, but the analysis can be applied to all the marker genes in the chromosome. The results of doing so are set out in Table 6, from which it is clear that the y^2 w^a 'locus' has associated with it chaeta genes in respect of which L and S differ from one another and also their average differs from T. L and S do not differ from one another, but they do jointly differ from T, in chaeta genes associated with the cv and f loci, while there is a suggestive difference just inside the 5% level of significance ($P = 0.07 − 0.06$), between L and S in the genes associated with the v locus. Evidently polygenic activity in respect of front coxal chaeta number is widely

TABLE 6
Distribution along the X chromosome of genetic differences between L and S inbred lines in respect of front coxal chaeta number (Mather and Hanks, 1978).

	Experiment 1					Experiment 2			
	y^2wa	cv	ct	v	f	y^2	wa	ec	'Rest'	
L − T	4·180	−1·050	0·235	−0·330	−1·000	0·748	2·219	0·834	−2·774	all
S − T	·0·295	−1·545	0·220	0·570	−0·630	−0·176	−0·116	−0·119	−1·412	±0·345
½(L − S)	1·943‡	0·248	0·008	−0·450*	−0·185	0·462*	1·168‡	0·477†	−0·681‡	all
½(L + S − 2T)	2·238‡	−1·298‡	0·228	0·120	−0·815‡	0·286	1·052‡	0·358	−2·093‡	±0·244

* $P = 0.06$ † $P = 0.05 − 0.01$ ‡ $P = 0.01$

Experiment	1	2
	y^2wa	$y^2 + wa + ec$
½(L − S)	1·943	2·106
½(L + S − 2T)	2·238	1·695
	cv + ct + v + f	'Rest'
½(L − S)	−0·380	−0·681
½(L + S − 2T)	−1·765	−2·093

distributed along the X chromosome, the genes giving the main differences between L and S being concentrated in the y^2w^a region, though there is a suggested indication of another gene, or others, near the v locus.

In a second experiment of the same kind Mather and Hanks investigated further the left end of the chromosome, this time using a tester with an additional marker, echinus eye (ec, 5.5), in it. They followed y^2, w^a, ec and cv as separate loci but effectively took the rest of the chromosome as a single unit tied to cv, by rejecting all recombinants to the right of cv. The results of this second experiment, analysed in the same way as the first, are also shown in Table 6. Both the w^a locus and the chromosome to the right of cv (denoted 'Rest' in the table) show evidence of chaeta genes in respect of which $\frac{1}{2}(L + S)$ and T differ; but chief interest must be in the evidence that the main difference between L and S lies in the region round the w^a locus, with evidence of smaller effects round y^2 and ec, that associated with y^2 being just subsignificant with $P = 0.06$. The L and S chromosomes also differ significantly in the long segment denoted as 'Rest', no doubt largely because of the gene(s) which the first experiment suggested as being associated with the v locus.

We may note a final point. The chromosome assay of the previous section revealed the X chromosomes of L and S as differing in genes which together caused a deviation of 1·450 in front coxal chaeta number. If we sum the values of $\frac{1}{2}(L - S)$ over all the regions in the first analysis of the X chromosome, so obtaining the amount by which this analysis indicates that the L and S chromosomes would cause deviations from the mean, we obtain a value of 1·562 chaetae and the same sum for the second analysis gives 1·425 chaetae. Evidently the two analytical experiments are indeed sorting out the genes whose combined effect was revealed by the chromosome assay.

We can take this point further. The difference $\frac{1}{2}(L - S)$ for the joint $y^2 w^a$ region was 1·943 in the first analysis, while the y^2, w^a and ec regions give a summed difference of 2·106 in the second analysis. Again the effects of the same genes are clearly being followed – as a pool in the first experiment and as three separate genes, or groups of genes, in the second. We may observe too that the deviation due to the combined genes in the y^2, w^a and ec regions is greater than that traced for the whole chromosome in the assay. This difference is clearly due to the balancing effects of genes at the right end of the chromosome, as revealed by $\frac{1}{2}(L - S)$ being positive in the y^2, w^a, ec part of the chromosome and indeed in the cv and ct regions, as revealed by the first experiment, but negative for the v and f regions in the first experiment and for the combined chromosome to the right of cv in the second.

Mather and Hanks did not take further the location of the chaetae genes by which the X chromosomes of L and S differed, particularly at the left end in the y^2, w^a, ec region. To do so they could have adopted the technique introduced by Thoday (1961) which was used by Wolstenholme and Thoday, some of whose data we have already taken for illustrative purposes. Thoday's technique uses progeny testing to ascertain the number of classes, genetically different in respect of chaeta number, to be found among the chromosomes recombinant for the two marker genes. In the experiment which we discussed earlier in this section, Wolstenholme and Thoday were able to recognize three such classes, which of course require two chaeta genes between their markers, cp and Sb. Since only three classes were distinguishable, not four, they were led to assume two chaeta genes whose allelic differences had approximately equal effects on the chaeta number. By employing further markers enabling them to delimit smaller segments within the cp-Sb region, they placed one of the two genes near 49 on the standard map and the other near 51. Our analysis carried out earlier in the section would, of course, locate the resultant point of action of these genes. Assuming the two chaeta gene-pairs to have equal effects on the character, i.e. $a_1 = a_2$, the resultant point would be expected to lie midway between the loci of the two gene pairs, i.e. at locus 50 on the standard map. The analysis of Wolstenholme and Thoday's experiment that we carried out above placed it at 51·2 which is near the rightmore of the two genes; but a second experiment that they described and which we have analysed in the same way elsewhere (Mather and Jinks, 1977) places the resultant at locus 50·3. This is close to expectation and indeed better overall agreement is hardly to be expected when a segment of no more than two map units long is involved: the deviation must be well within error variation.

It is obviously essential that the two chromosomes used in such experiments (in this case the $+/+$ and cp Sb chromosomes) should not differ in any genes affecting the metrical character outside the region delimited by the two markers. The results of both Wolstenholme and Thoday's analysis and our own would have been distorted by the resultant action of any such genes, as will indeed be made immediately apparent by consideration of our analysis of Mather and Hanks' experiment using multiple markers.

Marker genes have been used in a different way by Breese and Mather (1957, 1960) to locate along the chromosome genes contributing to continuous variation, and to analyse their effects. Breese and Mather's technique was too elaborate for detailed description here, but essentially it consisted of using a marked chromosome as an intermediary in building up new chromosomes compounding known segments from two unmarked chromosomes whose differences it was desired to investigate. No marker genes were left in the compound chromosomes thus built up, so that no differences were attributable to pleiotropic effects of the marker genes. Comparisons could thus be made directly between segments of the two unmarked chromosomes, whose effects were under test, without the necessity of using the marked chromosome as a common standard with which the unmarked chromosomes were first compared directly and through the medium of which they were compared indirectly with one another. Breese and Mather were themselves concerned to analyse the differences in effect on chaeta number (both abdominal and sternopleural) and viability of the chromosome III from two lines of *Drosophila*, one originally selected for high and the other for low numbers of chaetae. The two chromosomes showed differences in all six segments into which each was divided for the study of effects on chaeta numbers and in all four segments followed in the observations on viability. Once again there is a clear association of the chromosomes and their segments with polygenic activity. Once again it is clear that continuous variation is mediated by genes borne on the chromosomes, like the genes of major effect familiar from genetical experiments of the traditional kind, and each having its own locus within a chromosome, though in this case no attempt was made to locate them more precisely.

4. LINKAGE BETWEEN GENES AFFECTING DIFFERENT CHARACTERS

The observations reported by Breese and Mather show that polygenes affecting characters as different as chaeta number and viability are distributed along the whole length of chromosome III in *Drosophila melanogaster* and that they occur together in the segments into which the chromosome was divided for the purposes of the experiment. The different segments, however, contributed differently to the variation in chaeta number and viability in that their effects were of different magnitudes relative to one another, and that they showed different dominance relations and different properties of interaction in respect of the two characters. It is therefore highly unlikely, to say the least, that the effects on the two characters were traceable to the pleiotropic action of the same polygenes. Rather we must suppose that the two characters were affected each by its own polygenic system with the member genes of the two systems intermingled in their distribution along the chromosome. Indeed there is evidence that even such similar characters as number of sternopleural chaetae and number of abdominal chaetae are affected each by its own polygenic system, the two systems being intermingled along the chromosomes (Davies and Workman, 1971).

Now it has been established that while polygenes mutate like other genes, the effects of mutation on the variation of a character are far too small to account for the differences that can be built up by selection (Mather, 1941, 1956; Mather and Wigan, 1942; Paxman, 1957). Differences produced by selection in the overall action of chromosomes such as are shown in Table 3, must be the result of redistribution by recombination of the genes within the chromosomes. Selection will therefore pick out recombinant chromosomes and wherever the genes controlling a second character are intermingled along the chromosome with those controlling the operative character, recombination of the one set will mean recombination of the other. Fixation by selection of the redistributed gene combinations for the one character will then mean fixation of redistributed gene combinations for the other, with the consequent possibility of a change in phenotype (Fig. 4). The second character can thus show a correlated response to a selection which did not aim at altering

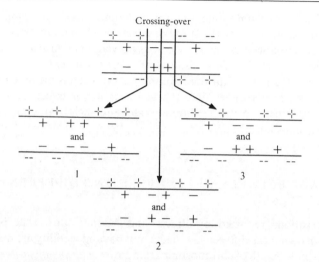

Figure 4 Correlated responses to selection. Recombination of the genes controlling the primary character (-+- and --) is accompanied by recombination of the genes controlling the secondary character for which no selection is practised (+ and −). This recombination may lead to no unbalance (2) or to unbalance in either direction (1 and 3) of the genes controlling the secondary character, according to the position of the crossing-over which gave rise to the recombination. The response of this character to selection for the primary one may therefore be in either direction or may be absent.

it, though the direction and magnitude of this correlated response may well be unpredictable (Wigan and Mather, 1942).

Correlated responses are indeed a commonplace of selection experiments. They are quite regularly revealed by falls in fertility of selected lines. These were recorded by Mather (1941) in his lines of *Drosophila* selected for chaeta number and have indeed proved to be a general accompaniment of selective change in chaeta number and other characters in the many later experiments undertaken with this fly. They have been found too in such diverse organisms as fishes (Svärdson, 1944), mice (see Falconer, 1960), poultry (Lerner and Dempster, 1951) and rye-grass (Cooper, 1961). The effect might, of course, be ascribed to pleiotropic action of the polygenes if the relation of the two characters proved to be constant, and indeed it would be difficult to assert that correlated response is never due to pleiotropy. Evidence is, however, available against pleiotropic action as a general explanation and the more thoroughly a case of correlated response is explored the less likely pleiotropy becomes as an explanation, except in cases where there is a close developmental relation between the characters.

The evidence against pleiotropy is clear from the results of Mather and Harrison (1949). They practised selection for increase in number of abdominal chaetae on flies from a cross between the Oregon and Samarkand stocks of

Drosophila (Fig. 5). The mean number of chaetae rose for 20 generations and, as usual, fertility fell at the same time. After these 20 generations of selection, fertility was so low that the line could not be maintained under selection and mass culture was then resorted to. Fertility immediately began to rise rapidly, doubtless as a result of natural selection for it within the culture bottles. This increase in fertility was accompanied by a correspondingly rapid fall in chaeta number, as indeed we would expect whether the effects on chaeta number and fertility were pleiotropically determined by the same genes or whether they were due to linkage of genes affecting fertility with those raising chaeta number.

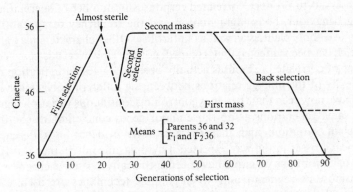

Figure 5 Selection for increased number of abdominal chaetae in *Drosophila melanogaster* (Mather and Harrison, 1949). In the first selection, increase in mean chaeta number was accompanied by a decrease in fertility. When selection for chaeta number was relaxed fertility took charge and chaeta number fell. The second selection for chaeta number gave no correlated response in fertility, and the second relaxation of selection was followed by no fall in chaeta number, even though the stock was not homozygous as shown by successful back selection. Solid lines indicate selection, and broken lines mass culture without selection.

After three generations of mass culture, selection was again practised for increased chaeta number and in four generations it restored the chaeta number to the peak value originally obtained at the time when mass culture became necessary. But this time there was not the same loss of fertility, and the new high line maintained the peak value for over 100 generations after selection was again replaced by mass culture. Furthermore, not only did this new mass culture maintain its chaeta number at the level where fertility after the first selection had been so low, but it did so even though it had within itself all the genic materials for a fall in chaeta number, as was shown when a downward selection was successfully attempted with it later. Thus lower fertility was now associated not with higher but with lower chaeta number.

The conclusion is clear. In the second selection for high chaeta number the association of low fertility with high chaeta number was broken, with the result that the second massed line had not the incubus of poor fertility combinations which had been present after the first selection. Such a re-association must often follow by recombination on the assumption of linkage. It is capable of no simple explanation on the alternative assumption of pleiotropic action. Polygenes controlling different characters must thus be linked and inter-mingled along the chromosome.

This is but one part of the extensive evidence Mather and Harrison obtained about the relation between chaeta number and fertility, all pointing to an interpretation in terms of linkage. Nor was the linkage confined to the polygenes controlling chaeta number and fertility. Selection for chaeta number was observed to produce correlated responses in number of spermathaecae, mating behaviour, body pigmentation and the frequency of appearance of certain eye abnormalities as well as in fertility. Other characters may also, of course, have been affected but escaped notice. However this may be, the number of correlated responses actually observed gives an indication of the complexity of the linkage relations between the different polygenic systems.

We have now seen that the genes controlling continuous variation segregate in the same generations as do those major genes controlling discontinuous Mendelian variation, and that there is little evidence of differences in reciprocal crosses in the way expected if cytoplasmic inheritance is involved. Genes of continuous variation have been specifically located in chromosomes, in chromosome segments and, when suitable techniques are used, even at particular loci. They can be manipulated by treatment of the chromosomes and their relations in response to selection are such as to be expected from determinants borne by the chromosomes. Furthermore, when in *Drosophila* a balance sheet was struck it accorded well with the view that all the heritable variation was accountable by genes carried on the chromosomes. The nucleus is apparently as potent, relative to the remainder of the cell, in its control of continuous variation as it is in its control of major discontinuities: we may confidently build our methods of analysis of continuous variation on the assumption of nuclear control.

5. POLYGENES AND MAJOR GENES

The use of major genes for markers has allowed us not merely to see that the determinants of continuous variation are borne on the chromosomes but that in *Drosophila* the three major chromosomes all carry members of the polygenic system determining variation in abdominal chaeta number. Furthermore, each of the six segments into which Breese and Mather divided chromosome III showed activity in determining differences in this character and Wigan (1949) has shown that activity in the determination of variation in a related character – number of sternopleural chaetae – is widely distributed along the X chromosome. The chromosome assays of Mather and Harrison's (1949) selection lines (Table 3) require for their interpretation at least two polygenes in chromosome II, at least three in chromosome III and probably at least three in the X – a minimum of eight taking all three chromosomes together. Breese and Mather have demonstrated at least six genes in chromosome III affecting this same character of abdominal chaeta number, using lines of which one was descended from Mather and Harrison's selection and the other was derived also from the cross between Or and Sk. The minimum number of genes in the system is thus 11, and if we take into account the evidence from other observations in this character (e.g. by Mather 1941, 1942) and the implications of Wigan's findings, we can hardly doubt that the minimum must be 18 or more and the true number may be much larger than this. The various investigations of sternopleural chaeta number, by Wigan, Cooke and Mather, and Thoday and his collaborators reveal a similar picture: indeed from a single set of experiments Davies (1971) reports evidence of genes at 15 loci, at least, affecting sternopleural chaetae number and 14 or 15 affecting the number of abdominals.

'Student' (1934) has estimated that at least 20 polygenes must be involved in the control of oil content in maize, and he expressed the opinion that the number of genes was in fact nearer to 200 than to 20. More recently, Falconer (1971) has reported evidence leading him to the view that some 80 gene differences affecting litter size must have been segregating in the base population of mice from which his selection for this character was started. Since except in special circumstances we cannot recognize polygenes as individuals we cannot count them as individuals, and for technical reasons which we shall see later, we must tend to recognize groups of them associated in a segment of chromosome rather than individual genes. It is, however, both clear and sufficient for our present purpose that the minimum is many rather than few.

Now with a dozen gene pairs segregating, or even half a dozen, many genotypes and many expressions of the character will be realized. Provided that no one member gene of the system has a grossly disproportionate effect on the character, each of them will contribute but a small proportion of the variation and none will produce such a discontinuity in the phenotypic distribution as to disrupt the gradation of expression of the character. Furthermore, in the case of abdominal chaeta number that we have been discussing, the non-heritable variance is of the order of 6 to 8 (Mather, 1941, 1949a) giving a non-heritable standard deviation of 2·5 to 2·8. Now the extreme lines shown in Table 3 differed by $26·7 + 7·5 = 34·2$ abdominal chaetae, and with no more than the minimum of 11 genes which we confirmed above from this material, the average effect of each gene substitution would be about three chaetae. The average effect is therefore of the same order of size as the non-heritable standard deviation, and if 11 is an underestimate of the number of polygenes, this average effect must be correspondingly smaller than the non-heritable component of variation. The non-heritable agencies can thus easily blur any minor discontinuities of expression of the character that the genic differences might produce. In other words such a polygenic system provides a fully adequate interpretation of continuous variation.

The different members of this polygenic system were distinguished only by their linkage relation. They had similar effects on the ultimate character of chaeta number and this similarity extended even to the gross actions of the whole chromosomes. Equally, the effects of the different genes must be supplementary as well as similar, since no one chromosome, and *a fortiori* no one gene, could itself account for the differences in chaeta number observed between Mather and Harrison's lines as shown in Table 3. All the chromosomes played their parts, and often more or less equal parts, in producing the differences. The same is true of the polygenic systems governing other continuously varying characters that have been analysed such as sternopleural chaeta number, coxal chaetae number, viability and the heterosis in egg production of *Drosophila* females discussed by Strauss and Gowen (1943). Supplementation of effect between genes acting in the same direction implies a balancing of effect between genes acting in opposite directions. Upon this balancing depends the great capacity of polygenic systems to hide or store variability behind the façade of phenotypic uniformity, a capacity which increases directly with the number of genes in the system (Mather, 1943, 1953a, 1960a, 1973).

The genes of quantitative variation and the major genes of Mendelian genetics are inferred to account for different types of variation, continuous and discontinuous. They have been found to be alike in their transmission from cell to cell and from generation to generation as constituents of the chromosomes. Yet in other respects they appear to be in contrast with each other: the genes of continuous variation appear as acting in polygenic systems whose members have small, similar and supplementary effects, while the effects of major genes appear as drastic and individually specific. To what extent is this a valid contrast? And if it is, from what properties of the genes does it arise?

The two inferences, of major genes and of polygenic systems, are generally derived from analyses which make use of statistically different techniques of comparison, the one by the familiar use of ratios and the other by consideration of the means, variances and covariances which can be obtained from continuous distributions. The distinction that we have drawn between them should not, however, be confused, as sometimes appears to be the case, with the differences in statistical technique. Genes producing discontinuities in the phenotypic distributions can indeed be handled by consideration of the variances and covariances to which they give rise and, in fact, might sometimes best be considered in this way; but usually the more familiar techniques are to be preferred as being more expeditious and more convenient. Equally, a gene of relatively small and non-specific effect can be handled by the familiar techniques of Mendelian genetics if the obscuring effects of segregation of other genes is removed by suitable breeding techniques, non-heritable variation reduced as far as possible by rigorous control of the environment, and recourse had to sufficiently extensive progeny testing, whose value has been so well shown by Thoday (1961) and Wolstenholme and Thoday (1963). But such techniques can be prohibitively demanding, and biometrical analysis of the kind we shall discuss must in effect be the only method generally open to us.

At the same time, the use of contrasting techniques can introduce a degree of artificial overemphasis into the comparison between the two types of gene that we are led to infer from the two types of variation. The biometrical technique covers all the genes contributing to the variation in the chosen character: the effects on the character of even the smallest gene difference will be brought into account as well as those of the largest, for the variation is dealt with as a whole (see Mather, 1971). The Mendelian approach, on the other hand, will emphasize major genic differences, no matter what character they affect, because the magnitude and constancy of their effects allow them to be recognized and followed in genetic studies with an ease that cannot be matched by gene differences of lesser effect on the phenotype. For this reason it is easy to be led into seeing a sharp contrast between the two ends of a spectrum which is in truth continuous. Observations such as those of Spickett and Thoday (1966) leave no doubt that gene differences (albeit perhaps compound differences acting as a unit in transmission) contributing to quantitative variation can differ in the magnitude of their effects. Equally, at the other end of the spectrum, series of alleles like that at the white locus in *Drosophila* make it clear that a genic structure, recognizable as major by some of the differences to which it gives rise, can also give rise to differences whose effects on the phenotype cannot be detected without recourse to special, sophisticated techniques (Muller, 1935). Thus in respect of the size of their effects, members of a polygenic system and major gene differences may indeed be no more than the ends of a continuous spectrum. Whether, however, gene differences of intermediate magnitude are in fact as common as those at the extremes is far from clear, and indeed Timoféeff-Ressovsky (1934, 1935, quoted by Dobzhansky, 1951) found that among induced mutations affecting the viability of *Drosophila*, intermediates are less common than either those of small effect or those whose effects are so drastic as to produce complete lethality.

Turning to other aspects of gene action, both dominance and the interaction of non-allelic genes have been recognized since early times as a commonplace of major genes (see Bateson, 1909). Equally, the first genetical analysis of quantitative variation by biometrical methods revealed that the genes mediating this variation displayed at least partial dominance (Fisher, 1918), and this has been established as a general property of such genes by innumerable biometrical studies since that time. Similarly, as we shall see later, the interaction of non-allelic genes has been revealed as a commonplace of polygenic systems in more recent years. Thus both dominance and non-allelic interaction are regular features of the genes mediating quantitative variation just as they are of major genes. There is, nevertheless, a difference. Both dominance and interaction are most usually complete, or virtually so, where major genes are concerned, whereas with quantitative variation, as pointed out by Mather (1979), they appear commonly to be only partial.

Again, with the exception of such traits as isozyme and antigenic characters, which are the immediate products of gene action and hence if detected at all must appear as strictly 'one or the other', all characters are subject to both discontinuous and continuous variation and both major genic and polygenic differences are to be inferred from them. To take an example, we have in man the dwarfs who, like achondroplasiacs, depart sharply from the normal because of specific major genic differences, while the variation in stature among 'normals' is one of the classic cases of continuous and, one assumes, polygenic variation. Abdominal chaeta number in *Drosophila*, too, may be drastically altered by such major genic differences as scute, while, as we have seen, showing polygenic variation among wild-type individuals – and also among scute individuals for that matter (Cocks, 1954). And even where isozymic or antigenic differences are concerned, variation in the ultimate level of the enzymic or antigenic activity displayed is subject to the influence of many more genes than the one whose alleles were determining the specific electrophoretic or immunological properties.

Similarity of the member of genes of a polygenic system in their effects on a character implies a certain lack of specificity, but this need not be a complete absence. A character such as chaeta number might be changed in a variety of ways. It might be raised, for example, by an increase in the average number of chaetae per unit area of surface; or the rise might be due to an increase in surface area with the density of chaetae remaining constant. The effects of genes bringing about changes of these two kinds would be the same on the ultimate character but would depend on different initial actions. We know that such differences in initial action do occur among members of the polygenic system governing variation in sternopleural chaetae number (Spickett, 1963). Thus at least some of the genes contributing to the continuous variation do so in different ways. Further evidence, from fungi, confirms that the similarity in effect of members of a polygenic system may be superficial in the sense that it vanishes when we look more closely into the ways in which they contribute to the variation shown by the character (Caten, 1979). This does not, of course, mean that all the genes contributing to quantitative variation do so in ways differing from those of their fellow members of the polygenic system. Indeed, when we consider how large even the minimal estimates of the number of genes in such a system may be, it is difficult not to conclude that although the similarity of action in some cases may not reach below the level of the character itself, in many others it must go deeper.

The difficulty in investigating similarities and differences in the action of the members of a polygenic system lies in isolating them as individual genes in order to examine and compare their effects. Though in principle this would be possible for any such genes using Thoday's (1961) technique, in practice even those with relatively large effects prove difficult enough to isolate and follow, and Spicketts' genes appear to have been units built up in Thoday and Boam's (1961) selection experiments. In pointing this out, Thoday (1973) views these results as suggesting that when genetic materials of unspecific effects are brought together by recombination it may give rise to genes showing more specific action. The rise of non-allelic interaction as a result of selection could accord with this view, though it is capable of other interpretations. Such a process of bringing together genetic elements of less specific action to produce more complex aggregates of greater specificity in action could be closely involved in the evolution of genes as we have come to know them (Mather, 1949a, 1954) and it accords too with the information about genetic structures that molecular genetics has given to us.

There is other evidence that the genes of quantitative variation in general fail to match the major genes, normally recognized by Mendelian means, in the specificity of their action. Every major gene appears to be highly specific in that, apart from the special cases of polyploidy or other form of duplication, the part it plays in development can rarely be replaced by other genes. Or to put it another way, if the gene is not playing its part in development, the shortcoming cannot be made good by other genes of the nucleus. The phenotype may be restored, at any rate in large part, by such other genes as suppressors where the shortcoming is due to mutation, but actual loss of even very small parts of the chromosome known to carry major genic loci is commonly lethal when both homologues are so affected. There are, however, chromosomes supernumerary to the basic complement that have long been known to occur (see Darlington, 1937, 1950) in individuals of a wide range of species, both plant and animal. These B chromosomes, as they are called, are smaller than the A chromosomes of the basic complement, and they vary in number from one individual to another without any effects on the phenotype comparable with those of extra chromosomes or chromosome parts from the basic complement. Nor are they essential to the organisms in the way that A chromosomes are, for individuals may lack them completely without any striking effect. Indeed they have given no evidence of any major genic action. Nevertheless, they are not without their effects on plant size, vigour and fertility, and perhaps most interesting of all, on the behaviour in pairing and crossing-over of the A chromosomes of the basic complement (see Rees and Jones, 1977 for review). The magnitudes of these effects vary with the number of B chromosomes in the nucleus and are thus quantitative rather than qualitative in nature. In short they are in a sense the cytological embodiment of genes of quantitative variation, or of aggregates of such genes; and in their contrast with the A chromosomes they illustrate the contrast of polygenes and major genes.

Though not always so, B chromosomes are commonly heterochromatic. Heterochromatin is also to be found in A chromosomes: in *Drosophila melanogaster* the Y chromosome and the proximal region of the X are heterochromatic, as indeed are the proximal regions of the large autosomes too. Apart from the bobbed gene (bb) and the male fertility genes of the distal half of the Y, both the Y and the heterochromatic region of the X are inert in the sense that they carry no major genic loci, so of course resembling B chromosomes. Nor is the Y essential for development in males where it normally occurs (although it is a requirement for the production of motile sperm), and the number of Y chromosomes can vary in both sexes without any disturbance of normal development comparable with aneuploidy of even the small chromosome IV. At the same time, however, there are reports of it contributing to continuous variation in the number of sternopleural chaetae (Mather, 1944) and to cell size (Barigozzi, 1951). Thus this heterochromatin, like B chromosomes, displays quantitative genetic activity without showing those properties of indispensibility, or near indispensibility, and need for numerical balance if development is to be normal, which are characteristic of the euchromatin where major genic structures are to be found. The lesser specificity of the genes of quantitative variation is thus unlikely always to be due to difficulties of tracing differences in initial action, even though such differences may be demonstrable in some cases.

These comparisons of B with A chromosomes, and of hetero- with euchromatin, show us that genetic materials exist in the nucleus whose changes can contribute to polygenic but not to major genic activity, alongside other materials which characteristically can show changes from whose effects major genic changes are to be inferred. This does not mean, however, that polygenic activity is confined to heterochromatin, that euchromatin cannot also contribute by its changes to quantitative variation. It might do so in two ways: first, by changes producing phenotypic differences from which major genes would be inferred while at the same time contributing to quantitative variation through pleiotropic action; and secondly through the genic structures being capable of two kinds of change on separate occasions, one leading to the inference of a major gene, but the other contributing to quantitative variation.

Pleiotropy of major genic differences is, of course, commonplace and Penrose (1951) and Grüneberg (1952) have emphasized the contribution that major genic differences can make to quantitative variation in this way. Indeed it is not unlikely that most, if not all, genic differences of major effect have simultaneous smaller side effects which could contribute to quantitative variation. Even if this is so, however, it does not follow that all minor effects are associated with major differences, and the evidence of observation leaves no doubt that polygenic variation exists in the phenotype independently of major genic differences: quantitative variation of the kind which experience has shown typically to be under polygenic control commonly occurs among individuals and strains which are not distinguished by any detectable major genic differences, even in the most intensively studied plants and animals. And the argument that in such cases the major genic differences are there but have not yet been found because they affect unrecognized characters, can only be regarded as evasion, at any rate until such time as the cryptic major differences have been exposed in at least one case.

In recent time it has been suggested that quantitative variation might be related to the polymorphisms for isozyme differences which are so commonly found in natural populations, and as we have seen earlier the difference between the two alleles of the Adh gene in *Drosophila* does contribute to the quantitative variation in overall activity of the enzyme. On the face of it this would be another example of a difference having a major effect, in that it produced two clearly and regularly distinguishable alleles, while simultaneously contributing to polygenic variation in overall activity; and it would thus be open to the objection we have just discussed, that much of the polygenic variation we observe is not relatable to polymorphisms of this kind. The similarity between isozyme polymorphism and a major genic difference, such as that for phenylketonuria in man which Penrose (1951) discussed, does not, however extend beyond their both contributing simultaneously to quantitative variation: the main effect of phenylketonuria on mental ability is indeed major and it cripples the fitness of phenylketonuriacs, whereas the sharp segregation for the isozyme alleles is not accompanied by any obvious major biologically recognizable effect but stems solely from the high discriminatory power of the electrophoretic technique which makes its recognition possible.

The second way in which a major genic structure may also contribute to polygenic variation is by it being capable of two kinds of change, one producing effects from which a major gene is inferred and the other contributing to quantitative variation. That this type of relationship occurs is indicated by surveys of polygenic activity in the X chromosome of *Drosophila*. Such activity is not confined to the heterochromatic right end of the X, which is devoid of major genes apart from bb, but has also been traced to the euchromatic parts of the chromosome. Wigan's (1949) survey of polygenic activity in respect of sternopleural chaeta number shows the left end of the chromosome, euchromatic in nature and plentifully supplied with major genes, to be at least as active as the heterochromatic right; and Mather and Hanks' (1978) survey in respect of coxal chaeta number, to which we have earlier referred, shows the same to an even more striking degree. True, some small heterochromatic inserts have been claimed in the euchromatin of the left end, but these would seem to be wholly insufficient to account for the activity that has been observed. Spickett and Thoday's (1966) observations on the location of quantitative genes having relatively large effects, built up by selection for sternopleural chaeta number, confirms this association with euchromatic areas carrying major genes; and Breese and Mather (1957, 1960) found all parts of chromosome III, euchromatic and heterochromatic alike, to display polygenic activity in respect of abdominal chaeta number. It thus appears that a given genetic structure may be capable on different occasions of the two kinds of change, one leading the greater and more specific change of action which leads to the inference of a major gene, and the other to the small and less specific action leading to the inference of a member of a polygenic system.

A genic structure which can vary in these different ways must have a corresponding complexity of parts. A complexity of parts would also appear to be implied by Mather and Hanks' (1978) study of the numbers of chaetae borne on the coxa of the three pairs of legs of *Drosophila*. These numbers are subject to polygenic variation on all the three pairs, but the front pair of legs (F) has a greater mean number than the middle (M), which in turn has a greater number than the rear (R). The relations of the numbers on the three pairs of legs in flies of different genotypes show that this rise from R through M to F cannot be due to a uniformly proportionate enhancement of the effects of the genes concerned, but suggest that three classes of genes are involved; α genes which are equally active on all three pairs of legs; β genes which are inactive on R but active on both M and F; and γ genes active only on F. This interpretation allows a variety of predictions to be made about the responses of the chaeta numbers on the three pairs of legs to different types of selection, and Hanks and Mather (1978) have shown that the results of selection do indeed agree with these expectations. While other explanations cannot be finally excluded, the most attractive interpretation of the different behaviours of α, β and γ genes is in terms of a regulatory system involving *inter alia* control and structural elements in the genes, similar to the tripartite system postulated by Britten and Davidson (1969) and Davidson and Britten (1973), so implying yet another similarity between major genes, such as that at the rosy locus in *Drosophila* (Chovnick *et al.*, 1976), and the genes of quantitative variation. It is of interest to note too that Mukai and Cockerham (1977) also came to the conclusion, from their study of induced mutation in genes mediating continuous variation in the viability of *Drosophila*, that much of the induced variation must have arisen from mutation not of the structural genes, but of controlling elements. Indeed they speculated that the viability polygenes must in the main have been such elements external to the structural genes themselves.

The evidence for the implication of regulatory action and controlling elements in the determination of the number of coxal chaetae comes from comparisons between the three pairs of legs. We have no direct evidence of them being involved in the variation in number on the individual pairs of legs, R, M and F. At the same time, *prima facie* we might expect the same mechanisms and the same kinds of genic structures to be involved, albeit in a finer and more localized way, in the intra-leg variation as in the inter-leg differences; and Mukai and Cockerham's findings might be held to point in the direction of regulation and controlling elements playing a greater part than structural genes in the mediation of continuous variation. Much more hard information will, however, be needed before we can arrive at firm conclusions about the relative parts played by regulatory and structural genes in the determination of quantitative variation as opposed to major gene differences. Birley *et al.* (1981) have evidence that the level of ADH activity in *Drosophila melanogaster* is determined partly by the intrinsic differences between the enzymes to which the fast and slow alleles at the Adh locus in chromosome II give rise, partly by the action of modifier genes regulating the rate of enzyme synthesis, and partly by the action of still further genes, borne on chromosome III, affecting catalytic efficiency *per se* through post-translational modification. Thus information on the variety of ways in which genes act to mediate quantitative variation is beginning to come along; but even so, generalizations could be no more than speculations in our present state of knowledge.

The similarities and contrasts that we have noted between major genic structures and the genes of quantitative variation in their actions, relations and distributions along the chromosomes should not, however, blind us to the different parts that they must play in the control and adjustment of the phenotype, especially in relation to the action of selection. The drastic differences produced by the changes from which we infer major genes, which cannot be made good by other genes in the nucleus, must generally be disastrous to Darwinian fitness. These genes are of major importance to the functioning of the organism. They seem seldom to have been concerned in the differentiation of species (Mather, 1943, 1953a) and although their deleterious mutant alleles often occur hidden under the cloak of dominance in wild populations they are seldom found in homozygous condition outside the laboratory. The homozygotes which will occur by the mating of heterozygotes in the wild must, in fact, be weeded out by natural selection. Such genes are the backbone of the genotype and being essential like a backbone, the normal allele of each has an unconditional advantage over its mutant alternatives.

The polygene, on the other hand, is one member of a system whose parts are interchangeable, at least up to a point, in development. Though their individual effects are not large, the members of a system may act together to produce big differences, as for example between the extreme selection lines of Mather and Harrison's experiment. Equally they may act against each other to give similar phenotypes from different genotypes, as in Mather and Harrison's parental lines. No allele of a polygene has therefore an unconditional advantage in selection over the other: its advantage or disadvantage depends on the alleles present at the other loci of the system. Great variability can lie hidden sheltered from selection in the form of balanced combinations, the proportion of the variability so hidden rising as $(k-1)/k$ where k is the number of genes in the system (Mather, 1960a), and great genetical diversity and change can occur behind the façade of phenotypic uniformity.

These special properties of polygenic variation in selection arise from the relatively small and interchangeable effects of the members of the polygenic system. Now natural selection acts on the phenotype as a whole and so the total action of a gene difference must be relatively small and interchangeable if it is to function as part of a polygenic system. A gene difference which, while contributing to polygenic variation in one character, simultaneously has a major and more specific effect on another, cannot have the properties of a member of a polygenic system in respect of its total action, because it cannot be balanced, reinforced or replaced in the necessary way. It will therefore respond to selection as a major genic effect. Polygenic systems in fact provide the variation of fine adjustment, clothing as it were the indispensable genetic skelton and moulding the whole into the fine shape demanded by natural selection. They are the systems of smooth adaptive change and of speciation. They are also the genes with which the plant and animal breeder is generally concerned in his endeavour to produce improved forms.

2 *Characters*

6. PHENOTYPE AND GENOTYPE

In genetics, the term 'character' is applied to any property of an organism in regard to which similarities or differences, especially those of a heritable nature, are recognizable between individuals. A great variety of characters is known to show heritable variation and as our knowledge of the properties of living organisms grows, the list of their characters having heritable variations grows with it. Gene and chromosome behaviour, cell shape and structure, gross morphology, physiological and biochemical properties, psychological and behavioural characteristics, mating capacity and propensity, fertility, resistance to disease and toxic agents, ability to infect a host, ability to act as a vector of disease, antigen production: all are known to show heritable variation. Indeed the proposition can hardly be questioned that no character of any organism would fail to show heritable variation were it subjected to adequate examination. Furthermore, the magnitude of the heritable differences shown in a character may range from the smallest that is detectable to the largest that is possible.

Since genes are generally inferable only from the effects of their differences in changing the expression of the characters observable in an organism, all genetical study requires some consideration of the relation between genes and characters and our discussion of continuous variation has already led us to touch on those relations. Three principles have emerged which we must now examine in more detail.

The first of these, and one which we owe to Johannsen, is the principle that the phenotype is the joint product of genotype and environment. Variation in a character may, therefore, result from variation in either genotype or environment and, as Johannsen showed, the two kinds of variation in the character, heritable and non-heritable, cannot be distinguished by mere inspection. A plant or animal may be small because of insufficient feeding or because of its ancestry, and only the breeding test can distinguish the one situation from the other. The production of certain specific substances such as the antigens of higher animals may constitute an exception to this principle; but it is worth

37

noting that the antigens of *Paramoecium* vary with temperature as well as genotype (Beale, 1954).

The second principle is that heritable variation in a character may be brought about by alteration in any of a number of genes. The genes causing differences in the same character may be related in various ways. They may, of course, all be members of a polygenic system, as in the case of the genes affecting the number of abdominal chaetae in the selection lines of *Drosophila*. They may, on the other hand, all be major gene changes whose effects on the character are neither small nor supplementary in a non-specific way. The alterations produced in the character by such major genic changes may be distinguishable by inspection as is the case with many genes affecting, for example, flower colour in plants or eye colour in *Drosophila*; or their effects may appear alike. Many chlorophyll-deficient barley plants, for example, look alike though traceable to mutation in different genes, and the same is true of 'minute' bristles which can be traced to any of 70 known gene changes in *Drosophila*. As the study of biosynthetic processes in micro-organisms has shown us, numbers of genes may be acting, and their actions may be related in a variety of ways, in the production even of such basic materials as amino-acids, and these genes may be varying individually or simultaneously. Finally, variation may at the same time be due partly to major gene differences and partly to a polygenic system. So not only is a breeding test necessary to show how far the difference between two individuals or strains is heritable, it is also necessary to reveal the nature and relations of the gene changes determining any heritable differences that may be found.

Furthermore, just as two individuals or strains may owe their differences to any of a number of genes, two which are alike may owe their similar phenotype to different genes. In cotton, for example, *Gossypium hirsutum* and *G. barbadense* both have forms with large coloured spots on the petals. They look alike, but breeding tests have shown that the genetical architecture of the character is different in the two cases (Harland, 1936). Similar differences may occur within species. Short tail on the hindwings of the butterfly *Papilio dardanus*, for example, is due in some races to a single gene, but in the Abyssinian and Madagascan types it is due to polygenic action (Clarke and Sheppard, 1960). That differences in the genetic architecture of similar characters has been important in evolution as Harland foresaw, is attested by the variety of ways in which the incompatibility reaction is determined in plants (Lewis, 1954) and even more so by the many mechanisms, cytological and genetical, on which sex-determination rests, especially in animals (White, 1954).

The genetical basis of any one of a range of phenotypes can be discovered only by breeding. Sometimes a single test will clarify the situation once and for all, because where a major genic difference is demonstrated, the genotype can subsequently be inferred from the phenotype within the range for which the difference has been established. This, however, is never possible where the variation is mediated by a polygenic system, because of the interchangeability in effect of the member genes of the system both with one another and with non-heritable agencies.

The third of our principles relating to gene and character is that just as one character can be influenced by more than one gene, one gene can influence more than one character. It has long been recognized that genes can have manifold effects in *Drosophila* (see Dobzhansky, 1927, 1930) and indeed pleiotropic action is quite common among genes. Sometimes the connection between the various effects appears fairly obvious. Thus the gene H in *Drosophila* not only removes certain of the major bristles of the head and thorax but also reduces the number of abdominal and sternopleural chaetae (Cocks, 1954). That the situation is not, however, as simple as it appears is shown by Cocks' further observation that the gene sc which removes major bristles from the scutellum and simultaneously reduces the number of abdominal chaetae even more than does H, has very little effect on the number of sternopleural chaetae, while Sp whose characteristic effect is nearly to double the number of sternopleurals, hardly affects the number of abdominal chaetae. Sometimes too the characters associated in the effect of a gene are surprising. A series of four alleles affect the size of the coloured 'eye' round the mouth of the corolla tube in *Primula sinensis*. So far as is known this is the only effect of three of them; but the fourth also shortens the style in such a way that flowers otherwise of the pin type become homostyled. This is all the more remarkable because the effect of this fourth allele on the eye is indistinguishable from that of one of the other three, which itself causes no shortening of the style.

The common developmental origin of a number of apparently unrelated changes caused by one gene is sometimes made clear by embryological studies. A recessive gene is known in the rat which kills the animals soon after birth. The affected animals show a wide range of peculiarities, which are especially marked in the circulatory and respiratory systems, but also appear in the form of the snout, the occlusion of incisor teeth, and the ability to suckle. When traced back, however, they all arise from an initial breakdown of the cartilage (Grüneberg, 1938). The complexity of the gene's pleiotropic effects is thus traceable to a single initial action, which changes the general course of development and so leads to the gene change expressing itself in a syndrome of varied abnormalities. This is likely to be true of many, if not all, of the varied patterns of effects which are consequent on single gene changes in rats and mice (Grüneberg, 1963) and indeed in any organism with a complexity of development and differentiation.

The longer the chain of events between the first action of the gene and its final expression in the phenotype, the greater the complexity of effects which can arise from a simple alteration. Conversely where a character, such as the production of a particular antigen, is a more immediate expression of gene action, a simpler correspondence would be expected and is in fact observed between gene change and character change. This simple correspondence is seen clearly in the variation of human haemoglobins. The difference between, for example, normal and sickle-cell haemoglobins is due to a single gene change which results in glutamic acid being replaced by valine in a specific place in the globin part of the molecule, and other abnormal haemoglobins show similar correspondences (Jukes, 1966). At this level the relation between gene and character is simple and precise, though we might observe in passing that the relation between the triad of nucleotides in nucleic acid and the amino-acid for which it codes is not unique in all cases (Crick, 1967). Yet when we look at the anaemics who owe their condition to being homozygous for the sickling gene we find them showing a syndrome of abnormalities, all reasonably attributable, however, to the consequences of the simple substitution of the amino-acids which is the primary effect of the gene change. The complexity of relations between gene and character is thus attributable to the multiplicity of stages between initial action and final expression. This multiplicity not only permits one gene change to produce a number of apparently different effects, but also ensures that, because of the interplay at these intervening stages with the products of other genes' action, the consequences of the gene's primary actions, and with them the final character itself, will be subject to modifications by other genes as well as by non-heritable agencies. The complexity will be all the greater if a single gene can, in fact, have more than one primary action and if, as not only polygenic systems but also the study of isozymes (Harris, 1969) suggests, the same effective primary action can be shared by a number of genes. We are coming to understand increasingly the way genes exert their control of protein synthesis from observations on specific molecular substances in micro-organisms. As Lwoff (1962) points out, however, there is a difference between the cells of the multicellular organisms and the microbial cell, from which most of our recent information has stemmed. It is a long step from these primary, or near primary, actions of the genes in a single cell to the complexity of their ultimate expression in a multicellular soma, where there is interplay between the differentiated cells themselves as well as between materials within each of them.

It is therefore clear that while a given phenotype, taken as a whole, can be related to a given genotype, acting as a whole, no similar correspondence can be assumed between the parts of the phenotype and the parts of the genotype: the genes of a nucleus must be related to one another in action and the characters must be related to one another in development.

This complexity of relations between genes and characters has consequences outside the scope of our present discussion. It has also, however, an immediate importance for us because it imposes certain restrictions on the lines to be followed in making the genetical analysis with which we are concerned.

7. GENETIC ANALYSIS AND SOMATIC ANALYSIS

The object of much genetical experimentation, especially in the earlier decades, has been to throw light on the organization and transmission of the genotype. In such experiments the relations between the genes which are used and their phenotypic expressions are of secondary interest. They are of importance only in so far as they limit recognition of the genotype, with which the experimenter is primarily concerned, through the changes they produce in the phenotype. An unfortunate choice of marker genes, whose effects are indistinguishable or overriding, results in the confusion of the phenotypes associated with certain of the genotypes. Genotypic classification is thereby made less adequate and information is lost. The most useful choice of genes is clearly that which, in an appropriately arranged experiment, leads to each of the possible genotypes giving rise to a unique and distinctive phenotype; all the genotypes can then be identified with confidence and analysis is complete. In this type of experiment, therefore, the nature of the changes brought about by the genes in the phenotype is of purely technical interest. Emphasis is on the analysis of the genotype rather than the adjustment of the character.

The emphasis is obviously different in studies, of growing importance in recent years, aimed towards the elucidation of the manner in which genes and their parts act and interact in controlling the biochemical processes and products of the cell. Yet here again genes, specially chosen, though by different criteria, must be utilized. No longer are genes taken chiefly because of their strategic locations along the chromosomes and because they show as little as possible obscuring interaction with one another. Rather they must be chosen for the appropriateness of their effects to the biochemical study in hand and for the ease and precision of recognition of their specific consequences in the associations and combinations with which the experimenter is concerned. Emphasis is now on the analysis of biochemical relations and consequences, but discrimination in the genetic differences taken for study is as essential as in the older type of study, albeit discrimination of a different kind.

In the study of the interplay of variation and natural selection in determining the structure and evolution of populations, and in the practical utilization of genetical knowledge for the improvement of domestic animals and plants, interest must centre on the phenotype whose origin it is desired to understand and whose manipulation it is desired to achieve. All the variation must be taken into account and the genetic architecture must be viewed as a whole. All genetic differences are prospective raw material for the action of natural selection and all genes are affected by its impact according to the nature of their effects on the phenotype. In the same way the applied geneticist is basically concerned with improving yield, quality, disease resistance or some other feature of his plants and animals and he seeks knowledge of the genetical architecture of the character with that end in view. He has but limited choice in the genes with which he will work; he is concerned with all that contribute to the variation in the character, whether they are of major effect or members of a polygenic system and whether they have pleiotropic effects or not. He therefore requires analytical methods capable of handling all types of genetical variation.

In so far as there are discontinuities in the variation of the character the breeder can use the methods of Mendelian genetics. The inheritance and interactions of any major genes distinguishable in this way can be ascertained and by appropriate breeding any desired type can be produced and recognized at will. The success of this approach derives from the identification and isolation of the genes involved so that they may be put together in any combination that is desired. It may be that a given phenotypic difference can be produced by more than one gene; but as Nilsson-Ehle showed us with his cereals, we can still isolate the genes one from another and put them together in the way we wish, provided they can be identified as units through their production of discontinuities in the variation of the character.

The application of Mendelian analysis can be extended by special methods, notably by progeny tests and by the use of inbred lines to permit the examination of particular gene differences on a uniform genetic background. In this way the obscuring effects of other gene segregations can be removed and genes recognized by their production of discontinuities in variation that would otherwise have been obscured by the mass of segregation. This method is powerful, but it is also laborious and expensive of time. Furthermore, even were we prepared to adopt these measures for the analysis of a polygenic system in which we were interested, the synthesis of any desired type by crossing these lines would bring back all the difficulties which arise from the obscuring effects of the genes on one another's segregation. A second and perhaps more important limitation is imposed by the obscuring effect of non-heritable variation on the segregation of polygenes. This is not, of course, eliminated by the use of inbred lines. In fact the inbreeding depression seen in organisms which normally cross-breed is commonly accompanied by an increase in non-heritable variation (Mather, 1946c, 1953b; Lerner 1954). Where the contribution of a particular locus, or closely placed group of loci, to the polygenic variation turns out to be especially large, perhaps as a result of earlier selection, the locus or group may be located by appropriate testing with a suitable tester stock (Thoday, 1961), or the genes may be located by the use of special chromosomes which have lost material or undergone structural change (Law, 1967). The use of such methods is, however, obviously dependent on the availability of the special tester or the special chromosomes and hence will not be generally possible.

While, therefore, we may conceive of all genes as capable of isolation and handling by Mendelian means, given an ideal uniformity of genetical background and elimination of non-heritable variation, we must in practice accept a situation falling short of this ideal. We can reduce non-heritable variation but we cannot at present wholly eliminate it; and indeed we have no certainty that it could ever be wholly eliminated even under the most rigorously controlled conditions, if only because a portion of it appears to spring from chance internal upsets of development rather than from differences in the environment external to the organism (Mather, 1953b). Equally, the most inbred stock still has a residuum of variation from mutation (Durrant and Mather, 1954) though it must be said that the increment added in each generation to continuous variation by polygenic mutation would appear to be small in *Drosophila* (Clayton and Robertson, 1955; Paxman, 1957 and see Mukai, 1979). Finally, even could we achieve the ideal state in which genes of small and similar effect could be separated and handled by Mendelian means, the cost in labour and facilities would make its general application to polygenic systems prohibitively uneconomic for practical purposes. We must then be prepared to accept the situation in which genes, though contributing as units to the variation, are not individually identifiable by their effects on the phenotype and therefore are not separable in analysis. The genetical analysis is thus limited by the complexity of relations between gene and character.

From time to time attempts have been made to simplify the genetical analysis by a prior somatic analysis of the character. Many characters such as yield in crop plant can be regarded as made up of a number of sub-characters. If each of these were under separate genetical control, the inheritance of its variation would be simpler than that of the whole character. A preliminary somatic analysis would then make the genetical analysis easier.

Consideration of the relations between genes and characters has shown us, however, that there is no ground for expecting the sub-characters we can recognize in the phenotype to be under completely separate genetical control. Rather we must expect that some genes will affect both of any pair of sub-characters while other genes affect only one of the pair, as has in fact been found to be the case in the control of the numbers of chaetae borne on the coxae of the three pairs of legs in *Drosophila melanogaster* (Mather and Hanks, 1978; Hanks and Mather, 1978). The variation of the two will then be correlated, but not completely so, and the degree of correlation cannot be predicted. It will depend on the sub-characters in question and on the genes which are contributing to the variation. The causes of the correlation may be several. In the production of inducible enzymes in micro-organisms it would appear that the action of structural genes mediating the production of different enzymes may be switched on and off simultaneously by the activity of a single operator gene, which is itself subject to repression by substances deriving partly from a still further repressor gene (Jacob and Monod, 1961). The system involved is thus one of some complexity. While it is known only from the study of enzyme production in micro-organisms there is no reason for regarding this form of control of gene action as being restricted to such cases, and indeed it has been used as the basis for a plausible, though as yet unproven, theory of the way in which genes act and interact in the determination of differentiation (Britten and Davidson, 1969; Davidson and

Britten, 1973). Such relationships afford every reason for expecting correlations to exist between different activities in the cell, even where these are directly mediated by different genes. Nor need we expect the correlations of activity so produced always to be complete.

Partial correlation of characters or sub-characters can also arise whenever two biosynthetic chains utilize a common precursor. Biochemical studies have shown that this must be a common situation, which we can illustrate by reference to one of the early objects of such study, the genetical control of flower colour in higher plants (see Beadle, 1945; Harborne, 1967). Flower colour depends on the presence of anthocyanins and anthoxanthins, among other pigments. Genes are known which affect the type and intensity of anthocyanin pigmentation while having relatively little effect on the anthoxanthins. Others in turn change the anthoxanthin pigmentation while leaving the anthocyanins relatively unaffected. The two kinds of pigment appear, however, to have a common precursor substance. Thus a negative correlation between variation in the two kinds of pigmentation can, and indeed does, arise from competition for this precursor where it is in restricted supply. Equally, a positive correlation may be produced by changes in the action of genes whose chief effect is on the supply of precursor. The degree of correlation observable in any group of individuals will depend therefore on the developmental relations between the two types of pigmentation and on the genes which happen to be contributing to the variation. Furthermore, where the variation is due to immediate segregation, the linkage relations of the genes must also have their effect on the correlation. Clearly not only must correlations be expected between sub-characters, they must also be expected to change with the genetical constitution of the individuals in which they are being observed.

In the case of variation determined by major gene differences the correlation between the sub-characters, here anthocyanic and anthoxanthic pigmentation, need cause us no concern. We can isolate the genes, determine the effects of their individual differences and also discover their interactions of effect. The partition of the character even helps us with the genetic analysis to the extent that it enables us to distinguish the various genes through their different effects on the sub-characters which we can establish by observation to be characteristic of each. The analysis is, however, primarily into units of inheritance. The somatic analysis is built on to the genetic analysis and must be justified at every stage by reference to it. Pleiotropic action and linkage can be detected, and the subsequent use of somatic analysis can be confined to effecting the distinction between those genes which it has shown itself in practice as capable of distinguishing.

The situation is quite different when the genes themselves cannot be isolated in experiment. Somatic analysis cannot then be founded on prior genetic analysis and in consequence its limitations cannot be gauged. Though the sub-characters may each have a simpler genetical basis than the whole character, we know neither how many genes have been separated by the analysis into subcharacters nor how far any failure to achieve separation depends on linkage as opposed to pleiotropic action. Furthermore, each subcharacter will still be showing the effects in its variation of an unknown number of individually unrecognizable genes. Such a somatic analysis does not simplify the genetical problem: after it has been made we still have to deal with a system of genes affecting each subcharacter and we do not know either how or how far these systems are interlocked.

It may be argued that genes which affect the same character, but do so by altering different subcharacters, cannot be regarded as members of the same polygenic system, because the similarity of their effects is only superficial. However justifiable this may be theoretically, we have no practical means of distinguishing them from genes which affect both subcharacters simultaneously, and it is with the practical problems of genetical analysis that we are concerned. Thus where we are dealing with continuous variation, due to genes which we cannot expect to be readily recognizable as individuals in segregation, we equally cannot expect automatically to overcome the intrinsic difficulty of the situation by attempting a prior somatic analysis. We must in any case use methods capable of dealing with systems of genes as wholes, and we may as well do so first as last. Somatic analysis is of use for genetical purposes only where experiment and observation have shown its application to be justifiable and helpful.

8. SUBCHARACTERS AND SUPERCHARACTERS

Although the somatic analysis of a character showing continuous variation into its subcharacters cannot, in general, be expected to simplify the genetical analysis of the character and its variation, it may nevertheless have its uses in predicting the average breeding behaviour in respect of the character. Yield in wheat plants may, for example, be regarded as built up from the subcharacters, number of ears, number of grains per ear and average weight per grain. The genetical variation of these sub-characters has been shown by Smith (1936) to be correlated in Australian varieties. The somatic analysis, therefore, does not simplify the methods which must be used in genetical analysis of the polygenic system mediating variation in yield. It does, however, aid us in partially disentangling the heritable and non-heritable variation. Plants may have the same yield but have it for different reasons: excellence in one subcharacter may be balanced by poor performance in another. In so far as these balancing effects are genetic the somatic analysis does not of itself help us to predict breeding behaviour in respect of yield as such (though, of course, it may help us to breed for, say, ear number as a desirable character in its own right, rather than as a mere subcharacter of yield, in which latter capacity its improvement is no more a desirable means of increasing yield than is the improvement of any other subcharacter). But where the genetical quality is being obscured by the effects of variation of non-heritable agencies or of the background genotype, somatic analysis may help us, because one or more of the subcharacters may show relatively less variation than the character does as a whole. In his wheat, Smith found that this was indeed the case. In particular, knowledge of the subcharacter, average weight of the grain, was helpful in lessening the error caused in assessment of the genotype from the phenotype by the intervention of non-heritable agencies. Prediction of breeding behaviour in respect of yield was therefore capable of improvement by the use of somatic analysis into subcharacters, information about which could be used to supplement that about yield as a whole.

Such a value of somatic analysis for the improvement of prediction about average breeding behaviour cannot, of course, be assumed without evidence, any more than can its value in aiding genetical analysis. It requires experimental justification every time. Smith supplied this by his observation that average weight of grain showed relatively less non-heritable variation than did yield as a whole. If somatic analysis is to have a genetical use its application must be based on adequate genetical evidence.

In Smith's wheat, its lesser, non-heritable component of variation relative to the other subcharacters put average weight per grain in a special place because effectively this implied that it was capable of more precise genetical assessment and more effective manipulation by the breeder. Subcharacters may command special attention for other reasons too. They may pose different problems and offer different possibilities, as Lerner's (1950) analysis of egg production in poultry shows well, and if one of them is particularly weak in its expression or limiting in its effect on the overall character, it will obviously demand special attention from the breeder. It becomes virtually a character in its own right, though in so far as it is negatively correlated in its expression with other subcharacters, to treat it in complete isolation would be self-defeating. Thus in sugar beet, where one main object of the breeder must be to raise the yield of sugar, it would be easy to raise the yield of root. But as this shows a strong negative correlation with sugar percentage in the root, to do so by indiscriminate selection would have as one of its first results the lowering of sugar percentage. The gain in weight of root would thus be offset, at any rate in large measure, by loss of sugar percentage and the breeder could end up in the position of having produced a strain which gave virtually the same yield of sugar but had to produce more root to do it.

A further use of somatic analysis has been suggested by Grafius (1961). He distinguishes between characters whose component subcharacters are combined multiplicatively to give the character (characters which, to use his term, 'have a geometry') and characters whose components do not combine in this way. The former, he considers, are likely to show greater effects of genic interaction in their determination and in particular may show a high degree of heterosis as a result of simple, or even incomplete, dominance. Thus the 'genetic' relation of the subcharacters would offer a guide to the prospects of successfully utilizing hybrid vigour in a breeding programme. The significance of multiplicative action is a matter to which we must return when we come to consider the choice of a scale on which to represent the degree of expression of a character, but in the meantime we may note that to be of value such a generalization must be firmly grounded in experience. Again, as in other connections, the value of somatic analysis must be justified by observation, not merely assumed.

Just as a character may be capable of resolution into a number of subcharacters, it may itself be only one of a number of characters in whose joint properties we are interested. Thus yield is only one of a number of features on which the general merit of a wheat plant depends. Baking quality, straw properties and resistance to unfavourable conditions and to various diseases are other characters which the breeder cannot ignore. These, together with yield, may be taken as components of a supercharacter which is itself the overall merit of the plant.

In the same way that two subcharacters may not be independent in their genetic control, whether for physiological or mechanical reasons, so we may not be in a position to treat any two characters themselves as independent. And just as the division into subcharacters does not simplify the problem posed by continuous variation in such a way as to permit the use of simpler genetical methods, so the aggregation of characters into a supercharacter need not necessitate the use of more complex genetical methods in its analysis. However, the two relations, of subcharacter to character and of character to supercharacter, differ in one respect. The former is a fixed relation since the character is itself definable and measurable; but the latter is not fixed because the supercharacter is not generally capable of final definition and measurement: the method of combining the characters is not clear. Thus the yield in grammes of a wheat plant, for example, is obviously the product of the number of ears, the average number of grains per ear and the average weight in grammes per grain. But how should yield, baking quality and so on be combined to arrive at a measure of the supercharacter, overall merit?

It is obvious that the relation of a supercharacter to its component characters cannot be delimited or even investigated until the supercharacter itself is defined. Such a definition must spring from the purpose of the experiment or analysis. Thus if we are concerned with the changes in the genetical constitution of a natural population of plants or animals, the constitution of the population can be understood ultimately only in terms of the Darwinian fitness of its constituent individuals, i.e. of their prospects relative to one another of leaving posterity; and change in any character that we choose to consider can be understood only in terms of the relationship this character bears to fitness, which for our present purpose may be regarded as a supercharacter to which, together with others, our character contributes. Yet such a definition does not get us very far unless it enables us to assess fitness sufficiently well for us to establish its relation to the character over the range of material and circumstances in which we are interested, so that once established, the implications of the character for fitness and of fitness for the character can be seen or predicted in any given instance.

Two difficulties then arise. The first is that of assessing fitness in such a way as to allow its relation to the character to be ascertained. The second is that this relation may change as the genetical constitution of the population changes. So formidable are these difficulties that the attempt to arrive at the relationship is seldom made in this way. Instead we generally try to assess the effect on fitness from consideration of the changes in the character itself, rather than seek to measure their relationship directly in the first place, or even to arrive at it by *a priori* argument rather than by empirical observation. In principle, the position is, of course, easier in artificial selection experiments where the character for which selection is practised is, by the nature of the case, the dominant component of fitness; though even here we know from such experimental experience as was discussed in Section 4 that other components, uncontrolled and often subsumed under the general heading of natural selection, must be taken into account if we are to see the implications of our observations. Thus fitness may be definable in principle as the capital element in our consideration of evolutionary and selective changes, but it is extremely difficult to measure and often far from easy to define in usable terms. Its use as a supercharacter is therefore greatly restricted, and not least by the difficulty of listing its components in such a way that its relations to them all can be usefully investigated.

The applied geneticist meets similar difficulties. Returning to the wheat breeder's problem of combining yield, baking quality, disease resistance and so on into a single supercharacter: if some measure of overall merit were available the problem could be solved by treating this as a single character with yield, etc. as subcharacters. Then for any range of material the multiple regression of overall merit on its subcharacters could be found and used as a means of predicting or estimating merit from measurements of its components. The problem then becomes that of finding a measure of overall merit. It would be possible in this case to use the market value of the produce from a variety. The objections to such a course – artificial intervention in controlling prices, fluctuations in price even when taken relative to a standard and the undesirability of confining attention to the features which preponderantly determine market prices at any given moment – are too obvious to require stressing. Nevertheless this means of evaluating the emphasis placed by commerce on the various qualities of a biological product may well have its uses, albeit limited ones. One of the main limitations is likely to be its use over time, for the advent of a new race of a pest, whether by introduction or genetic change, could completely upset the value of some hitherto desirable quality of disease resistance, just as the advent of a new technique of cultivation or the introduction of a new chemical treatment or additive for the product could change the emphasis on field or quality characters. Any definition of merit must be subject to change with the biological situation and the technological requirements.

In some circumstances an alternative approach is open to the definition of the supercharacter where its aim can be seen to be essentially discriminatory. The characters can then be combined into a single measure by use of Fisher's discriminant functions. These bear a formal resemblance to multiple regression equations, although their aim is a different one. They may be illustrated by reference to Smith's consideration of yield in wheat, mentioned earlier. The phenotypic relations of the various subcharacters of yield to yield itself are clear and fixed. Since, however, yield and its subcharacters are subject to non-heritable variation, the relations of genotypic value in respect of yield to the various phenotypic measurements are not so obvious. The problem is that of how best to predict the genotypic value in respect of yield (as distinct from yield itself) from the phenotypic observations of the subcharacters; or in other words how best to combine the subcharacters so as to discriminate among genotypes. This Smith approached by finding a function of the phenotypic measurements, number of ears, mean number of grains per ear and mean weight per grain, selection for which would give the maximum advance in yield. In this way coefficients could be calculated for the combination of the subcharacters with one another in the way most useful for the purpose in hand. The function so obtained as the best for discriminating among genotypes is a discriminant function.

Discriminant functions are functions, usually linear, of the various measurements which are available and which may themselves be correlated with one another. The coefficients are chosen so as to maximize the differences between two or more classes of objects or individuals relative to the variation within the classes. They thus afford the best available means of discriminating between the classes. The theory of these functions was first described by Fisher (1936, 1938), and a great literature has grown up about them (see Kendall, 1957; Rao, 1952). They have nevertheless been put to little use in genetics. In addition to Smith's use, just described, a discriminant function was calculated by Mather and Philip (see Mather, 1949a) for defining ear-conformation in barley. The characters measured were overall length of ear, maximum width of ear and internode length between the spikelets as represented by the combined length of the central six internodes. The material was an F_2 between the two varieties Spratt and Goldthorpe and two ears were measured from each of 170 plants of the F_2. The variation within plants is obviously non-heritable, apart from possible somatic mutation, and the linear function of the three measurements calculated as maximizing variation between plants relative to variation within plants should go a long way towards giving a measure of ear-conformation whose genetical variation in this material is at a maximum compared with at least one important kind of non-heritable variation. Ear-conformation is not a character which can be measured directly. In fact before it could be measured it had to be defined as that compound of ear length, ear width and internode length which maximized the inter-plant variation, and which contains a large genetic element, with respect to intra-plant variation, which is almost wholly non-genetic. Once the supercharacter was so defined the three characters could be combined to give it in a way specially suited to the genetical needs of the experiment.

The method is open to use whenever it is possible to represent our needs in the form of maximization of differences between recognizable classes: in Mather and Philip's barley, the F_2 plants. It may therefore be applied to the definition of characters which are either undefinable or not capable of measurement by other means. Thus flavour in fruit is generally incapable of measurement. We can, however, measure such characters as pH and sugar concentration. If we can classify a range of fruits into good and poor, it is possible to calculate the discriminant function of pH, sugar concentration and such other measurable characters as are available, which maximizes the difference between the flavour classes. In this way a usable definition can be given to the flavour classes. Once again, of course, the definition cannot be a final one, since some new measurable property of the fruit may be observed which supplements or even partially replaces one or more properties of which our earlier discriminant has been compounded.

There are thus various ways in which the definition of a supercharacter may be attempted. Once usable criteria have been set and the difficulties of defining the supercharacter satisfactorily overcome, it may be handled in genetical analysis in just the same way as are characters, or subcharacters, themselves capable of direct measurement. Most applied geneticists must have super-characters, albeit empirical ones, in mind in their breeding work, for improvement in one character loses much of its value if it is accompanied by regression in others. Nor do the precise details of the way in which the characters are weighted and combined count for so much if the super-character is broadly correct in conception. The validity of the supercharacter, whether conceived intuitively or established statistically, will depend on the breeder's judgement not only of present but also of future needs. Although, as in Smith's wheat, the supercharacter may be defined with genetical purposes in mind, genetics cannot in general set the breeder's aims: the task of genetical analysis is to help him in achieving them.

3 Sources of variation: scales

9. GENIC VARIATION

Continuous variation is, as Johannsen showed, caused by partly heritable and partly non-heritable factors. By far the greater part of the heritable variation can be traced to differences in nuclear genes. These genic differences may give rise to variation of three kinds. The main and basic type of variation is that reflecting the effects of substituting one allele for another at any of the loci in the genic system mediating the heritable variation of the character. In a diploid organism with two alleles at a locus, three genotypes can be distinguished in respect of that locus. Denoting the alleles as A and a, where A indicates the allele making for greater manifestation of the character and a the allele making for lesser manifestation, the three genotypes are aa, Aa and AA. Two substitutions of A for a are required to convert aa into AA. If the effects on the manifestation of the character of these two substitutions are equal, Aa being midway between the two homozygotes in its phenotype, and if the effect of substituting A for a is the same no matter which alleles are present at the other loci in a polygenic system, the variation associated with the A–a gene pair is entirely of this main or basic type and is completely described by the magnitude of effect of the substitution. Such variation may be said to be fixable in that its effects can be fixed as the difference between two true breeding lines.

Not all genic variation is, however, of this kind. If it were, the hybrid between any pair of true breeding lines would always lie exactly midway between its parents in the manifestation of the character. That this is not so is attested by the many examples of heterosis to be observed in outbreeding species, where in respect of a wide range of characters the F_1 between two inbred, and true breeding, lines displays the character to a degree exceeding that of either parent. There must, therefore, be interactions between the genes in the determination of their effects.

This interaction may be of two kinds. It may be between alleles or it may be between non-allelic genes. In the former case the effect on the character of substituting A for a in one chromosome depends on the allele, A or a, present in the other chromosome. The difference in phenotype between aa and Aa is then not the same as that between Aa and AA. Aa is not midway between aa and AA

51

and the allele characterizing the homozygote to which Aa approaches more closely is dominant. It should be observed that since A is defined as the allele making for greater manifestation and a as that making for lesser manifestation of the character, either A or a may be dominant, according to whether Aa approaches more closely to AA or aa in phenotype: the use of the capital A does not here indicate the dominant allele as in the conventional usage of Mendelian genetics.

Dominance is said to be present if Aa falls other than midway between aa and AA in phenotype expression. This implies a scale on which the expression of the character is properly to be measured, and the choice of scale can determine not merely the degree of dominance, but even its occurrence or its direction. Let the expression of the character associated with the genotypes aa, Aa and AA be 1, 4 and 9 as measured on some scale. The mid-point between 1 and 9, the expressions associated with aa and AA, is 5. The expression of Aa falls short of this and a is judged to show dominance over its allele A. But suppose that, for some reason, the logarithm of this scale was regarded as more appropriate for the representation of the character. Then the measurements associated with aa, Aa and AA would be $\log 1 = 0 \cdot 000$, $\log 4 = 0 \cdot 602$ and $\log 9 = 0 \cdot 954$ respectively. The point midway between $0 \cdot 000$ and $0 \cdot 954$ is $0 \cdot 477$ which is exceeded by $0 \cdot 602$, the manifestation of Aa. A must now be judged to show dominance over its allele a: the change of scale has reversed the direction of dominance. Finally, suppose that neither the direct nor the logarithmic scale is acceptable but that a square root scale is taken as the most appropriate for representing the manifestation of the character. Then the measurements associated with aa, Aa and AA become $\sqrt{1} = 1$, $\sqrt{4} = 2$ and $\sqrt{9} = 3$. The phenotype associated with Aa is now midway between those of aa and AA, and dominance must be judged to be absent. Thus dominance as we have defined it, is not an unconditional property of the genes, in the way that it so often appears in Mendelian genetics: its occurrence, direction and degree may depend on the scale used in representing the degree of expression of the character.

The second kind of interaction is that between non-allelic genes. The effect of substituting A for a then depends on the alleles present at other loci in the system. Such non-allelic interactions between the genes at two loci may themselves be of three kinds. In the first the difference in effect between aa and AA depends on whether the constitution at the second locus is bb or BB. The variation due to an interaction of this kind is fixable, like the main effect of the gene substitution itself, and it can contribute to the differences between true breeding lines. The second is where the dominance shown by the gene pair A–a differs according to the constitution, bb or BB, at a second locus; or equally, of course, whether the dominance of B–b depends on whether the constitution is aa or AA at the first locus. Finally the manifestation of the heterozygote Aa may depend on whether there is heterozygosity, Bb, at the second locus. Neither of the second or third types of non-allelic interactions is fixable but they may play their part in the determination of heterosis. Where genes at three or more loci are involved a correspondingly greater range of types of interaction is possible, but the increased complexity is entirely in the homozygous/heterozygous group of interactions.

Non-allelic interactions in continuous variation are often referred to as a group under the general designation of epistasis. This usage of the term epistasis should not be confused with the earlier usage in Mendelian genetics where it was confined to the relationship between genes which gave either a $9:3:4$ or $12:3:1$ ratio in F_2 (see Bateson 1909). The classical types of epistasis as well as the other classical relationships, complementary and duplicate interactions, are all capable of description in terms of the type of interactions that we have been describing (see Section 16).

Like dominance, non-allelic interaction may depend for its magnitude and even for its direction and occurrence, on the scale used in representing the manifestation of the character. The situation shown in Table 7 is an extension of that already used to illustrate the dependence of dominance properties on scale. In the direct scale (top figures in the table) not only is the effect of substituting A for a less in the presence of a than in that of A, it is also less in the presence of bb than of Bb, and greatest of all in BB. The relations of A–a and B–b are reciprocal; there is non-allelic interaction in the same direction as the dominance it accompanies. In taking logarithms (middle figures in the table) the dominance is reversed and so is the interaction, for the effect of substituting A for a is now greatest with bb, intermediate with Bb and least with BB. Using a square root scale (bottom figures in the Table) both dominance and non-allelic interactions are absent.

Nor need the significance of the scale be confined to determining the degree, direction and occurrence of dominance on the one hand and non-allelic interaction on the other. In the situation shown in Table 8, neither A–a nor B–b

TABLE 7
The effect on non-allelic interaction of change in scale

	aa	Aa	AA
bb	1	4	9
	0·000	0.602	0.954
	1	2	3
Bb	4	9	16
	0·602	0·954	1·204
	2	3	4
BB	9	16	25
	0·954	1·204	1.398
	3	4	5

In each cell the top figure is the manifestation on the direct scale, the middle figure is the manifestation on the log scale and the bottom figure is the manifestation on the square root scale.

TABLE 8
Interchange of dominance and non-allelic interaction by change in scale

	aa	Aa	AA
bb	1 0·000	2 0·301	3 0·477
Bb	2 0·301	4 0·602	6 0·778
BB	3 0·477	6 0.778	9 0.954

In each cell the upper figure is the minifestation on the direct scale and the lower figure is the manifestation on the log scale

shows dominance on the direct scale (upper figures in the table): at each of their loci the heterozygote is exactly intermediate between the two homozygotes on this scale. There is, however, non-allelic interaction, the effect of substituting A for a being least with bb, intermediate with Bb and greatest with BB, and of course reciprocally. But if we represent the manifestation of the character on a logarithmic scale (lower figures in the table) dominance appears, with A and B showing dominance over a and b, while the non-allelic interaction vanishes, the effect of substituting A for a being the same with all genotypes in respect of B–b, and reciprocally. The change of scale has resulted in a change from apparently non-allelic interaction to dominance, i.e. allelic interaction. Not only can the scale chosen for representing the expression of the character affect the relative amount of the heritable variation appearing in a given category; it may even determine the category in which the heritable variation appears.

10. PLASMATIC VARIATION

By far the greater amount of heritable variation can be traced to differences in the nuclear genotype and indeed such variation is the chief subject of this book. In general, variation tracing to differences in the cytoplasm, to the plasmatype as we may say, is insignificant by comparison with the genotypic variation. The determination of differences caused by the cytoplasm passed on from parent to offspring will be revealed by differences between the F_1s from reciprocal crosses, since in general the cytoplasm is virtually entirely maternal in origin. Such reciprocal differences have been recorded but they are rare. The plasmatype appears to play as small a part in determining continuous variation as it does in determining the major differences with which we are familiar from Mendelian analysis.

In two circumstances, however, the plasmatype may reveal itself of importance: in the determination of differences in the juvenile characters of sexual offspring and in the determination of differences among the members of a clone. It has long been known that juvenile characters may be maternal in expression. Thus the rate of early cleavage in the eggs produced by hybridization between the echinoderms *Dendraster* and *Strongylocentrotus* is characteristic of the mother, and indeed in most echinoderm hybrids the paternal characters begin to show only at the gastrula stage (see Waddington 1939). The effects of this early domination by the maternally derived cytoplasm may persist in the adult, as in the determination of the direction of coiling in the shell of *Limnea* (see Darlington and Mather, 1949). These snails show us, however, that the cytoplasm is acting only as the agent of the mother's nuclear genes, for the direction of coiling in the offspring reflects the mother's genotype. Such determination by the cytoplasm acting as the agent of a pre-existing nucleus is by no means uncommon (see, for example, Mather, 1948) and we might well expect it to appear in continuous variation too.

Several cases of maternal effects are known in seedling characters of plants. Radiosensitivity is under polygenic control in the tomato and the effects of radiation as they appear in the cotyledons, first true leaves and second leaves show pronounced maternal effects, though these are not always consistent (Davies, 1962). The length of the first seedling leaf in the rye-grass *Lolium perenne* shows strong maternal influence (Beddows *et al.*, 1962), and other seedling characters in this grass have been found to show evidence of residual differences between reciprocal crosses, even though constant maternal effects could not be observed (Hayward and Breese, 1966). Possibly, since Hayward and Breese were observing somewhat older plants than Beddows *et al.*, their results indicate a diminution of the maternal effect with increasing age, as we might indeed expect from the general evidence cited above. More persistent differences between reciprocal crosses which are still detectable at flowering time and in final plant height have, however, been reported in one cross in *Nicotiana rustica* (Jinks *et al.*, 1972). Furthermore, it persisted, albeit at a reduced level, for two further generations of selfing and back-crossing of the reciprocal crosses. For plant height the results fitted maternal and grandmaternal determination.

Before leaving these observations, it should be noted that in no case could the differences be ascribed unequivocally to the cytoplasms derived from the mother; differences in maternal nutrition of the seed could not be ruled out as their cause. They nevertheless supply a *prima facie* case for plasmatic differences contributing to continuous variation in the *Lolium* plants. There is some evidence too that the plasmatype contributes to differences in the growth of mouse embryos, though here again the case is by no means a clear one (Brumby, 1960).

The evidence relating variation within clonally propagated plants to differences in the plasmatype does not suffer from this residual ambiguity. Jinks (1957) has shown that selection can be practised successfully for a number of characters, including rate of growth and production of perithecia, in clones of the fungi *Aspergillus nidulans* and *A. glaucus*, propagated either by asexual spores (conidia) or by hyphal tips. He was able to show by appropriate tests that the nuclear genotype of the clone was unchanged by the selection, the altered phenotype thus springing from change in the cytoplasm. An example of the responses he obtained to selection are illustrated in Fig. 6. It is clear from such results, and from the study of mutations induced in the cytoplasm by radiation and chemical mutagens, that the transmissible determinants of the cytoplasm must be particulate. They interact with the nuclear genes, at any rate to the extent that the limits within which cytoplasmic selection can change the phenotype depend on the nuclear constitution also. These studies also reveal that the plasmatype will play little part in the determination of variation among the sexual progeny, for a single propagation through the sexual ascospores serves virtually to eliminate even the greatest changes that have been produced in the cytoplasm either by artificial selection or by the normal processes of differentiation (Mather and Jinks, 1958).

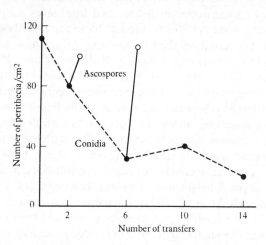

Figure 6 The effects of propagation by conidia and by ascospores on the production of perithecia in *Aspergillus glaucus*. Continued propagation by conidia (dotted line) results in a gradual decline in perithecial production, which is however immediately restored to its original level by a single propagation through an ascospore. The decline and restoration of perithecial production result from changes in the cytoplasm, the nucleus remaining unaltered. (Reproduced by permission from Mather and Jinks 1958.)

There is evidence of similar lability of the plasmatype from *Lolium perenne* (Breese *et al.*, 1965). The rate of tillering can be altered by selection within young clones raised from seedlings. Again, as in the fungus, there is evidence of an interaction between nucleus and cytoplasm, perhaps even of a domination of the cytoplasm by the nucleus, for the success of the somatic selection depended on the genotype of the clone even though the result was produced by alteration of the plasmatype.

These results serve to tell us that in certain circumstances, particularly where juvenile characters or differences within clones are being considered, we cannot afford to leave the plasmatype out of account, though in the cases we have considered the nuclear genotype is usually, if not always, in ultimate control and the plasmatype is of relatively small importance in determining variation among all but the most juvenile progeny arising from sexual reproduction. Major mutants whose hereditary determinants are located in the cytoplasm, usually in the DNA of plastids or mitochondria, are however known and indeed are now becoming relatively commonplace. But apart from the fact that they show preferential maternal transmission we know too little of cytoplasmic elements for us to formulate an analytical theory of any variation for which they may be responsible. We do, however, consider a general treatment of maternal effects, to which the plasmatype would contribute (Sections 51, 52).

11. INTERACTION OF GENOTYPE AND ENVIRONMENT

The plasmatype need only be brought into account in the analysis of continuous variation in certain special cases. The environment, on the other hand, must be brought into account in all but perhaps a few special cases, for virtually all characters show non-heritable variation. The only exceptions are such things as antigen production, and even here the quantity if not the specificity of the antigens may be subject to non-heritable influences. The heritable and the non-heritable components of continuous variation can be distinguished only by a breeding test. The effect of the environment is therefore a distraction in the genetical analysis, and our aim will thus be to isolate it and set it on one side in the analysis rather than to make it the subject of analysis in its own right, except of course where genotype and environment interact in producing their effects.

Genotype and environment may interact in a variety of ways. First, the environment may affect the genetic constitution of a population by the pressure of selection it exercises on the population. In the longer term this, of course, may lead to evolutionary changes with which we are not concerned; but in the short term, too, the pressure of selection from the environment may distort the segregations and recombinations that genetic theory leads us to expect in populations under experiment or observation. This impact of the environment on the genotype must then be taken into account, together with any other forces of selection, in the genetic analysis.

The environment also may alter the genetic picture by stimulating permanent changes in the genetic materials themselves. These may be of two kinds. One is the familiar and general mutagenic effect of radiations and various chemical substances. This need not detain us since the mutagenic agencies will seldom be encountered unless they are deliberately introduced as part of the experimental plan.

The second effect of the environment on the genetic materials is of a different kind, discovered by Durrant (1958, 1962). He found that flax plants of certain varieties when grown in soil carrying combinations of nitrogenous, phosphatic and potassic fertilizers not merely reflected the effects of their treatments in their own growth but also transmitted them to their offspring in the next and later generations. Similar conditioning changes have been observed in *Nicotiana rustica* (Hill, 1967; Hill and Perkins, 1969) and perhaps also in peas (Highkin, 1958), but they have not always been found where they have been sought and indeed are not even shown by all varieties of flax. They may be exceptional in their occurrence. The changes are transmitted equally by pollen and egg, and by being both inevitable in occurrence and determinate in type, resemble Brink's (1960) paramutation in maize rather than the more familiar mutation induced by radiations and chemical mutagens (see Mather, 1961).

There can be little doubt that the changes produced by 'conditioning', as Durrant calls it, are seated in the nucleus (Jinks and Towey, 1976). Indeed in flax when newly induced they are associated with detectable changes in the amount of DNA (Evans, 1968), which may, however, be subsequently lost by reversion (Durrant and Jones, 1971). Both in flax and in *Nicotiana* the induced changes do not behave entirely like normal gene changes in their transmission from parent to offspring. They sometimes segregate in the F_1 of crosses between lines which are apparently true breeding for induced changes (Durrant and Tyson, 1964; Perkins *et al.*, 1971) although in later generations of the same crosses there may be little to distinguish them from normal gene changes (Eglington and Moore, 1973). For these various reasons the changes produced by conditioning appear unlikely to disturb genetical analysis of continuous variation in normal experiments and we shall therefore pursue them no further.

Finally, and by far the most important for our present consideration, genotype and environment may interact in the production of differences among the individuals and families under direct observation by virtue of their interplay in development to produce non-transmissible effects.

These interactions commonly appear in one or other of two ways according to whether the differences in the environment are defined or not. They may be revealed by the non-heritable variance of a character, in individual or family, changing with the genotype. Or where use is made of two or more environments whose differences are defined, as for example two or more controlled temperatures, the interaction may appear as a change in the difference of mean expression of the character between two genotypes when the organisms are moved from one environment to another.

The non-heritable variance of an individual or a family may reflect differences of the micro-environment in which different members of a family are raised or in which different parts of a given individual develop. It may also, especially where the variation is between repetitive parts of the same individual, reflect the accidents and chances of development. What ever its causation the variance may show changes related to the mean expression of the character in a·fairly direct way, rising, for example, as the mean.rises. In such a case the differences in variance can commonly be removed by a change of scale, say for example a logarithmic transformation, or by using as a measure of the variation not the variance itself but the coefficient of variation in which the variance or its square root, the standard deviation, is divided by the mean (see Table 9).

TABLE 9
Example of the use of the coefficient of variation to equalize the variances of inbred and F_1 families taken from Powers (1942). The character is fruit weight in the cross between two tomato varieties

Variety	Mean weight	Standard error	Standard error / Mean
Danmark	51·17	15·40	0·30
F_1	5·48	1·50	0·27
Red currant	0·92	0·23	0·25

Not all apparent interactions of genotype and environment are direct metrical relations of this kind, removable by a simple change of scale. Rees and Thompson (1956) report a study of meiosis in four inbred lines of rye and their F_1s in which were recorded the frequencies with which chiasmata were observed. The chiasmata were recorded from 20 pollen mother cells of each plant, the number of plants from each line and F_1 ranging between 5 and 10. Table 10 shows for each line and F_1 the mean frequencies of chiasma formation per bivalent, the variance of chiasma frequency between the plants of the line or F_1, and the average variance of chiasma frequency between the pollen mother cells of a single plant. The variances between individual pollen

TABLE 10
Means and variances of chiasma frequencies in rye (Rees and Thompson, 1956)

Inbred line	Mean	Plant variance	Cell variance	F_1	Mean	Plant variance	Cell variance
3	1·85	8	7	3 × 6	2·03	1	7
6	1·71	48	15	3 × 12	2·07	12	7
12	1·70	42	11	3 × 13	2·09	1	4
13	1·73	167	13	6 × 12	2·04	7	7
				6 × 13	2·04	6	5
				12 × 13	2·07	5	6

All variances are given in units of 0·0001: for example the plant variance of line 6 shown as 48 is 0·0048 mother cells recorded by Rees and Thompson have been divided by 20 to make them comparable to the variances between plants, which are derived from plant totals each reflecting the mean behaviour of 20 pollen mother cells.

We first note that while the F_1s all have higher means than the parent lines, they also, with a single exception, have lower variances between plants and very much lower variances than lines 6, 12 and 13. The differences between the cell variances are less striking but again quite clearly the F_1s in general yield lower values than the parent lines. Since the F_1 means are higher, these differences in variance obviously cannot be accommodated by taking some coefficient of variation in place of variance itself. Rather we must regard the lower variance of the F_1s as reflecting their genotypes' ability to determine more stable development than can the genotypes of the parent lines. This greater stability is reflected partly in a nearer approach to uniformity among the repetitive parts (the 20 pollen mother cells) within a plant and partly by a greater ability to accommodate the vagaries of the environment as revealed by the lower variation between different plants within the family. Here is a true interaction of genotype and environment which no simple change of scale could always eliminate. The stability of development and the ability to accommodate the vagaries of the environment are under genetic control and it is a general observation that in normally outbreeding species inbred lines are less stable than their F_1s (Mather, 1946c, 1953b; Lerner, 1954) though in normally inbreeding species the F_1s have no major advantage in this respect (Jinks and Mather, 1955).

Turning to the situation where interactions are revealed by comparison of the behaviour of genotypes in a limited number of defined environments, we may take for illustration the example discussed by Hogben (1933). He quotes data obtained by Krafka for the average number of ocelli in eyes of two strains of *Drosophila*, referred to as 'Low Bar' (L) and 'Ultra Bar' (U), at two temperatures, 15 and 25 °C. The averages are set out in Table 11. As Hogben points out, the effects of the genetic and the environmental differences cannot be regarded as additive since the differences in phenotype of L and U are not even approximately the same at the two temperatures, or to put the same thing in other words, the differences due to temperature are not even approximately the same in the two strains. In statistical terms the data are completely described by three sources of deviation from the mean of the observations, d measuring the average effects of the difference in genotype, e measuring that of the difference in environment, and g measuring their interactions. We can then find:

$$d = \tfrac{1}{4}(198 - 52 + 74 - 25) = 48 \cdot 75;$$
$$e = \tfrac{1}{4}(198 + 52 - 74 - 25) = 37 \cdot 75;$$
$$g = \tfrac{1}{4}(198 - 52 - 74 + 25) = 24 \cdot 25.$$

The interaction item is almost half as large as the effect ascribable to genotype and not much less than two-thirds that of the environment.

Yet here again, as with genic interaction, the size of the interaction between

TABLE 11
Number of ocelli in eyes of two strains of
Drosophila at two temperatures (Hogben,
1933)
Direct measure

Temperature (°C)	Strain	
	L	U
15	198	52
25	74	25

Logarithmic measure

Temperature (°C)	Strain	
	L	U
15	2·297	1·716
25	1·869	1·398

genotype and environment depends on the scale used for representing the manifestation of the character. If we transform the data by taking logarithms we obtain the figures set out in the lower part of Table 11. Now we find $d = 0·263$, $e = 0·187$ and $g = 0·028$. The value of g is now little more than one-tenth that of d and nearer one-seventh than one-sixth that of e. The interaction has been very much reduced and may be regarded as almost negligible, for taking the logarithms of means, as we have had to do, does not give the means of the logarithms of the original observations, which are really needed. The logarithm of a mean in fact exceeds the mean of the logarithms by an amount related to the variance of the observations, and since this variance was probably higher the higher the mean, the upward bias is most likely greater in the larger figures of Table 10. The estimate of $g = 0·028$ is thus very likely to be spuriously high. In any case there is every reason to believe that the interaction could be removed entirely by a transformation which would foreshorten the upper end of the scale more than does the taking of logarithms. So not only does this example show that the apparent interaction can be reduced by using an appropriate scale, it suggests that it could be removed entirely, Thus, despite Hogben's doubts, the effects of genotype and environment can, on a log scale, be treated as substantially additive, and on an even more appropriate scale might be wholly additive. Here, as with genic interactions, choice of scale affects our picture of the structure of the variation.

12. THE PRINCIPLES OF SCALING

With discontinuous variation the scale used in representing a character is unimportant. Individuals are assignable to one or other of a number of distinct classes and the resulting data consist of the frequencies with which the individuals fall into their various classes. The characters by which the classes are recognized may be expressed in terms of some convenient metric (though they seldom are); but this is not essential for the analysis of the data. As an example, the tall peas of a segregating family such as Mendel considered may average 6 ft in height and their short companions only 2 ft; but it would make no difference to the analysis of the data or to the conclusions reached if the talls were 10 ft high and the shorts but 1 ft. Nor would it matter if the talls were 4 ft and the shorts 3 ft, or 3 ft 6 in, provided that every individual fell unambiguously into one class or the other. The analysis requires no assumption about the metrical relations of the classes because the genetical situation is completely described by the frequencies of talls and shorts. The only essential is that the shortest tall be unambiguously taller than the tallest short; that in fact the distribution of heights be discontinuous. The precise metrical relation of the character in the two classes is a matter of indifference so far as the treatment of frequency data goes. The result would be just the same if plant height was expressed in log feet or antilog feet, or indeed on any reasonable scale, for the discontinuity would still exist and the regular separation of the classes would still be possible.

With continuous variation the situation is different. Each observation is a measurement which must be regarded as potentially unique even where, by reason of the shortcomings of our measuring instruments or our system of recording, two or more individuals appear as having the same measurement. Without regular discontinuities each datum can have no significance other than as a measurement of character expression. Thus the common variation of human stature cannot be described adequately for genetical purposes by saying that there exist so many talls and so many shorts, for such a classification has no genetical foundation. Even if the variation were to be expressed in the form of frequencies of individuals with heights lying within successive ranges of, say, one inch, the representation would still be inadequate because the classification into these ranges would have no genetical significance. The grouping would be artificial. Nor where the grouping is imposed naturally need it have a genetical significance. The distribution of chaeta number in *Drosophila* obviously cannot be continuous in the strict sense for one cannot have fractional chaetae; but the minor discontinuities so inferred can have no genetical meaning and the data can be legitimately and even desirably treated for genetical purposes as though the discontinuities did not exist. In fact we assume an underlying continuous distribution of chaeta-producing potential, upon which the discontinuities are imposed by the meristic nature of the character. We treat the variation as 'quasi-continuous', to use Gruneberg's (1952) term. The same assumption may properly be made even where the discontinuities are greater and the number of classes smaller, as for example with variation in the number of scutellar bristles in *Drosophila*, and in the limit it may be necessary for a character which shows only two levels of expression, because a single threshold is operating. In such extreme cases, special considerations may apply and special methods be needed in interpreting the variation as for example in Rendel's (1979) use of the probit transformation in interpreting the variation in the number of scutellar bristles of *Drosophila*; but the chief point at issue is that unless the discontinuities can be shown to have a genetical significance, by relating them to the segregation of identifiable genes, the frequencies to which they lead cannot be made the basis of Mendelian analysis. Rather they must be related to an underlying continuous distribution, using such transformations as probits if necessary.

Thus with continuous and even quasi-continuous variation we must necessarily use biometrical quantities, means, variances, etc., to replace frequencies in describing and analysing the variation. The resulting description and analysis can be valid only in terms of the scale on which the measurements are taken. A change of scale will change the values of the biometrical quantities and it will change them unequally for measurements of different magnitudes. The description of such a body of data and *a fortiori* the comparisons we make within it or between it and other similar bodies of data must therefore depend on the scale used in measurement, and indeed we have already seen how interactions among genes themselves and between genes and environmental agencies can change, and even appear or disappear, as the scale is altered. Clearly the choice of an appropriate scale is the first step in the analysis of polygenic variation.

The scales of the instruments which we employ in measuring our plants and animals are those which experience has shown to be convenient to us. We have no reason to suppose that they are specially appropriate to the representation of the characters of a living organism for the purposes of genetical analysis. Nor have we any reason to believe that a single scale can reflect equally the idiosyncrasies of all the genes affecting a single character, or that the agencies determining the discontinuities in a meristically varying character bear any scalar relation to the actions of the genes mediating the underlying continuous distribution of potential manifestation of the character. We cannot even assume without evidence that a scale appropriate to the representation of variation of a character in one set of individuals under one set of conditions will be equally appropriate to the representation of that same character either in a different set of individuals, which may be heterogenic for different genes, or under different conditions. It may well, therefore, never be possible to construct an *a priori* scale for the representation of variation in a character. Certainly with no more than our present knowledge of gene action the construction of such a scale is impossible. Available observations, such as those of Powers (1941) discussed in the next chapter, serve merely to emphasize this conclusion.

The scale on which the measurements are expressed for the purposes of genetical analysis must therefore be reached by empirical means. Obviously it should be one which facilitates both the analysis of the data and the interpretation and use of the resulting statistics. Now sums, sums of squares and sums of cross-products, on the partitioning of which the analysis must depend, may be regarded most conveniently as composed of various components determined by the various genes and by the non-heritable agents, which are summed to give the quantity in question. As we shall see, components can be included in such sums to cover the interactions of the genes with one another and with non-heritable agents, but the more components there are the more complex the analysis becomes and the more statistics are necessary to provide the means of separating these components. The scale should preferably be one on which the analysis is as simple as possible, which means one on which interactions among the genes and between genotype and environment are absent, or at any rate as small as they can reasonably be made. Interaction may, of course, occur between alleles, but such dominance covers a more limited range of relations than does non-allelic interaction and so presents relatively little difficulty in analysis, interpretation and prediction. Our aim in choosing a scale should therefore be to remove as far as possible the effects of non-allelic interactions and genotype × environment interactions, allowing dominance to take its own value on the scale so reached. So far as possible the non-allelic genes and non-heritable agents should all be additive in action.

In practice, as we shall see, such a scale may be impossible to find. Each gene and each non-heritable agent may be acting on its own unique scale, or at least all may not be acting on the same scale. Since, however, the genetical analysis cannot pretend to ascertain the individual effects and properties of genes which are not individually distinguished, we may take the more limited aim of seeking a scale on which the genic and non-heritable effects are additive on the average, as far as the data go. As we have already seen from Rees and Thompson's rye, even this limited aim is not always attainable. The search should however always be made, for a scale satisfying these criteria facilitates analysis so very much. But even where such a scale can be found its limitations must not be forgotten. In particular, discovery of an empirically satisfactory scale cannot of itself be used to justify theoretical conclusions concerning the physiology of gene action, though it is of course legitimate to test the agreement of any empirical scale with one expected theoretically from other considerations.

4 Components of means: additive and dominance effects

13. ADDITIVE AND DOMINANCE COMPONENTS

In diploid organisms the individual can fall into any one of three genetic classes (AA, Aa and aa) in respect of a gene for which there exists two alleles (A and a). The effect of the gene difference on the phenotype can be described by two parameters and specified completely if the values of these two parameters are known. In the system adopted by Fisher *et al.* (1932) and Mather (1949a) one parameter (d) is used to represent the phenotypic difference between the two homozygotes AA and aa and the other (h) to represent the departure in phenotype of the heterozygote Aa from the mid-point between AA and aa (m). Taking this mid-point as the origin the effects on the magnitude of the characters in question are then (Fig. 7)

$$\text{AA } d_a; \text{ Aa } h_a; \text{ aa} - d_a$$

so that the gene's contribution to the fixable or additive genetic variation is proportional to d, while h reflects the dominance properties of the gene and represents the contribution to the unfixable heritable variation. It should be noted that the definition of the additive variation used here is strictly the genetically additive variation and so differs from the widely used definition which is synonymous with the statistically additive variation. It will be recalled that the designation of one allele by A and the other by a does not here have the conventional implication with regard to dominance: it merely implies an increasing and decreasing effect on the phenotype, respectively.

Similarly for gene B–b the three effects will be BB d_b; Bb h_b; bb $- d_b$ and so on for any number of genes.

In all cases d represents an increment in a constant direction along the scale of measurement from the origin, termed for convenience the $+$ direction, while h may be an increment in either direction, $+$ or $-$, according to whether the increasing or the decreasing allele is dominant. With an adequate scale, d and h will have constant average values over the whole length of the scale, i.e. they will be independent of m.

The relative values of d_a and h_a will depend on the degree of dominance. If there is no dominance $h_a = 0$, if A is dominant h_a is positive, if a is dominant h_a

Figure 7 The *d* and *h* increment of the gene A–a. Deviations are measured from the point 0, midway between the homozygotes. Aa may lie on either side of this point and the sign of *h* will vary accordingly.

is negative. If dominance is complete h_a is equal to $+d_a$ or $-d_a$. The degree of dominance is h_a/d_a.

A true breeding strain will have an average measurement of

$$m + S(d_+) - S(d_-)$$

where $S(d_+)$ indicates the summed effects of those genes under consideration which are represented by their + alleles, i.e. those adding increments in the + direction, $S(d_-)$ indicates the corresponding sum of effects of genes represented by their − alleles, and *m* is a constant depending on the action of genes not under consideration and of non-heritable agencies. If two such strains are grown under comparable conditions their mean measurements will differ by

$$2[S(d_+) - S(d_-)]$$

where only the genes by which they differ are taken into account and the smaller mean measurement is subtracted from the larger. The mean of the two strains is *m*, and is independent of the distribution of the genes between the strains. This mid-point value is thus the natural zero point from which measurements can be expressed as deviations.

The difference between the mean measurements of the two strains depends on the distribution of the genes between them. Where all the *k* genes by which the strains differ are represented in one strain by their + alleles and in the other by their − alleles (i.e. are isodirectionally distributed, all like alleles being associated in the strains) the difference between the strain means becomes

$$2[S(d_+)]$$

which equals $2kd$ in the special case where $d_a = d_b = d_c \ldots = d_k = d$.

In general, however, of the *k* genes for which the two strains differ, k' alleles of + (increasing) effect will be present in one strain along with $k - k'$ alleles of − (decreasing) effect and *vice versa* for the other strain. The difference between the strain mean becomes

$$2\left[\underset{k}{S}(d) - 2\underset{k'}{S}(d_-)\right];$$

if we now write

$$r_d = \frac{\underset{k}{S}(d) - 2\underset{k'}{S}(d_-)}{\underset{k}{S}(d)}$$

this becomes $2r_d\underset{k}{S}(d)$

which equals $2rkd$ in the special case where $d_a = d_b = d_c \ldots = d_k = d$.

The value of r_d is a measure of the gene distribution between the two strains. If all genes of + (increasing) effect are present in one strain then $r_d = 1$. If, on the other hand, they are equally shared between the strains, i.e. they are dispersed, then $r_d = 0$. The value of r_d is thus a measure of the degree of association of genes of like effect.

For ease of presentation we shall write

$$2r_d \underset{k}{S}(d) = 2[d].$$

The F_1 between the two strains discussed above must be heterozygous for all k genes, irrespective of their distribution in the strains, and when grown under comparable conditions its mean will deviate from the mid-parent by $S(h)$ or kh in the special case where $h_a = h_b = \ldots = h$ taking sign into account. For uniformity of presentation we shall write

$$\underset{k}{S}(h) = [h].$$

In such a cross the ratio which the deviation of F_1 from the mid-parent bears to half the parental difference is frequently regarded as a measure of average dominance of the genes concerned. We now see that the measure is in fact

$$\frac{[h]}{[d]} = \frac{S(h)}{r_d S(d)}$$

which in the special case of equal effects of the genes becomes

$$\frac{kh}{rkd}.$$

Its use as a measure of overall dominance is, therefore, tantamount to assuming:

(i) that $r_d = 1$, that is the genes of like effect are completely associated in the parental strains; and

(ii) that all the h increments have the same sign, that is, dominance is unidirectional at all loci.

Neither assumption can be generally true and so it is better to speak of it as the potence ratio measuring the relative potence of the parental gene sets (Mather, 1946b; Wigan, 1944). This is made the more desirable by the possibility of measuring the average dominance of the genes in a way to be discussed later. It should be noted that assumptions (i) and (ii) hold by definition when only one gene is responsible for the difference between the parental strains and hence the potence ratio is identical with the dominance ratio in this limiting case.

The potence ratio can in fact theoretically take any value between 0 and ∞. This is true even with the restriction that $h \leqslant |d|$ for each gene, that is in respect of any individual gene the heterozygote never falls outside the range delimited by the two homozygotes. If the h increments are balanced, in the sense that the sum of h increments of genes whose heterozygotes resemble the $+$ homozygote more than the $-$ homozygote ($+$ allele dominant) equals the summed h increments of the genes to which the opposite applies ($-$ allele dominant) then

$$S(h) = [h] = 0$$

and the F_1 will fall on the mid-parent (m). Potence is zero, no matter what the average dominance may be. Similarly, where the parents do not differ in mean measurement because the genes are dispersed, i.e. $r_d = 0$ so that $r_d S(d) = [d] = 0$, potence must be ∞ with any condition other than $[h] = 0$.

In short, observable potence of sets of genes indicates dominance of the individual genes preponderantly in the same direction; but zero potence does not of necessity indicate absence of dominance.

If the relative frequencies of the three genetic classes in respect of each gene are known for any population of individuals the expected deviation of the mean of the population can be specified in terms of the two parameters d and h. For example, in an F_2 we expect half the individuals to be heterozygous and half homozygous in respect of each gene pair which is segregating. Since among the homozygotes half will be homozygous for the increasing allele and half homozygous for the decreasing allele their contributions to the generation means, which will be $+d$ and $-d$ for each gene difference respectively, will cancel out. Heterozygotes, on the other hand, will contribute to the generation mean. Hence summing over all segregating genes, the F_2 generation mean ($\bar{F}_2$) will equal $m + \frac{1}{2}[h]$. Thus the $\bar{F}_2$ will deviate from the mid-parent value by an amount which is equal to half the deviation of the $\bar{F}_1$ from the mid-parent.

This result can be generalized to any generation derived by selfing the successive generations derived from an F_1. The F_n generation will consist of a proportion $(\frac{1}{2})^{n-1}$ individuals which will be heterozygous in respect of each gene segregating in the cross and a proportion $\frac{1}{2}[1 - (\frac{1}{2})^{n-1}]$ individuals of each of the corresponding homozygotes. The generation mean will therefore be

$$\bar{F}_n = m + (\tfrac{1}{2})^{n-1}[h].$$

If instead of selfing we sib-mate the F_1, the expected generation mean $\bar{S}_2$ is the same as $\bar{F}_2$. However, in subsequent generations obtained by sib-mating the expectations are not the same as for the selfing series. In the sib-mating series the frequency of heterozygosity in the nth generation can be obtained by substituting the nth term of the Fibonacci series (fn) for the $(\frac{1}{2})^{n-1}$ of the selfing series. The successive terms in the Fibonacci series starting from $n = 2$ (the $F_2 = S_2$) are $fn = \frac{1}{2}, \frac{2}{4}, \frac{3}{8}, \frac{5}{16}, \ldots$, etc.

The frequency of homozygosity in respect of each gene is of course $(1 - fn)$ and this will be made up equally of the increasing homozygote ($+ +$) and the decreasing homozygote ($- -$) whose contributions to the generation mean will be equal and opposite in sign and hence cancel out. All we need concern ourselves with, therefore, is the contribution of the heterozygote. The expected generation mean will then be

$$\bar{S}_n = m + fn[h].$$

The expected generation means for a backcrossing series started from an F_1 can be obtained similarly. In backcrosses, however, only two of the three possible genotypes in respect of a single gene difference are present in each generation, namely, the heterozygote and one of the two homozygotes. In the nth generation of repeated backcrossing to one of the parents (B_n) the frequency of heterozygous individuals in respect of any gene is simply $(\frac{1}{2})^n$ and the frequency of the single homozygote is $1 - (\frac{1}{2})^n$. Let us now consider the result of backcrossing to the larger parent whose mean is $m + S(d_+) - S(d_-)$ where $S(d_+) > S(d_-)$ and which, therefore, equals

$$m + [d].$$

In the nth generation of repeated backcrossing to the larger parent B_{n1}, the expected mean is:

$$\bar{B}_{n1} = m + (1 - (\tfrac{1}{2})^n)[d] + (\tfrac{1}{2})^n[h].$$

The comparable series of backcrosses to the smaller parent whose mean is $m - [d]$ will be

$$\bar{B}_{ns} = m - (1 - (\tfrac{1}{2})^n)[d] + (\tfrac{1}{2})^n[h].$$

While the expectations we have given are of the commonest generations we can arrive at comparable expectations for any generation, no matter how unusual, providing that the relative frequencies of the two types of homozygotes and the heterozygote in respect of any gene difference can be predicted from the breeding programme. For any generation the expected mean is

$$(u - v)[d] + \beta[h]$$

where u is the frequency of the increasing allele at all loci, $v (= 1 - u)$ is the frequency of the decreasing allele at all loci and β is the frequency of heterozygosity at all loci.

If the generation means depend only on the additive and dominance effects of the genes, i.e. there are no interactions between non-allelic genes and there are no other disturbing factors such as differential viability or differential fertility, then it is clearly possible to find simple relationships between the expected values of different generation means. We have already noted that the F_2 generation mean under these conditions is expected to deviate from the mid-parent value by half the amount that the F_1 generation mean deviates from the mid-parent value. That is

$$\bar{F}_2 - \tfrac{1}{2}(\bar{P}_1 + \bar{P}_2) = \tfrac{1}{2}[\bar{F}_1 - \tfrac{1}{2}(\bar{P}_1 + \bar{P}_2)]$$

which can be rewritten in a number of ways, one of the more useful being

$$\bar{F}_2 = \tfrac{1}{4}\bar{P}_1 + \tfrac{1}{4}\bar{P}_2 + \tfrac{1}{2}\bar{F}_1.$$

Similar relationships can readily be obtained for the expected value of the backcross generations mean. Reference to the expected means of the generations produced by backcrossing an F_1 to each of its parents shows that

$$\bar{B}_1 - \bar{P}_1 = \tfrac{1}{2}(\bar{F}_1 - \bar{P}_1) \text{ and } \bar{B}_2 - \bar{P}_2 = \tfrac{1}{2}(\bar{F}_1 - \bar{P}_2)$$

which are equivalent to

$$\bar{B}_1 = \tfrac{1}{2}\bar{P}_1 + \tfrac{1}{2}\bar{F}_1 \text{ and } \bar{B}_2 = \tfrac{1}{2}\bar{P}_2 + \tfrac{1}{2}\bar{F}_1.$$

From which it is not difficult to see that

$$\bar{F}_2 = \tfrac{1}{4}(\bar{P}_1 + \bar{P}_2 + 2\bar{F}_1) = \tfrac{1}{2}\bar{B}_1 + \tfrac{1}{2}\bar{B}_2.$$

Corresponding expressions can be found for the means of other types of generations (Table 12).

TABLE 12
Some simple relationships between the means of generations derived from a cross between two true breeding lines assuming that an additive-dominance model is adequate on the scale used

$$\bar{S}_2 = \bar{F}_2 = \tfrac{1}{4}\bar{P}_1 + \tfrac{1}{4}\bar{P}_2 + \tfrac{1}{2}\bar{F}_1$$

$$\bar{F}_3 = \tfrac{1}{4}\bar{P}_1 + \tfrac{1}{4}\bar{P}_2 + \tfrac{1}{2}\bar{F}_2 = \tfrac{3}{8}\bar{P}_1 + \tfrac{3}{8}\bar{P}_2 + \tfrac{1}{4}\bar{F}_1$$

$$\bar{F}_n = \tfrac{1}{4}\bar{P}_1 + \tfrac{1}{4}\bar{P}_2 + \tfrac{1}{2}\bar{F}_{n-1} = (\tfrac{1}{2} - \tfrac{1}{2^n})(\bar{P}_1 + \bar{P}_2) + (\tfrac{1}{2})^{n-1}\bar{F}_1$$

$$\bar{B}_1 = \tfrac{1}{2}\bar{P}_1 + \tfrac{1}{2}\bar{F}_1$$

$$\bar{B}_2 = \tfrac{1}{2}\bar{P}_2 + \tfrac{1}{2}\bar{F}_1$$

$$\bar{B}_{11} = \tfrac{3}{4}\bar{P}_1 + \tfrac{1}{4}\bar{F}_1 \qquad \text{where } B_{11} = B_1 \times P_1$$

$$\bar{B}_{12} = \tfrac{1}{4}\bar{P}_2 + \tfrac{3}{4}\bar{F}_1 \qquad \text{where } B_{12} = B_1 \times P_2$$

$$\bar{B}_{21} = \tfrac{1}{4}\bar{P}_1 + \tfrac{3}{4}\bar{F}_1 \qquad \text{where } B_{21} = B_2 \times P_1$$

$$\bar{B}_{22} = \tfrac{3}{4}\bar{P}_2 + \tfrac{1}{4}\bar{F}_1 \qquad \text{where } B_{22} = B_2 \times P_2$$

$$\bar{B}_{1S} = \tfrac{5}{8}\bar{P}_1 + \tfrac{1}{8}\bar{P}_2 + \tfrac{1}{4}\bar{F}_1 \qquad \text{where } B_{1S} = B_1 \text{ selfed}$$

$$\bar{B}_{2S} = \tfrac{5}{8}\bar{P}_2 + \tfrac{1}{8}\bar{P}_1 + \tfrac{1}{4}\bar{F}_1 \qquad \text{where } B_{2S} = B_2 \text{ selfed}$$

$$\bar{F}_2 = \tfrac{1}{2}\bar{B}_1 + \tfrac{1}{2}\bar{B}_2 = \tfrac{1}{4}\bar{B}_{11} + \tfrac{1}{4}\bar{B}_{12} + \tfrac{1}{4}\bar{B}_{21} + \tfrac{1}{4}\bar{B}_{22}$$

$$\bar{F}_3 = \tfrac{1}{2}\bar{B}_{1S} + \tfrac{1}{2}\bar{B}_{2S} = \tfrac{1}{2}\bar{B}_{11} + \tfrac{1}{2}\bar{B}_{22}$$

We can test these predicted relationships between generation means and hence test the expectation that they depend only on the additive and dominance effects of the genes, in a number of ways. One of these is the scaling tests of Mather (1949a).

14. SCALING TESTS

In practice the relationships between the generation means can hold only within the limits of accuracy set by the sampling errors to which the various mean measurements are subject. When using the relationships as tests of conformity with the additive-dominance model, regard must therefore be paid to the variances of the generation means. This can be achieved by putting, for example,

$$A = 2\bar{B}_1 - \bar{P}_1 - \bar{F}_1,$$

whereupon
$$V_A = 4V_{\bar{B}_1} + V_{\bar{P}_1} + V_{\bar{F}_1},$$

with $V_{\bar{B}_1}$ as the variance of the mean measurement of the B_1 generation etc. The standard error of A is then obtained as $\sqrt{V_A}$ and a test of significance is applied by the customary methods.

This approach can be illustrated by reference to the mean measurements of the two parents, the F_1 and F_2 generations and the backcrosses of the F_1 to the parents in a number of tomato crosses (Powers, 1941, 1942). The mean number of loculi per fruit in the generations derived from four crosses — one of which, Danmark × Johannisfeuer, was grown in two years 1938 and 1939 – are given in Table 13. The quantities, A, B and C and their variances have been calculated to test the adequacy of the additive-dominance model in each case, using Mather's formulae

$$A = 2\bar{B}_1 - \bar{P}_1 - \bar{F}_1 \qquad\qquad V_A = 4V_{\bar{B}_1} + V_{\bar{P}_1} + V_{\bar{F}_1}$$
$$B = 2\bar{B}_2 - \bar{P}_2 - \bar{F}_1 \qquad \text{and} \qquad V_B = 4V_{\bar{B}_2} + V_{\bar{P}_2} + V_{\bar{F}_1}$$
$$C = 4\bar{F}_2 - 2\bar{F}_1 - \bar{P}_1 - \bar{P}_2 \qquad\quad V_C = 16V_{\bar{F}_2} + 4V_{\bar{F}_2} + V_{\bar{P}_1} + V_{\bar{P}_2}.$$

TABLE 13
Mean number of loculi per fruit in tomato (Powers, 1941)

	J × D 1938	J × D 1939	J × BB 1939	D × RC 1938	J × RC 1939
$\bar{P}_1$	$9·125 \pm 0·091$	$9·028 \pm 0·084$	$9·028 \pm 0·084$	$5·405 \pm 0·068$	$9·028 \pm 0·084$
$\bar{B}_1$	$7·500 \pm 0·100$	$7·344 \pm 0·136$	$7·034 \pm 0·162$	$3·473 \pm 0·037$	$4·356 \pm 0·140$
$\bar{F}_1$	$5·500 \pm 0·086$	$6·051 \pm 0·096$	$6·329 \pm 0·111$	$2·395 \pm 0·018$	$2·517 \pm 0·029$
$\bar{F}_2$	$6·595 \pm 0·118$	$6·826 \pm 0·150$	$6·781 \pm 0·208$	$2·570 \pm 0·022$	$2·886 \pm 0·078$
$\bar{B}_2$	$5·575 \pm 0·064$	$5·898 \pm 0·097$	$5·452 \pm 0·077$	$2·205 \pm 0·015$	$2·183 \pm 0·016$
$\bar{P}_2$	$5·475 \pm 0·057$	$6·183 \pm 0·065$	$6·318 \pm 0·069$	$2·050 \pm 0·014$	$2·034 \pm 0·004$
A	$0·375 \pm 0·236$	$-0·391 \pm 0·300$	$-1·289 \pm 0·353$	$-0·854 \pm 0·102$	$-2·833 \pm 0·294$
B	$0·175 \pm 0·164$	$-0·438 \pm 0·266$	$-1·743 \pm 0·202$	$-0·035 \pm 0·038$	$-0·185 \pm 0·043$
C	$0·780 \pm 0·581$	$-0·009 \pm 0·639$	$-0·880 \pm 0·887$	$-1·965 \pm 0·118$	$-4·552 \pm 0·328$

In each case P_1 is the larger and P_2 the smaller parent
  D = Danmark
  J = Johannisfeuer
  BB = Bonny Best
  RC = Red Currant
  P_1 corresponds with the first variety in each column heading

If the model is adequate these quantities A, B and C will each equal zero within the limits of sampling error. Powers gives the standard errors of his means and by squaring these the corresponding variances can be found. V_A, V_B and V_C are then determined from the formulae given above. The standard errors of A, B and C, entered in Table 13, have been found as the square roots of the corresponding variances. For example, in the cross Danmark × Johannisfeuer as grown in 1938,

$$B = 2 \times 5\cdot575 - 5\cdot500 - 5\cdot475 = 11\cdot150 - 10\cdot975 = 0\cdot175$$

and $\quad V_B = (4 \times 0\cdot064^2) + (0\cdot086^2) + (0\cdot057^2) = 0\cdot0270$

giving a standard error of B of $\sqrt{0\cdot0270}$ or $0\cdot164$. The values of A, B and C together with their standard errors are also entered for the various crosses in Table 12.

The data from the cross Danmark × Johannisfeuer, whether grown in 1938 or 1939, give insignificant values of A, B and C. Hence the additive-dominance model is perfectly adequate for the analysis of the variation in these single sets of data. The cross Johannisfeuer × Bonny Best, however, shows significant deviations from zero for both A and B and in consequence the model must be judged inadequate for this set of data. It may be observed that this difference between crosses is not due to the second one including a greater range of loculus numbers than the first. The two crosses which include Red Currant both indicate the inadequacy of the model, the values of A and C being especially large in the cross Johannisfeuer × Red Currant where the range in loculus numbers is greatest. Thus a model which is adequate for one cross may not be adequate for another which covers the same range of variation, and still less adequate for other crosses covering wider ranges of variation.

Further analysis of the data for which the model is inadequate must proceed along one of two lines. Either the model must be extended to include those components such as non-allelic interactions which were excluded from the simple model (see Chapter 5) or alternatively, a scale must be sought on which the simple model is adequate. However, before discussing these alternatives further in relation to the present data, a more general method of testing the expected relationship between generation means on the additive-dominance model will be described.

A procedure known as the joint scaling test was proposed by Cavalli (1952). It consists of estimating the parameters, m, $[d]$ and $[h]$ from means of the available types of generations followed by a comparison of the observed generation means with expected values derived from the estimates of the three parameters. A minimum of three types of families is, of course, required to estimate the three parameters. With only three, however, there must be a perfect agreement between the observed family means and those expected and this provides no test of the goodness of fit of the model. To provide such a test four or more types of families must be raised. The three parameters must then be estimated by weighted least squares, taking as weights the reciprocals of the squared standard errors of each mean. The comparison between observed and expected means can then be effected by assuming the sum of squares minimized in the fitting process to be distributed as a χ^2 with degrees of freedom three less than the number of family means available (three less because three parameters have been fitted).

This is a standard procedure (Fisher, 1946; Mather, 1946a; Searle, 1966) and may be illustrated by reference to the cross Danmark × Johannisfeuer grown in 1938. Six equations are available for estimating m, $[d]$ and $[h]$ and these are obtained by equating the observed family means to their expectations, in terms of these three parameters (Table 14). Since there are more equations than unknowns the estimation must be by a least squares technique. The various generation means are not known with equal precision, for example the standard error of $\bar{P}_2$ is half that of the $\bar{F}_2$. The generation means and their expectations must, therefore, be weighted, the appropriate weights being the reciprocal of the squared standard errors. For $\bar{P}_2$ this is equal to

$$\frac{1}{(0.057)^2} = 307.7870,$$

and so on for the other families (Table 14).

TABLE 14
Joint scaling tests on the cross Danmark × Johannisfeuer 1938

Generation	Weight	m	$[d]$	$[h]$		Observed generation mean	Expected generation mean
P_1	120.7584	1	1	0	=	9.125	9.188
B_1	100.0000	1	0.5	0.5	=	7.500	7.393
F_1	135.2082	1	0	1	=	5.500	5.599
F_2	71.8184	1	0	0.5	=	6.595	6.470
B_2	244.1406	1	−0.5	0.5	=	5.575	5.546
P_2	307.7870	1	−1	0	=	5.475	5.494

The six equations and their weights may be combined to give three equations yielding weighted least squares estimates of the three parameters as follows. Each of the equations is multiplied through by the coefficient of m which it contains, and by its weight, and the six are then summed. Then we have

m		$[d]$		$[h]$		
120.7584	+	120.7584			=	1101.920 400
100.0000	+	50.0000	+	50.0000	=	750.000 000
135.2082			+	135.2082	=	743.645 100
71.8184			+	35.9092	=	473.642 348
244.1406	−	122.0703	+	122.0703	=	1361.083 845
307.7870	−	307.7870			=	1685.133 825
979.7126	−	259.0989	+	343.1877	=	6115.425 518

The two further equations are found in the same way using the coefficients of $[d]$ and $[h]$ in turn and the weights as multipliers so that we have three simultaneous equations that may be solved in a variety of ways to yield estimates of m, $[d]$ and $[h]$.

A general approach to the solution is by way of matrix inversion. The three equations can be rewritten in the form

$$\underbrace{\begin{bmatrix} 979{\cdot}712\,60 & -259{\cdot}098\,90 & 343{\cdot}187\,70 \\ -259{\cdot}098\,90 & 514{\cdot}580\,55 & -36{\cdot}035\,15 \\ 343{\cdot}187\,70 & -36{\cdot}035\,15 & 239{\cdot}197\,95 \end{bmatrix}}_{\mathbf{J}} \underbrace{\begin{bmatrix} \hat{m} \\ [\hat{d}] \\ [\hat{h}] \end{bmatrix}}_{\hat{\mathbf{M}}} = \underbrace{\begin{bmatrix} 6115{\cdot}425\,518 \\ -888{\cdot}755\,348 \\ 2036{\cdot}008\,197 \end{bmatrix}}_{\mathbf{S}}$$

where $\mathbf{J}$ is the information matrix, $\hat{\mathbf{M}}$ is the estimate of the parameters and $\mathbf{S}$ is the matrix of scores.

The solution then takes the general form

$$\hat{\mathbf{M}} = \mathbf{J}^{-1}\mathbf{S}$$

where $\mathbf{J}^{-1}$ is the inverse of the information matrix and is a variance-covariance matrix. The inversion may be achieved by any one of a number of standard procedures (Fisher, 1946; Searle, 1966) and one method is illustrated in Section 27, where the procedure of least squares estimation is set out in greater arithmetic detail in relation to the components of variation. For the present data, inversion leads to the following solution.

$$\begin{bmatrix} \hat{m} \\ [\hat{d}] \\ [\hat{h}] \end{bmatrix} = \begin{bmatrix} 0{\cdot}002\,482\,34 & 0{\cdot}001\,011\,16 & -0{\cdot}003\,409\,20 \\ 0{\cdot}001\,011\,16 & 0{\cdot}002\,375\,94 & -0{\cdot}001\,092\,82 \\ -0{\cdot}003\,409\,20 & -0{\cdot}001\,092\,82 & 0{\cdot}008\,907\,33 \end{bmatrix} \begin{bmatrix} 6115{\cdot}425\,518 \\ -888{\cdot}755\,348 \\ 2036{\cdot}008\,197 \end{bmatrix}$$

The estimate of m is then

$$\hat{m} = 0{\cdot}002\,482\,34 \times 6115{\cdot}425\,518 - 0{\cdot}001\,011\,16 \times 888{\cdot}755\,348$$
$$- 0{\cdot}003\,409\,20 \times 2036{\cdot}008\,197 = 7{\cdot}340\,732 = 7{\cdot}341$$

to the accuracy required, and the standard error of $\hat{m}$ is

$$\sqrt{0{\cdot}002\,482\,34} = \pm 0{\cdot}049\,823\,2 = \pm 0{\cdot}050$$

to the accuracy required. In a similar way

$$[\hat{d}] = 1{\cdot}847 \pm 0{\cdot}049$$
and
$$[\hat{h}] = -1{\cdot}742 \pm 0{\cdot}094.$$

All are highly significantly different from zero on an approximate c test.

The adequacy of the additive-dominance model may next be tested by predicting the six family means from the estimates of the three parameters. For example, $\bar{B}_2$ is expected to be

$$7{\cdot}341 - \tfrac{1}{2}(1{\cdot}847) + \tfrac{1}{2}(-1{\cdot}742) = 5{\cdot}546.$$

This expectation along with those for the other families is listed in Table 14. The agreement with the observed values appears to be very close and in no case is the deviation more than 2 % of the observed value. The goodness of fit can be tested by squaring the deviation of the observed from the expected value for each type of family, multiplying by the corresponding weight and summing the products over all six types of families. The contribution made to this sum by $\bar{P}_2$, for example, is

$$(5\cdot494 - 5\cdot475)^2 \times 307\cdot7870 = 0\cdot1111.$$

Summing these contributions over the six types of families gives 4·3702 which is a χ^2 for $(6-3)$ degrees of freedom. This χ^2 has a probability between 0·30 and 0·20 and the model must be judged satisfactory. This conclusion is in complete agreement with the outcome of the A, B and C scaling tests (Table 13).

A summary of the estimates of the three parameters and the test of goodness of fit of the model for the remaining four crosses analysed earlier is given in Table 15. The outcome of these analyses is in complete agreement with the A, B and C scaling tests (Table 13) in showing that the additive-dominance model is adequate for the cross Danmark × Johannisfeuer in both seasons but inadequate for the remaining three crosses. For the latter three crosses the χ^2s, which are the residual error sum of squares in a weighted regression analysis, are significant. In these circumstances the variances on the leading diagonal of the variance-covariance matrix $(\mathbf{J}^{-1})$ must be premultiplied by the χ^2 divided by its degrees of freedom before they can be used to obtain the standard errors of the estimates of m, $[d]$ and $[h]$. Estimates and their standard errors are, however, of limited value when the model is inadequate. They are biased to an unknown extent by effects not attributable to the additive and dominance action of the genes and their standard errors, being derived from the residual error, are based on very few degrees of freedom, in the present case three only. One would rarely find a use for such estimates particularly when, as in these crosses, adequate models can be found (see p. 90 and Tables 23 and 24).

For the cross Danmark × Johannisfeuer the model is adequate, the estimates of $[d]$ and $[h]$ are significantly different from zero, $[h]$ is negative and the

TABLE 15
The results of the joint scaling test for mean number of loculi per fruit in the tomato (Powers, 1941)

Cross	J × D 1938	J × D 1939	J × BB 1939	D × RC 1938	J × RC 1939
m	$7\cdot341 \pm 0\cdot050$	$7\cdot579 \pm 0\cdot051$	$7\cdot556 \pm 0\cdot268$	$3\cdot365 \pm 0\cdot234$	$5\cdot292 \pm 0\cdot368$
$[d]$	$1\cdot847 \pm 0\cdot049$	$1\cdot428 \pm 0\cdot051$	$1\cdot431 \pm 0\cdot263$	$1\cdot331 \pm 0\cdot234$	$3\cdot260 \pm 0\cdot368$
$[h]$	$-1\cdot742 \pm 0\cdot094$	$-1\cdot625 \pm 0\cdot103$	$-1\cdot798 \pm 0\cdot577$	$-1\cdot045 \pm 0\cdot312$	$-2\cdot907 \pm 0\cdot435$
$\chi^2_{(3)}$	$4\cdot370$	$5\cdot027$	$79\cdot659$	$285\cdot178$	$267\cdot866$
P	$0\cdot30-0\cdot20$	$0\cdot20-0\cdot10$	$<0\cdot001$	$<0\cdot001$	$<0\cdot001$

relative values of $[d]$ and $[h]$ more or less equal. For this cross we can therefore conclude that dominance is present and that the genes producing fewer loculi per fruit are in general dominant to their alleles which produce more loculi per fruit.

15. TRANSFORMATION OF THE SCALE

It is possible that a scale could be found which would make the additive-dominance model adequate for all the tomato crosses. This has, however, not yet been obtained, and the possibility cannot be tested further without resort to the original data, which have not been published. The preponderantly negative values of A, B and C suggest a transformation which would foreshorten the upper end of the scale and Powers tried an approximation to the logarithmic transformation on the data. This, though giving a better fit with the model than the untransformed scale, was, he thought, still not fully adequate.

Powers has, however, published the weight per loculus of fruit for the cross Danmark × Red Currant grown in 1938 on both the original scale and on a logarithmic scale (Table 16). The scaling tests on the original scale provide overwhelming evidence of the failure of the additive-dominance model (Table 16). The same tests carried out on the log transformed data provide equally clear evidence of the adequacy of this model on the new scale (Table 16). While we cannot therefore successfully predict the F_2, B_1 and B_2 means from those of P_1, P_2 and F_1 on the original scale we can do so on the log scale. And having made the correct predictions on the log scale we can, by taking antilogs, obtain the expected means of F_2, B_1 and B_2 on the original scale. For example, the expected F_2 mean on the original scale obtained as $\frac{1}{4}\overline{P_1} + \frac{1}{4}\overline{P_2} + \frac{1}{2}\overline{F_1}$ equals 3.86 which differs significantly from the observed mean of 2.12. The antilog of the corresponding prediction made on the log scale is 2.09.

TABLE 16
Danmark × Red Currant 1938 weight per loculus of fruit (Powers, 1951)

Generation	Original scale mean and standard error	Logarithmic scale mean and standard error
P_1	$10\cdot36 \pm 0\cdot581$	$0\cdot9769 \pm 0\cdot02661$
B_1	$4\cdot82 \pm 0\cdot253$	$0\cdot6357 \pm 0\cdot01706$
F_1	$2\cdot33 \pm 0\cdot130$	$0\cdot3346 \pm 0\cdot02673$
F_2	$2\cdot12 \pm 0\cdot105$	$0\cdot2726 \pm 0\cdot01465$
B_2	$0\cdot97 \pm 0\cdot045$	$-0\cdot0512 \pm 0\cdot01467$
P_2	$0\cdot45 \pm 0\cdot017$	$-0\cdot3643 \pm 0\cdot01836$
Scaling test		
A	$-3\cdot05 \pm 0\cdot791$	$-0\cdot0401 \pm 0\cdot05085$
B	$-0\cdot85 \pm 0\cdot159$	$-0\cdot0727 \pm 0\cdot04373$
C	$-6\cdot99 \pm 0\cdot763$	$-0\cdot1914 \pm 0\cdot08565$
Joint	$\chi^2_{(3)} = 96\cdot59$	$\chi^2_{(3)} = 5\cdot66$

The consequences of using an undesirable transformation can also be illustrated by the data of Powers. We have already seen that for the mean number of loculi per fruit in the cross Danmark × Johannisfeuer grown in 1939, the additive-dominance model is adequate on the original scale (Table 12 and 13). Nevertheless, Powers (1951) has examined these data on a logarithmic scale. Application of the scaling tests to the transformed data shows that on this scale the additive-dominance model is inadequate. For example, the $\chi^2_{(3)}$ which tests the goodness of fit of the model on this scale has a value of 11·93 ($P = 0·01$–$0·001$). Hence, the transformation to the log scale to satisfy some other criterion, in this case an attempt to normalize the distributions in the non-segregating parental and F_1 generations, has led to the failure of the additive-dominance model.

The tests of the adequacy of an additive-dominance model may be rendered nugatory by differential viability or fertility in segregating generations (see page 57). Coarser tests then become necessary. One general rule may be mentioned as useful in at least some cases. Reference to the general expectations for the generation means in the selfing series (page 68) shows that $\bar{F}_1, \bar{F}_2, \bar{F}_3$ etc. must all, apart from error variance, lie on the same side of m though with diminishing deviations from it. This holds, as we shall see, even in the presence of interactions between non-allelic genes (Chapter 5) unless they are of a very unusual and extreme form. If, therefore, a scale is obtained upon which $\bar{F}_1$ and $\bar{F}_2$ etc. fall reasonably close to, but on differing sides of, m their departure must be due largely to error variation, which differential viability and fertility may be inflating beyond the level assumed in the precise tests of the adequacy of an additive-dominance model. In such a case the scale must be regarded as appropriate for fitting an additive-dominance model within the limits of the system even though the precise tests show significant discrepancies, for these discrepancies will generally be caused by failure of the genetic assumptions rather than by the use of an inappropriate scale for the analysis.

A case in which the use of such an approximate scale is unavoidable is afforded by the data on corolla length in a species cross *Petunia axillaris* × *P. violaceae* taken from Mather (1949a) and set out in Table 17. The measurements were originally taken in millimetres, but it is quite clear from the C scaling test that an additive-dominance model is inadequate on this scale. Backcrosses are available and they give values of the A and B scaling tests which confirm this inadequacy, though these tests are less trustworthy than the C scaling test because of the possibility of high selective elimination of certain classes of gamete (Mather and Edwards, 1943).

On transforming the data into log measure the agreement with an additive-dominance model is more nearly satisfactory. C is still significantly negative, and this might be taken as suggesting the need for a still stronger transformation. The F_1 and F_2 means are, however, now on opposite sides of the mid-parent value ($m = 5·270 \pm 0·0496$) so showing that, in regard to this

TABLE 17
Corolla length in *Petunia* (Mather, 1949a)

Generation	Original scale mean and standard error	Logarithmic scale mean and standard error
P_1	$61 \cdot 28 \pm 0 \cdot 745$	$7 \cdot 871 \pm 0 \cdot 0585$
F_1	$35 \cdot 20 \pm 0 \cdot 122$	$5 \cdot 467 \pm 0 \cdot 0152$
F_2	$29 \cdot 90 \pm 0 \cdot 366$	$4 \cdot 953 \pm 0 \cdot 1315$
P_2	$18 \cdot 51 \pm 0 \cdot 333$	$2 \cdot 669 \pm 0 \cdot 0800$
C	$-30 \cdot 60 \pm 0 \cdot 408$	$-1 \cdot 662 \pm 0 \cdot 536$

P_1 refers to *P. axillaris*, which is the parent with the larger mean

test at least, the scale is as satisfactory as can be expected. It therefore appears that some agency other than the scale is responsible for the failure of the additive-dominance model. Some differential viability and selective fertilization are to be expected in the offspring of a species cross, and indeed there is evidence of such from the F_2 data themselves (though not to the extent observed in backcrosses). The significant value of C is, therefore, not to be taken as evidence so much that the scale is wrong for the model that is being fitted as of the existence of sources of error in the test which the standard errors of the generation means do not cover. The log scale is consequently an approximation sufficiently good for the purpose of analysis.

Scaling difficulties of this kind, arising from selective fertilization or differential viability, are likely to be especially common with species crosses, but disturbances of the kind produced by these agencies also occur in varietal crosses. Quisenberry (1926) has recorded grain length in two varieties of oats, Sparrowbill and Victor, and in the F_2 and F_3 generations of reciprocal crosses between them, all plants being grown in 1924 (Table 18). From his data

TABLE 18
Grain length in oats (Quisenberry, 1926)

Generation		Mean
P_1		$16 \cdot 36$
P_2		$11 \cdot 48$
F_2	$(P_1 \times P_2)$	$14 \cdot 20$
	$(P_2 \times P_1)$	$14 \cdot 29$
F_3	$(P_1 \times P_2)$	$13 \cdot 77$
	$(P_2 \times P_1)$	$13 \cdot 90$
Scaling test	$(P_1 \times P_2)$	$-1 \cdot 16 \pm 0 \cdot 467$
	$(P_2 \times P_1)$	$-0 \cdot 82 \pm 0 \cdot 449$

P_1 is the larger parent, Victor

summarized in Table 18 we can calculate the scaling test $\bar{F}_3 - \frac{1}{2}\bar{F}_2 - \frac{1}{4}\bar{P}_1 - \frac{1}{4}\bar{P}_2$. The deviation is significant for one reciprocal at least and taking reciprocals together the joint departure from expectation is also significant. Thus the scale is apparently unsuitable for fitting an additive-dominance model. However, the mid-parent value $m = 13\cdot92$, which lies between the F_2 and F_3 means. As in the case of the *Petunia* cross it must therefore be concluded that the scale is satisfactory, the discrepancy revealed by the more exacting test being due to slight departures from simple Mendelian expectations in F_2 and F_3. Such small departures are, however, not likely to prejudice the conclusions obtained from an analysis of the data as they stand.

Sufficient cases have been examined over the years to obtain some idea of the frequency with which the scale chosen for making the measurements proved to be unsatisfactory for fitting an additive-dominance model. In the tomato crosses of Powers we have seen that for one character, number of loculi per fruit, the original scale is satisfactory for only one out of the four crosses investigated. But of the remaining five characters he recorded, four gave a satisfactory fit with an additive-dominance model on the original scale.

In *Nicotiana rustica*, 28 different crosses have been examined from this viewpoint over a number of successive seasons for two characters, time of flowering and final height. When grown in 1952 the original scale was adequate for 18 crosses for final height and for all 28 crosses for flowering-time. When grown in 1953 the corresponding number of crosses was 16 and 14, respectively. A further cross has been examined in 18 different seasons during which time it has been grown in three different localities. For nine of these seasons the original scale was adequate for final height. It would appear, therefore, that rescaling is frequently necessary if we wish to fit the simple additive-dominance model.

Transforming a set of data by taking logarithms is equivalent to measuring the character on a scale graduated like a slide rule. In fact, any transformation is equivalent to the use of a measuring instrument graduated in some appropriate way. It is, however, clearly more expedient in the great majority of cases to take the measurements on a conventional scale and to transform them later, than it is to use a specially graduated instrument. The only drawback is a distortion of the errors of measurement, for equal errors on, say, a scale of millimetres will become unequal when the data are transformed into log millimetres. This is not likely to be a serious consideration, however, unless the range of variation is relatively large.

It is essential when changing the scale that the original measurements themselves be transformed individually, before the means, variances and other statistics are calculated. Otherwise the statistics will be distorted, because, to take an example, the log of a mean is not also the mean of the logs (Mather, 1946b). Nor can we test the adequacy of a log scale by finding geometric means of the statistics based on untransformed measurements. The statistics must be recalculated from the individually transformed measurements.

The transformation to be used in order to adjust the data to a more satisfactory scale must, in the absence of any theoretical considerations, be a matter of expediency. The log transformation is an obvious choice when the upper end of the scale needs foreshortening and an antilog transformation when it needs extending. The log transformation has also been regarded as especially appropriate on the grounds that growth is more likely to proceed on a log scale than on any other. There is no evidence to support this view from the analyses presented here. The decision as to whether or not a log scale should be used must rest on the outcome of the appropriate scaling tests and not on *a priori* arguments derived from mechanical, physiological or biochemical considerations.

Our primary concern must be to secure a scale permitting the type of analysis which is envisaged. In so far as it is desirable to be able to neglect non-allelic interactions, the scale used must be one upon which this neglect will not vitiate conclusions reached about the additive and dominance components. If all the genes bear the same relation to each other the choice of an appropriate scale will free all their individual effects from interactions. Where, however the interactions of one pair of genes differ from those of others, the scale can at best merely be one on which interactions on average are absent. Such residual, though balanced, genic interactions must constitute a source of error variance when fitting an additive-dominance model; but, as Fisher *et al.* (1932) point out, there is no reason to suppose that they will introduce any bias similar to that which arises from the use of an inadequate scale. In other words, residual genic interaction, like non-heritable variation and sampling variation, may lower the precision of the genetic analysis in which an additive-dominance model is fitted but it will not be expected to falsify interpretation where valid estimates of error are calculated.

There are occasions when transforming data to conform to an additive-dominance model is in conflict with the requirement of the data to conform with some different criterion. We have examined such an example in the tomato cross Danmark × Johannisfeuer grown in 1939 (see page 77). Thus for the number of loculi per fruit there is a conflict between a scale which is satisfactory for fitting an additive-dominance model and one which will normalize the distributions within the non-segregating generations. The literature contains many examples of this kind.

In this situation we must decide whether or not it is more important that one requirement rather than the other should be satisfied by the data for the type of analysis that is envisaged. Where the intention is to analyse the data assuming an additive-dominance model there can be no question that the requirement for a scale on which this model fits the data must override all other considerations.

If complementary non-allelic interactions of the classical kind occur in poly-genic systems they are expected to have effects which cannot wholly be removed though they may be reduced in magnitude and perhaps balanced one against another by rescaling. Attempts to find a satisfactory scale for fitting a simple model may therefore be unsuccessful for this reason. We must then consider an alternative to rescaling which is always available, namely the fitting of a model which allows for non-allelic interactions of classical and non-classical kinds. The appropriate procedures are described in the next chapter. This alternative, however, is worthwhile and indeed is possible only if the data are sufficiently extensive to allow estimation of the additional parameters that are necessary to specify the effects of interactions.

5 Components of means: interaction and heterosis

16. NON-ALLELIC INTERACTIONS

If we consider two different genes each with two alleles A–a and B–b, nine genotypes are possible in a diploid organism and eight parameters must be used to give a complete description of the differences among phenotypes. Four of these are the ds and hs appropriate to the two genes, i.e. d_a, d_b, h_a and h_b. The other four may then be derived conveniently to correspond to the interaction comparisons between non-allelic genes. The distribution of these four parameters among the nine genotypes using the Robson definitions of the interactions described by Van der Veen (1959) are shown in Table 19.

TABLE 19
The interactions of two gene pairs

	BB	Bb	bb	F_2 Mean
AA	$d_a + d_b + i_{ab}$	$d_a + h_b + j_{ab}$	$d_a - d_b - i_{ab}$	$d_a + \frac{1}{2}h_b + \frac{1}{2}j_{ab}$
Aa	$h_a + d_b + j_{ba}$	$h_a + h_b + l_{ab}$	$h_a - d_b - j_{ba}$	$h_a + \frac{1}{2}h_b + \frac{1}{2}l_{ab}$
aa	$-d_a + d_b - i_{ab}$	$-d_a + h_b - j_{ab}$	$-d_a - d_b + i_{ab}$	$-d_a + \frac{1}{2}h_b - \frac{1}{2}j_{ab}$
F_2 Mean	$\frac{1}{2}h_a + d_b + \frac{1}{2}j_{ba}$	$\frac{1}{2}h_a + h_b + \frac{1}{2}l_{ab}$	$\frac{1}{2}h_a - d_b - \frac{1}{2}j_{ba}$	$\frac{1}{2}h_a + \frac{1}{2}h_b + \frac{1}{4}l_{ab}$

The parameters fall into three classes. One of these i_{ab} (or $d_a \times d_b$) is the interaction of d_a and d_b and may be termed the homozygote × homozygote interaction. Two others j_{ab} (or $d_a \times h_b$) and j_{ba} (or $d_b \times h_a$) are the homozygote × heterozygote interactions, respectively, of d_a and h_b and d_b and h_a. The last l_{ab} (or $h_a \times h_b$) is the heterozygote × heterozygote interaction of h_a and h_b (see Hayman and Mather, 1955, who however use different definitions of i, j and l).

The coefficients of the interaction parameters are in all cases derived as the products of the coefficients of the corresponding pair of ds and hs taking sign into account. For example:

	d_a	d_b	i_{ab}
AABB	+1	+1	+1
AAbb	+1	−1	−1

83

The four interactions, as defined in this way, have clear genetical meanings, though they do not appear at first sight to be related to the conventional genetical classification of interactions between non-allelic genes. We can, however, cast all the classical types of interactions in terms of i, j and l (Hayman and Mather, 1955; Mather, 1967b). This can be illustrated by reference to an F_2 in which generation all the types of interaction show characteristic segregation ratios.

In an F_2 the nine genotypes of Table 18 occur with the following frequencies:

	AA	Aa	aa
BB	1/16	2/16	1/16
Bb	2/16	4/16	2/16
bb	1/16	2/16	1/16

The standard F_2 segregation into four phenotypic classes with frequencies 9:3:3:1 occur when $d_a = h_a$, $d_b = h_b$ and $i_{ab} = j_{ab} = j_{ba} = l_{ab}$. Thus although this type of F_2 is classically regarded as showing no interaction of the genes, interactions may, in fact, be present within certain restrictions. This and the following relationships which lead to the classical interactions differ from those given by Hayman and Mather (1955) because of the change in the definitions of the interaction parameters to those of the Robson system. If we now add the restriction that $d_a = d_b$ and $d_a = i_{ab}$, i.e. $d_a = d_b = h_a = h_b = i_{ab}$ $= j_{ab} = j_{ba} = l_{ab}$, this gives the 9:7 ratio of classical complementary genes. Equally, we can obtain this ratio when $d_a = d_b = -h_a = -h_b = -i_{ab} = j_{ab}$ $= j_{ba} = -l_{ab}$. Similarly, if we retain the condition that $d_a = d_b$, we obtain the 15:1 ratio of classical duplicate genes by adding the conditions that $d_a = d_b$ $= h_a = h_b = -i_{ab} = -j_{ab} = -j_{ba} = -l_{ab}$ or $d_a = d_b = -h_a = -h_b = i_{ab}$ $= -j_{ab} = -j_{ba} = l_{ab}$. When $d_a \neq d_b$ the complementary interaction becomes the classical recessive epistasis with 9:3:4 in the F_2 given $d_a = h_a = i_{ab} = j_{ab}$ $= j_{ba} = l_{ab}$ and $d_b = h_b$ or $d_a = -h_a = -i_{ab} = j_{ab} = j_{ba} = -l_{ab}$ and $d_b = -h_b$. Similarly, the duplicate relation becomes the classical dominant epistasis, with 12:3:1 in the F_2, given that $d_a = h_a = -i_{ab} = -j_{ab} = -j_{ba} = -l_{ab}$ and $d_b = h_b$ or $d_a = -h_a = i_{ab} = -j_{ab} = -j_{ba} = l_{ab}$ and $d_b = -h_b$. Indeed, any interaction of two genes can be achieved by imposing appropriate conditions.

This approach to the specification of interactions can be extended to cover interactions between three or more genes. With trigenic interactions we should recognize four categories, homozygote × homozygote × homozygote ($d \times d$ $\times d$), homozygote × homozygote × heterozygote ($d \times d \times h$), homozygote × heterozygote × heterozygote ($d \times h \times h$) and heterozygote × heterozygote × heterozygote ($h \times h \times h$). So four new types of parameters would come into the analysis, although two of the types, hom × hom × het and hom × het × het, would each include three individual parameters, making eight in all. To describe the phenotypes of the 27 genotypes produced by three genes requires 26 parameters. Of these 18 are readily available being 3 ds, 3 hs and the three sets of is, js and ls required to describe the digenic interactions among the three possible pairs with three genes. The remaining eight parameters are those required for the trigenic interactions. The coefficients and signs of these eight parameters will follow the same rule as for the digenic interaction, being the products of the coefficients of the corresponding set of 3 ds and hs taking sign into account; for example:

TABLE 20

Coefficients of interaction parameters

Genotype	d_a	d_b	d_c	i_{abc}
AABBCC	+1	+1	+1	+1
AABBcc	+1	+1	−1	−1
AAbbcc	+1	−1	−1	+1 etc.

The phenotype is found as the algebraic sum of all the parameters associated with the genotype in question. The sum of d and h gives a first approximation to the phenotype – one which neglects all interactions. At the next level of approximation we admit the digenic interactions, i, j and l. With a polygenic system the third approximation is obtained by bringing in the eight parameters for trigenic interactions and so on. The successive approximations might, however, be expected to confer less and less advantage. Most of the differences in phenotype will generally (though not, of course, necessarily) be accounted for by d and h, most of the rest by the parameters for digenic interactions and so on. There is, therefore, little justification for considering the more complex situations until a simpler model has been fully explored and has failed to provide an adequate representation of the differences in phenotype under investigation.

The contribution of two interacting genes to the mean expression of a character in the various types of family derivable from a cross between two true breeding lines are shown in Table 21. Two kinds of cross are possible, in respect of the two genes, one where the increasing alleles of the two genes are *associated* in one parent and the decreasing alleles in the other (AABB × aabb); and another where each parental line carries the increasing allele of one gene and the decreasing allele of the other (AAbb × aaBB), that is, the genes are *dispersed*.

Thus the mean expressions of the parental and backcross generations vary with the genic distribution in the parents but the means of the selfing (F) and biparental generations (S) are independent of distribution in the absence of linkage. Free recombination of the genes is assumed in all these formulae (see Section 18).

TABLE 21
Interactions in digenic crosses

(i) Associated AABB × aabb

	m	d_a	d_b	h_a	h_b	i_{ab}	j_{ab}	j_{ba}	l_{ab}
$\bar{P}_1$	1	1	1			1			
$\bar{P}_2$	1	-1	-1			1			
$\bar{F}_1$	1			1	1				1
$\bar{F}_2 = \bar{S}_2$	1			$\frac{1}{2}$	$\frac{1}{2}$				$\frac{1}{4}$
$\bar{F}_3$	1			$\frac{1}{4}$	$\frac{1}{4}$				$\frac{1}{16}$
$\bar{F}_{n+1}$	1			$(\frac{1}{2})^n$	$(\frac{1}{2})^n$				$(\frac{1}{2})^{2n}$
$\bar{S}_3$	1			$\frac{1}{2}$	$\frac{1}{2}$				$\frac{1}{4}$
$\bar{B}_1$	1	$\frac{1}{2}$	$\frac{1}{2}$	$\frac{1}{2}$	$\frac{1}{2}$	$\frac{1}{4}$	$\frac{1}{4}$	$\frac{1}{4}$	$\frac{1}{4}$
$\bar{B}_2$	1	$-\frac{1}{2}$	$-\frac{1}{2}$	$\frac{1}{2}$	$\frac{1}{2}$	$\frac{1}{4}$	$-\frac{1}{4}$	$-\frac{1}{4}$	$\frac{1}{4}$

(ii) Dispersed AAbb × aaBB

	m	d_a	d_b	h_a	h_b	i_{ab}	j_{ab}	j_{ba}	l_{ab}
$\bar{P}_1$	1	1	-1			-1			
$\bar{P}_2$	1	-1	1			-1			
$\bar{B}_1$	1	$\frac{1}{2}$	$-\frac{1}{2}$	$\frac{1}{2}$	$\frac{1}{2}$	$-\frac{1}{4}$	$\frac{1}{4}$	$-\frac{1}{4}$	$\frac{1}{4}$
$\bar{B}_2$	1	$-\frac{1}{2}$	$\frac{1}{2}$	$\frac{1}{2}$	$\frac{1}{2}$	$-\frac{1}{4}$	$-\frac{1}{4}$	$\frac{1}{4}$	$\frac{1}{4}$

The rest are the same as for association

The expectations in Table 21 may be obtained by combining the expectations of the nine phenotypes in respect of two gene differences given in Table 19 in the proportions in which they occur in any one generation. These expectations also follow directly from the expectations of the generation means in terms of ds and hs following the procedures for designating the interactions. For example, the expected generation mean for a first backcross of the F_1 to P_1 with dispersed genes in the absence of interaction is

$$\bar{B}_1 = \tfrac{1}{2}d_a - \tfrac{1}{2}d_b + \tfrac{1}{2}h_a + \tfrac{1}{2}h_b$$

from which it follows that in the presence of interactions this generation mean will contain

$$i_{ab} \text{ with a coefficient of } \tfrac{1}{2} \times -\tfrac{1}{2} = -\tfrac{1}{4}$$

since the coefficients of d_a and d_b are $\tfrac{1}{2}$ and $-\tfrac{1}{2}$ respectively, and similarly

$$j_{ab} \text{ with a coefficient of } \tfrac{1}{2} \times \tfrac{1}{2} = \tfrac{1}{4}$$
$$j_{ba} \text{ with a coefficient of } -\tfrac{1}{2} \times \tfrac{1}{2} = -\tfrac{1}{4}$$

and $\quad l_{ab} \text{ with a coefficient of } \tfrac{1}{2} \times \tfrac{1}{2} = \tfrac{1}{4}.$

Hence the interaction terms for any generation mean can be written down once the expectations of the non-interaction terms are known.

If we consider two true breeding lines that differ in a number of pairs of interacting genes, for every pair of interacting genes that are associated there will be a $+i$ increment to the mean phenotype and for every pair of dispersed genes there will be a $-i$ increment. The net effect of the interaction on the mean measurement will, therefore, be $S(i_+) - S(i_-)$ where $S(i_+)$ is summed over all pairs of associated genes and $S(i_-)$ over all pairs of dispersed genes. The general treatment for any number of pairs of genes arbitrarily distributed between two parental strains has been given by Jinks and Jones (1958).

For the case of two true breeding lines which differ for k genes of which k' alleles of increasing effect will be present in one line along with $k - k'$ alleles of decreasing effect, then of the $\frac{1}{2}k(k-1)$ pairs of genes $k'(k-k')$ are dispersed and will contribute a $-i$ to the mean and the remaining pairs $\frac{1}{2}k(k-1) - k'(k-k')$ will be associated and will contribute a $+i$. The contribution of i type interactions to the line means will, therefore, be

$$S(i) \quad - \quad 2S(i)$$
$$\scriptstyle\frac{1}{2}k(k-1) \qquad k'(k-k')$$

if we now make use of

$$r_i = \frac{\underset{\frac{1}{2}k(k-1)}{S(i)} - \underset{k'(k-k')}{2S(i)}}{\underset{\frac{1}{2}k(k-1)}{S(i)}}$$

the coefficient of association of interacting pairs of genes (page 66), this expression reduces to

$$r_i S(i)$$
$$\scriptstyle\frac{1}{2}k(k-1)$$

which equals $\frac{1}{2}rk(k-1)i$ in the special case where $i_{ab} = i_{bc} = i_{ac} \ldots = i$.

For many pairs of interacting genes we can now write

$$\bar{P}_1 = m + [d] + [i]$$
$$\bar{P}_2 = m - [d] + [i],$$

where for ease of presentation $[i]$ is

$$r_i S(i),$$
$$\scriptstyle\frac{1}{2}k(k-1)$$

irrespective of the distribution of the alleles in the parental lines since these are allowed for in the specifications of $[d]$ and $[i]$.

The F_1 produced by crossing this pair of parental lines will be heterozygous for all k genes and each of the $\frac{1}{2}k(k-1)$ pairs of genes, irrespective of their distribution between the parental lines, will contribute an l increment to the mean measurement. The expected F_1 mean will, therefore, be

$$\bar{F}_1 = m + [h] + [l]$$

where $[l]$ is the balance of all the l increments allowing for the signs of the individual increments of all pairs.

With many pairs of interacting genes the distinction between hom × het (j_{ab}) and het × hom (j_{ba}) interactions can no longer usefully be retained. By the same arguments as for the i type interactions we find that the contribution of these interactions, to, for example, the B_1 generation means is

$$\tfrac{1}{4} r_j \underset{k(k-1)}{S(j)}$$

since we are summing over $k(k-1)$ pairs of genes. This equals $\tfrac{1}{4} rk(k-1)j$ in the special case where $j_{ab} = j_{ba} = j_{bc} = j_{cb} \ldots = j$.

The backcross generation means are then

$$\bar{B}_1 = m + \tfrac{1}{2}[d] + \tfrac{1}{2}[h] + \tfrac{1}{4}[i] + \tfrac{1}{4}[j] + \tfrac{1}{4}[l]$$

and

$$\bar{B}_2 = m - \tfrac{1}{2}[d] + \tfrac{1}{2}[h] + \tfrac{1}{4}[i] - \tfrac{1}{4}[j] + \tfrac{1}{4}[l].$$

where for ease of presentation

$$[j] = r_j \underset{k(k-1)}{S(j)} .$$

All generation means can now be rewritten for an arbitrary number of pairs of genes with arbitrary distribution between the parents by using the $[d]$, $[h]$, $[i]$, $[j]$ and $[l]$ forms of the parameters as defined. There is, however, no need to rederive these expectations since the coefficients of $[i]$, $[j]$ and $[l]$ for any generation follow the same rules as for a single pair of interacting genes, namely, they are the products of the coefficients of the corresponding $[d]$s and $[h]$s taking sign into account.

17. DETECTION OF NON-ALLELIC INTERACTION

We have already seen (Section 14) that the presence of non-allelic interactions can be inferred from the failure to observe the relationships between generation means that are expected on an additive-dominance model either by the individual scaling tests or by the joint scaling test. The expectations for the scaling tests for a pair and for many pairs of interacting genes are set out in Table 22.

These expectations all reduce to zero when no interaction is present, but each type of test depends on characteristic combinations of interactions for its departure from zero. Hence each type of test is capable of detecting its own characteristic constellation of interactions. Thus D provides a test largely of the i type interaction. Test C depends to a greater extent on the l type interaction and C and D combined provide a means of assessing the relative importance of i and l interactions. The j type interactions have no effect on tests C and D but will affect the outcome of the backcross tests A and B.

TABLE 22

Expectation for the scaling tests in the presence of interactions

	Single pair of genes		Many pairs of genes
	Associated	Dispersed	
$A = 2\bar{B}_1 - \bar{P}_1 - \bar{F}_1$	$\frac{1}{2}(-i_{ab} + j_{ab} + j_{ba} - l_{ab})$	$\frac{1}{2}(i_{ab} + j_{ab} - j_{ba} - l_{ab})$	$\frac{1}{2}\{-[i] - [l] + [j]\}$
$B = 2\bar{B}_2 - \bar{P}_2 - \bar{F}_1$	$\frac{1}{2}(-i_{ab} - j_{ab} - j_{ba} - l_{ab})$	$\frac{1}{2}(i_{ab} - j_{ab} + j_{ba} - l_{ab})$	$\frac{1}{2}\{-[i] - [l] - [j]\}$
$C = 4\bar{F}_2 - 2\bar{F}_1 - \bar{P}_1 - \bar{P}_2$	$-2i_{ab} - l_{ab}$	$2i_{ab} - l_{ab}$	$-2[i] - [l]$
$D = 4\bar{F}_3 - 2\bar{F}_2 - \bar{P}_1 - \bar{P}_2$	$-2i_{ab} - \frac{1}{4}l_{ab}$	$2i_{ab} - \frac{1}{4}l_{ab}$	$-2[i] - \frac{1}{4}[l]$

There are obviously conditions under which one or more of these tests will fail to detect non-allelic interactions. Thus with a dispersed pair of genes the two j interactions and the i and l interactions may in part cancel out. With more than two interacting genes cancellation can arise not only from dispersion but also because the direction (sign) of the is, js and ls may differ from one pair of interacting genes to another. This balancing action introduced by differences in sign is always likely to be encountered in the contributions made to means and comparisons between them. It is partly or wholly overcome by turning to second degree statistics (see Chapters 6 and 7).

Provided that sufficient generation means are available we can not only detect the effects of non-allelic interactions on these means but we can also estimate their magnitude. Since there are six parameters, m, $[d]$, $[h]$, $[i]$, $[j]$ and $[l]$, a minimum of six family means are required for their estimation. These can be provided by the parents, F_1, F_2 and first backcross generations of a cross between two true breeding lines. The perfect fit solution is then given by the formulae of Jinks and Jones (1958),

$$m = \tfrac{1}{2}\bar{P}_1 + \tfrac{1}{2}\bar{P}_2 + 4\bar{F}_2 - 2\bar{B}_1 - 2\bar{B}_2$$

$$[d] = \tfrac{1}{2}\bar{P}_1 - \tfrac{1}{2}\bar{P}_2$$

$$[h] = 6\bar{B}_1 + 6\bar{B}_2 - 8\bar{F}_2 - \bar{F}_1 - 1\tfrac{1}{2}\bar{P}_1 - 1\tfrac{1}{2}\bar{P}_2$$

$$[i] = 2\bar{B}_1 + 2\bar{B}_2 - 4\bar{F}_2$$

$$[j] = 2\bar{B}_1 - \bar{P}_1 - 2\bar{B}_2 + \bar{P}_2$$

$$[l] = \bar{P}_1 + \bar{P}_2 + 2\bar{F}_1 + 4\bar{F}_2 - 4\bar{B}_1 - 4\bar{B}_2$$

The standard errors of these estimates can be obtained in the usual way. For example,

$$V_{[d]} = \tfrac{1}{4}V_{\bar{P}_1} + \tfrac{1}{4}V_{\bar{P}_2}$$

$$s_{[d]} = \sqrt{V_{[d]}}$$

and the significance of $[d]$ can be tested by

$$t = \frac{[d]}{s_{[d]}}$$

Clearly, finding [i], [j] or [l] significant by this test is equivalent to finding a significant derivation from zero in the scaling tests. It has the additional advantage, however, of providing estimates of the parameters. This approach to the detection of non-allelic interactions may be illustrated by referring once more to the tomato crosses of Powers (1941). The estimates of the six parameters and their standard errors for the five tomato crosses are given in Table 23. From the cross Danmark × Johannisfeuer grown in both seasons the estimates of the interaction parameters are either smaller than their standard errors or not significantly larger than them. There is, therefore, no evidence of non-allelic interactions in this cross which agrees with the conclusions from the individual and joint scaling tests (Tables 13 and 15). For each of the remaining three crosses, at least two of the estimates of the three interaction parameters are significantly different from zero. This is again in complete agreement with the scaling tests (Tables 13 and 15). For the cross Johannisfeuer × Bonny Best most of the significant interaction can be ascribed to [i] and [l]; for the cross Danmark × Red Currant [i] and [j] are mainly responsible and for Johannisfeuer × Red Currant all three types of interaction are making a contribution.

TABLE 23
Estimates of the additive, dominance and interaction parameters for the mean number of loculi per fruit in the tomato (Powers, 1941)

Cross	D × J 1938	D × J 1939	J × BB 1939	D × RC 1938	J × RC 1939
m	$7\cdot530\pm0\cdot531$	$8\cdot425\pm0\cdot689$	$9\cdot825\pm0\cdot908$	$2\cdot652\pm0\cdot124$	$3\cdot997\pm0\cdot423$
[d]	$1\cdot825\pm0\cdot054$	$1\cdot423\pm0\cdot053$	$1\cdot355\pm0\cdot054$	$1\cdot678\pm0\cdot035$	$3\cdot497\pm0\cdot042$
[h]	$-1\cdot710\pm1\cdot197$	$-4\cdot023\pm1\cdot575$	$-8\cdot680\pm1\cdot991$	$-0\cdot070\pm0\cdot315$	$-2\cdot964\pm1\cdot059$
[i]	$-0\cdot230\pm0\cdot528$	$-0\cdot820\pm0\cdot687$	$-2\cdot152\pm0\cdot906$	$1\cdot076\pm0\cdot119$	$1\cdot534\pm0\cdot420$
[j]	$0\cdot200\pm0\cdot261$	$0\cdot047\pm0\cdot351$	$0\cdot454\pm0\cdot375$	$-0\cdot819\pm0\cdot106$	$-2\cdot648\pm0\cdot294$
[l]	$-0\cdot320\pm0\cdot700$	$1\cdot649\pm0\cdot925$	$5\cdot184\pm1\cdot126$	$-0\cdot187\pm0\cdot198$	$1\cdot484\pm0\cdot644$

Before considering the estimates further we must examine more fully the meaning of some of the parameters. Reference to the equation for estimating m (page 89) shows quite clearly that it no longer corresponds with the original definition of m as the mid-parent value (page 65). Thus once we allow for the presence of interactions the mid-parent is no longer the origin. What then is the origin? Let us return to our consideration of a single pair of interacting genes. The four possible true breeding lines in respect of two gene differences and their expectations are,

$$AABB \quad m + d_a + d_b + i_{ab}$$
$$AAbb \quad m + d_a - d_b - i_{ab}$$
$$aaBB \quad m - d_a + d_b - i_{ab}$$
$$aabb \quad m - d_a - d_b + i_{ab}$$

and the mean of these four true breeding lines is m. However, a cross capable of segregating for both gene differences can be initiated only from one of the two pairs AABB × aabb and AAbb × aaBB whose means are $m + i_{ab}$ and $m - i_{ab}$, respectively. On the other hand, both crosses will, in the absence of linkage, produce all four true breeding lines in equal numbers by prolonged inbreeding. Hence m can be regarded as the mean of the inbred population (F_∞) derived from a cross between two true breeding lines. For this reason this definition of the parameters has been termed by Van der Veen (1959) the F_∞-metric. This definition of m has the advantage that it holds whether or not interactions are present, and if present, irrespective of the number of interacting genes.

For a single pair of interacting genes we can classify the type of interaction on the basis of two criteria (Section 16).

(i) The sign of the hs relative to those of i, and l. For as we have seen certain combinations of sign produce complementary or recessive epistasis, while others produce duplicate behaviour and dominant epistasis (see page 84).

(ii) The relative magnitudes and signs of d_a and d_b and h_a and h_b. For example, differences in magnitude and signs distinguish complementary genes and recessive epistasis, and duplicate genes and dominant epistasis.

With many pairs of interacting genes, however, these criteria do not always hold for $[d]$, $[h]$, $[i]$, $[j]$ and $[l]$. For example, let us consider a situation in which the is for all pairs of interacting genes are positive, i.e. $S(i)$ is positive. Then:

$$[i] = \underset{\frac{1}{2}k(k-1)}{r_i S(i)}$$

will be positive only when r_i is positive and since

$$r_i = \dfrac{\underset{\frac{1}{2}k(k-1)}{S(i)} - \underset{k'(k-k')}{2S(i)}}{\underset{\frac{1}{2}k(k-1)}{S(i)}}$$

this will occur only when

$$\underset{\frac{1}{2}k(k-1)}{S(i)} > \underset{k'(k-k')}{2S(i)}$$

i.e. when the sum of the interactions from dispersed pairs of genes is less than half the sum of all interactions. When the contribution from dispersed pairs is more than half the total r_i will be negative and hence $[i]$ will have the opposite sign to the individual is. The same argument applies when all the is are negative. It also applies when some is are positive and some negative when there is a positive or negative balance. In view of these conditions we cannot safely infer the sign of the is or the sign of the balance of the is from the sign of $[i]$.

Similar considerations show that when there are many pairs of interacting genes the magnitude and sign of the js or of the balance of the js, if they differ in sign, cannot in general be inferred from the magnitude and sign of $[j]$. Indeed, only the two components which are independent of r, namely, $[h]$ and $[l]$, will reflect the magnitude and sign of the individual hs and ls or of their balance. Any classification of interactions which depends on relative magnitudes and signs of the estimates of the six parameters must, therefore, largely depend on the magnitudes and signs of the estimates of $[h]$ and $[l]$. Thus if the hs and ls are predominantly of the same sign then $[h]$ and $[l]$ will be of the same sign and it can be concluded that the interactions are, on balance, mainly of a complementary or recessive epistatic kind. We cannot, however, distinguish between these two alternatives, because this requires knowledge of the relative magnitudes and signs of the individual ds of the interacting genes.

Similarly, if the hs and ls are predominantly of opposite sign then $[h]$ and $[l]$ will be of opposite sign and it can be concluded that the interactions are on balance mainly of the duplicate, dominant epistatic or recessive suppressor kind. Again we cannot distinguish between these alternatives because it requires knowledge of the relative magnitudes and signs of the individual ds and hs of the interacting genes.

In practice, therefore, we can classify interactions into two types only. Those in which $[h]$ and $[l]$ have the same sign – which we will refer to as complementary types; and those in which $[h]$ and $[l]$ have opposite signs – which we will refer to as duplicate types. We should, however, be quite clear that this classification refers only to the predominant type of interaction which is present in cases where different pairs of genes are showing different types of interactions.

Returning to the estimates of the six parameters for the three tomato crosses which show non-allelic interactions (Table 23) we find that for two of them, Johannisfeuer × Bonny Best and Johannisfeuer × Red Currant, $[h]$ is significantly negative and $[l]$ is significantly positive. The interaction appears, therefore, to be predominantly of a duplicate type. For the third cross, Danmark × Red Currant, neither $[h]$ nor $[l]$ is significantly different from zero and we cannot classify the type of interaction.

Reference to Table 23 shows that if we fit a model which allows for digenic interactions to a cross such as Danmark × Johannisfeuer where an additive-dominance model is adequate, the resulting estimates of the additive and dominance components have larger standard errors. For this cross the estimates of the parameters obtained by assuming that interactions are absent provide the better estimates of m, $[d]$ and $[h]$.

Examination of the estimates of the six parameters for the crosses where an interaction model is required shows that some of the interaction parameters are not significant. For example, it appears that a model in which $[j] = 0$ would be adequate for the cross Johannisfeuer × Bonny Best. Fitting a five-parameter model by omitting $[j]$ has two advantages. First it should increase the precision with which the remaining parameters are estimated and second it provides a test for one degree of freedom of the five-parameter model.

The results of fitting the five-parameter model are summarized in Table 24 along with the estimates obtained by fitting the full model for comparison. The five-parameter model is, as expected, adequate, the $\chi^2_{(1)}$ testing the goodness of fit having a probability of 0·30 to 0·20. There has also been an improvement, although a small one, in the precision with which the parameters are estimated as shown by their lower standard errors.

TABLE 24
Estimates of the components of the generation means for the cross Johannisfeuer × Bonny Best fitting a six- and a five-parameter model

	Six-parameter model	Five-parameter model
m	9·83 ± 0·91	10·09 ± 0·88
$[d]$	1·36 ± 0·05	1·37 ± 0·05
$[h]$	−8·68 ± 1·99	−9·48 ± 1·88
$[i]$	−2·15 ± 0·91	−2·42 ± 0·88
$[j]$	0·45 ± 0·37	—
$[l]$	5·18 ± 1·13	5·72 ± 1·04
$\chi^2_{(1)}$	—	1·47

While the outcome of this re-analysis is satisfactory it does not provide a general method for testing the adequacy of a model which allows for digenic interactions but excludes trigenic and higher order interactions. Such a method requires experiments which provide more family means than the minimum number that are necessary for fitting a full digenic interaction model. A suitable experiment for this purpose has been reported by Hill (1966) consisting of the usual parent, F_1, F_2, B_1 and B_2 generations of a cross between two inbred lines (1 and 5) of *Nicotiana rustica* and, in addition, the generations produced by backcrossing the first backcross generations (B_1 and B_2) to each of the two parents, i.e. B_{11} ($B_1 \times P_1$), B_{12} ($B_1 \times P_2$), B_{21} ($B_2 \times P_1$) and B_{22} ($B_2 \times P_2$) and by selfing the first backcrosses, i.e. B_{1S} and B_{2S}. This provides a total of 12 means from different types of family for testing the adequacy of a model which allows for digenic interactions but excludes higher order interactions. The experiment was grown in three successive years, 1960, 1961 and 1962, and final height and flowering-time recorded. For illustration we will first look at the final height of the 12 generations grown in 1961. Table 25 contains the generation means averaged over replications, their weights and their expectations on a model which allows for digenic interactions.

Five of the individual scaling tests that are available for use on these data (Section 14) were significant as was also the joint scaling test; an additive-dominance model is clearly inadequate on the scale of measurement. A model

TABLE 25
Final height in Nicotiana rustica grown in 1961 (Hill, 1966)

Generation	Mean	Weight	Model					
			m	$[d]$	$[h]$	$[i]$	$[j]$	$[l]$
P_1	146·53	0·0605	1	1	0	1	0	0
P_2	120·17	0·1792	1	-1	0	1	0	0
F_1	152·62	0·1490	1	0	1	0	0	1
F_2	141·40	0·1490	1	0	$\frac{1}{2}$	0	0	$\frac{1}{4}$
B_1	141·33	0·1957	1	$\frac{1}{2}$	$\frac{1}{2}$	$\frac{1}{4}$	$\frac{1}{4}$	$\frac{1}{4}$
B_2	132·97	0·3256	1	$-\frac{1}{2}$	$\frac{1}{2}$	$\frac{1}{4}$	$-\frac{1}{4}$	$\frac{1}{4}$
B_{11}	144·37	0·8787	1	$\frac{3}{4}$	$\frac{1}{4}$	$\frac{9}{16}$	$\frac{3}{16}$	$\frac{1}{16}$
B_{12}	144·45	1·3408	1	$-\frac{1}{4}$	$\frac{3}{4}$	$\frac{1}{16}$	$-\frac{3}{16}$	$\frac{9}{16}$
B_{21}	147·57	1·0734	1	$\frac{1}{4}$	$\frac{3}{4}$	$\frac{1}{16}$	$\frac{3}{16}$	$\frac{9}{16}$
B_{22}	128·62	1·6129	1	$-\frac{3}{4}$	$\frac{1}{4}$	$\frac{9}{16}$	$-\frac{3}{16}$	$\frac{1}{16}$
B_{1S}	142·32	1·8430	1	$\frac{1}{2}$	$\frac{1}{4}$	$\frac{1}{4}$	$\frac{1}{8}$	$\frac{1}{16}$
B_{2S}	132·59	2·1262	1	$-\frac{1}{2}$	$\frac{1}{4}$	$\frac{1}{4}$	$-\frac{1}{8}$	$\frac{1}{16}$

Component	
m	$139·00 \pm 2·56$
$[d]$	$12·49 \pm 1·30$
$[h]$	$-6·66 \pm 9·63$
$[i]$	$-4·78 \pm 2·44$
$[j]$	$-8·80 \pm 4·23$
$[l]$	$21·32 \pm 8·80$
$\chi^2_{(6)}$	$5·5560 \quad P = 0·50 - 0·30$

allowing for digenic interactions has therefore been fitted and its adequacy tested by a χ^2 for six degrees of freedom. The estimates of the six parameters in this model and their standard errors are given in Table 25. The estimates of all the parameters except $[h]$ are significant ($P \leqslant 0·05$) but the $\chi^2_{(6)}$ is not significant. Thus the digenic interaction model is adequate, there being no evidence of higher order interactions. This, however, is not always the case. The same analysis carried out for flowering-time shows that a digenic interaction model is inadequate in each of the three years in which the cross was grown.

While the failure of an additive-dominance model is usually a definite indication of non-allelic interactions, failure of a digenic interaction model may indicate either trigenic interactions or linkage of interacting genes or both. These two causes of failure will be examined in the following section.

18. LINKAGE OF INTERACTING GENES

The effects of linkage may be explored by considering a pair of interacting genes A–a and B–b that are linked with a recombination value between them of

$p(=1-q)$. Since linkage has no effect on the means of non-segregating famiIes such as those of true breeding strains or of the F_1 produced by crosses between them, we must begin by considering an F_2. Starting from an initial cross in the coupling phase AABB × aabb the ten genotypes of the F_2 are expected with the frequencies and mean phenotypes shown in Table 26, from which it can be seen that the F_2 mean is

$$\bar{F}_2 = m + \tfrac{1}{2}h_a + \tfrac{1}{2}h_b + \tfrac{1}{2}(1-2p)i_{ab} + \tfrac{1}{2}(1-2pq)l_{ab}.$$

The corresponding mean for an initial cross in the repulsion phase AAbb × aaBB can be obtained by substituting $q(=1-p)$ for p and $p(=1-q)$ for q in the expectation for the coupling phase. Thus

$$\bar{F}_2 = m + \tfrac{1}{2}h_a + \tfrac{1}{2}h_b \pm \tfrac{1}{2}(1-2p)i_{ab} + \tfrac{1}{2}(1-2pq)l_{ab}$$

where the $\pm$ term should be read as $+$ for coupling and $-$ for repulsion.

TABLE 26
Frequencies and mean phenotypes of the 10 genotypic classes in an F_2 for two interacting genes linked in coupling

	AA	Aa	aa
BB	q^2 $d_a + d_b + i_{ab}$	$2pq$ $h_a + d_b + j_{ba}$	p^2 $-d_a + d_b - i_{ab}$
Bb	$2pq$ $d_a + h_b + j_{ab}$	$2q^2$ $h_a + h_b$ $+ l_{ab}$　/　$h_a + h_b$ $+ l_{ab}$ $2p^2$	$2pq$ $-d_a + h_b - j_{ab}$
bb	p^2 $d_a - d_b - i_{ab}$	$2pq$ $h_a - d_b - j_{ba}$	q^2 $-d_a - d_b + i_{ab}$

All the frequencies should be divided by 4.

A similar treatment of the means of other segregating generations with respect to two interacting, linked genes leads to the expectations given in Table 27a. From these expectations we can draw the following conclusions.

(i) In the absence of interactions ($i_{ab} = j_{ab} = j_{ba} = l_{ab} = 0$) linkage has no effect on the generation means. This has been implicitly assumed throughout the discussion of non-interacting models without formal proof although it is easy to see that this must be so. Thus linkage affects neither the frequencies with which the alleles of a gene (A:a) are recovered in segregating generations, nor the frequencies with which the three genetic classes in respect of each gene (AA:Aa:aa) occur. Linkage only affects the frequencies with which the alleles of one gene and the genetic classes in respect of this gene occur in combination with the alleles and genetic classes of another gene to which it is linked. Hence where the increments added to the phenotype by the different genes are independent of one another, i.e. there is no interaction between them, linkage clearly cannot of itself have any effect on the mean measurement of segregating

TABLE 27(a)

The generation means with respect to two linked, interacting genes (after Van der Veen, 1959)

Generation	Coupling phase	Repulsion phase
P_1	$m + d_a + d_b + i_{ab}$	$m + d_a - d_b - i_{ab}$
P_2	$m - d_a - d_b + i_{ab}$	$m - d_a + d_b - i_{ab}$
F_1	$m + h_a + h_b + l_{ab}$	$m + h_a + h_b + l_{ab}$
F_2	$m + \frac{1}{2}h_a + \frac{1}{2}h_b + \frac{1}{2}(1-2p)i_{ab} + \frac{1}{4}(1-2pq)l_{ab}$	$m + \frac{1}{2}h_a + \frac{1}{2}h_b - \frac{1}{2}(1-2p)i_{ab} + \frac{1}{4}(1-2pq)l_{ab}$
F_3	$m + \frac{1}{4}h_a + \frac{1}{4}h_b + [\frac{1}{2}(1-2p)^2]i_{ab} + \frac{1}{4}(1-2pq)^2 l_{ab}$	$m + \frac{1}{4}h_a + \frac{1}{4}h_b - [\frac{1}{2}(1-2p)^2]i_{ab} + \frac{1}{4}(1-2pq)^2 l_{ab}$
B_1	$m + \frac{1}{2}d_a + \frac{1}{2}d_b + \frac{1}{2}h_a + \frac{1}{2}h_b + \frac{1}{2}(1-p)i_{ab} + \frac{1}{2}p j_{ab} + \frac{1}{2}p j_{ba} + \frac{1}{2}(1-p)l_{ab}$	$m + \frac{1}{2}d_a + \frac{1}{2}h_b + \frac{1}{2}h_a + \frac{1}{2}h_b - \frac{1}{2}(1-2p)i_{ab} + \frac{1}{4}(1-2p)^2 l_{ab} + \frac{1}{4}(1-2pq)^2 l_{ab} - \frac{1}{2}p j_{ba} + \frac{1}{2}(1-p)l_{ab}$
B_2	$m - \frac{1}{2}d_a - \frac{1}{2}d_b + \frac{1}{2}h_a + \frac{1}{2}h_b + \frac{1}{2}(1-p)i_{ab} - \frac{1}{2}p j_{ba} + \frac{1}{2}(1-p)l_{ab}$	$m - \frac{1}{2}d_a + \frac{1}{2}d_b + \frac{1}{2}h_a + \frac{1}{2}h_b - \frac{1}{2}(1-p)i_{ab} - \frac{1}{2}p j_{ab} + \frac{1}{2}p j_{ba} + \frac{1}{2}(1-p)l_{ab}$
B_{11}	$m + \frac{3}{4}d_a + \frac{3}{4}d_b + \frac{1}{4}h_a + \frac{1}{4}h_b + \frac{1}{4}(2+q^2)i_{ab} + \frac{1}{4}(1-q^2)l_{ab} + \frac{1}{4}(1-q^2)j_{ba} + \frac{1}{4}q^2 l_{ab}$	$m + \frac{3}{4}d_a - \frac{3}{4}d_b + \frac{1}{4}h_a + \frac{1}{4}h_b - \frac{1}{4}(2-q^2)i_{ab} + \frac{1}{4}(1-q^2)j_{ab} - \frac{1}{4}(1-q^2)j_{ba} + \frac{1}{4}q^2 l_{ab}$
B_{12}	$m - \frac{1}{4}d_a - \frac{1}{4}d_b + \frac{3}{4}h_a + \frac{3}{4}h_b + \frac{1}{4}(2+q^2)l_{ab} - \frac{1}{4}(1-q^2)j_{ba} + \frac{1}{4}(2+q^2)l_{ab}$	$m - \frac{1}{4}d_a + \frac{1}{4}d_b + \frac{3}{4}h_a + \frac{3}{4}h_b - \frac{1}{4}q^2 i_{ab} - \frac{1}{4}q^2 l_{ab} + \frac{1}{4}(1-q^2)j_{ba} + \frac{1}{4}(2+q^2)l_{ab}$
B_{21}	$m + \frac{1}{4}d_a + \frac{1}{4}d_b + \frac{3}{4}h_a + \frac{3}{4}h_b + \frac{1}{4}(2+q^2)l_{ab} + \frac{1}{4}(1-q^2)j_{ba} + \frac{1}{4}(2+q^2)l_{ab}$	$m + \frac{1}{4}d_a - \frac{1}{4}d_b + \frac{3}{4}h_a + \frac{3}{4}h_b - \frac{1}{4}q^2 i_{ab} - \frac{1}{4}q^2 l_{ab} + \frac{1}{4}(1-q^2)j_{ba} + \frac{1}{4}(2+q^2)l_{ab}$
B_{22}	$m - \frac{3}{4}d_a - \frac{3}{4}d_b + \frac{1}{4}h_a + \frac{1}{4}h_b + \frac{1}{4}(2+q^2)i_{ab} - \frac{1}{4}(1-q^2)l_{ab} - \frac{1}{4}(1-q^2)j_{ba} + \frac{1}{4}q^2 l_{ab}$	$m - \frac{3}{4}d_a + \frac{3}{4}d_b + \frac{1}{4}h_a + \frac{1}{4}h_b - \frac{1}{4}(2-q^2)i_{ab} - \frac{1}{4}(1-q^2)l_{ab} + \frac{1}{4}(1-q^2)j_{ba} + \frac{1}{4}q^2 l_{ab}$

TABLE 27(b)

Expected generation means in the presence of linked digenic interactions (for simplicity $[pi_{ab}]$ is written as pi etc.)

Generation	Components										
	d	$m+h+l$	$m+i$	pi	p^2i	pj	p^2j	pl	p^2l	p^3l	p^4l
P_1	1		1								
P_2	-1		1								
F_1		1									
F_2		$\frac{1}{2}$	$\frac{1}{2}$	-1				-1	1		
B_1	$\frac{1}{2}$	$\frac{1}{2}$	$\frac{1}{2}$	$-\frac{1}{2}$		$\frac{1}{2}$		$-\frac{1}{2}$			
B_2	$-\frac{1}{2}$	$\frac{1}{2}$	$\frac{1}{2}$	$-\frac{1}{2}$			$-\frac{1}{2}$	$-\frac{1}{2}$			
F_3		$\frac{1}{4}$	$\frac{3}{4}$	-2	1			-1	2	-2	1
$F_2 \times P_1 (L_1)$	$\frac{1}{2}$	$\frac{1}{2}$	$\frac{1}{2}$	$-\frac{3}{4}$	$\frac{1}{2}$	$\frac{3}{4}$	$-\frac{1}{2}$	$-\frac{3}{4}$	$\frac{1}{2}$		
$F_2 \times P_2 (L_2)$	$-\frac{1}{2}$	$\frac{1}{2}$	$\frac{1}{2}$	$-\frac{3}{4}$	$\frac{1}{2}$	$-\frac{3}{4}$	$\frac{1}{2}$	$-\frac{3}{4}$	$\frac{1}{2}$		
$F_2 \times F_1 (L_3)$		$\frac{1}{2}$	$\frac{1}{2}$	$-\frac{5}{4}$	$\frac{1}{2}$			$-\frac{5}{4}$	2	-1	
F_{2bip}		$\frac{1}{2}$	$\frac{1}{2}$	$-\frac{3}{2}$	1			$-\frac{3}{2}$	$\frac{13}{4}$	-3	1
B_{11}	$\frac{3}{4}$	$\frac{1}{4}$	$\frac{3}{4}$	$-\frac{1}{2}$	$\frac{1}{4}$	$\frac{1}{2}$	$-\frac{1}{4}$	$-\frac{1}{2}$	$\frac{1}{4}$		
B_{12}	$-\frac{1}{4}$	$\frac{3}{4}$	$\frac{1}{4}$	$-\frac{1}{2}$	$\frac{1}{4}$	$-\frac{1}{2}$	$\frac{1}{4}$	$-\frac{1}{2}$	$\frac{1}{4}$		
B_{22}	$-\frac{3}{4}$	$\frac{1}{4}$	$\frac{3}{4}$	$-\frac{1}{2}$	$\frac{1}{4}$	$-\frac{1}{2}$	$\frac{1}{4}$	$-\frac{1}{2}$	$\frac{1}{4}$		
B_{21}	$\frac{1}{4}$	$\frac{3}{4}$	$\frac{1}{4}$	$-\frac{1}{2}$	$\frac{1}{4}$	$\frac{1}{2}$	$-\frac{1}{4}$	$-\frac{1}{2}$	$\frac{1}{4}$		
$B_1 \times F_1$	$\frac{1}{4}$	$\frac{1}{2}$	$\frac{1}{2}$	-1	$\frac{1}{4}$	$\frac{1}{4}$		-1	$\frac{5}{4}$	$-\frac{1}{2}$	
$B_2 \times F_1$	$-\frac{1}{4}$	$\frac{1}{2}$	$\frac{1}{2}$	-1	$\frac{1}{4}$	$-\frac{1}{4}$		-1	$\frac{5}{4}$	$-\frac{1}{2}$	
B_{1bip}	$\frac{1}{2}$	$\frac{3}{8}$	$\frac{5}{8}$	-1	$\frac{1}{2}$	$\frac{1}{2}$	$-\frac{1}{4}$	-1	$\frac{3}{2}$	-1	$\frac{1}{4}$
B_{2bip}	$-\frac{1}{2}$	$\frac{3}{8}$	$\frac{5}{8}$	-1	$\frac{1}{2}$	$-\frac{1}{2}$	$\frac{1}{4}$	-1	$\frac{3}{2}$	-1	$\frac{1}{4}$
B_{1S}	$\frac{1}{2}$	$\frac{1}{4}$	$\frac{3}{4}$	$-\frac{5}{4}$	$\frac{1}{2}$	$\frac{1}{4}$		$-\frac{3}{4}$	1	$-\frac{1}{2}$	
B_{2S}	$-\frac{1}{2}$	$\frac{1}{4}$	$\frac{3}{4}$	$-\frac{5}{4}$	$\frac{1}{2}$	$-\frac{1}{4}$		$-\frac{3}{4}$	1	$-\frac{1}{2}$	

families. Linkage, therefore, does not vitiate the scaling tests nor does it bias the estimates of m, $[d]$ and $[h]$ in the absence of interactions.

(ii) In the presence of interactions the contribution the interacting genes make to the means of segregating generations depends on the magnitude of the linkage.

(iii) In the form given in Table 27a, the number of new parameters required to specify the contributions of linked interacting genes to the generation means always exceeds the number of generation means available for their estimation. However, Jinks and Perkins (1969) have shown that these expectations can be rewritten in a form in which all the interaction parameters are ascending powers of p. For example, the contribution of digenic interactions between homozygous combinations can be expressed in terms of the parameters i_{ab}, $p_{ab}i_{ab}$, $p_{ab}^2 i_{ab}$ and $p_{ab}^3 i_{ab}$ etc. and similarly for the j and l type interactions.

The expectations for the generation means expressed in this way for an arbitrary number of pairs of linked genes showing digenic interactions are given in Table 27b. In these expectations m, $[h]$ and $[l]$ are completely correlated as are also m and $[i]$. Hence $(m+[h]+[l])$ and $(m+[i])$ must be treated as two parameters only in estimating the parameters in these expectations. Thus by expressing the linkage components as ascending powers of p and recognizing these correlations, the number of parameters in the expectations is considerably reduced and they can be estimated from an appropriately designed experiment.

The minimum experiment for detecting linkage of genes showing digenic interactions consists of the six types of family P_1, P_2, F_1, F_2, B_1 and B_2 required to estimate digenic interactions in the absence of linkage plus the four double backcrosses $B_1 \times P_1 (B_{11})$, $B_1 \times P_2 (B_{12})$, $B_2 \times P_1 (B_{21})$ and $B_2 \times P_2 (B_{22})$. Their expected generation means require eight parameters for their specification leaving two degrees of freedom for testing the goodness of fit of the model.

A line of enquiry that can be profitably pursued further is the extent to which the estimates of the six parameters of a digenic interaction model which assumes no linkage are biased when linkage is present. This can be pursued by estimating the six parameters from the expectation of the parents, F_1, F_2 and backcross generations in Table 27 using the equations appropriate for no linkage. The outcome for two genes linked in the coupling and in the repulsion phase is summarized in Table 28.

Linkage clearly biases the estimates of all parameters except $d_a \pm d_b$ ($= [d]$ for many genes) which is the only parameter estimated entirely from the means of non-segregating generations. It is, however, possible to obtain estimates of certain combinations of parameters that are unbiased by linkage by using the means of the three non-segregating generations. For example, from the parental and F_1 generations means we can estimate

$$\tfrac{1}{2}(\bar{P}_1 - \bar{P}_2) = d_a \pm d_b = [d] \text{ for any number of genes}$$
$$\tfrac{1}{2}(\bar{P}_1 + \bar{P}_2) = m \pm i_{ab} = m + [i] \text{ for any number of genes}$$
$$\bar{F}_1 - \tfrac{1}{2}(\bar{P}_1 + \bar{P}_2) = h_a + h_b + l_{ab} \mp i_{ab} = [h] + [l] - [i].$$

These estimates have no value for separating the effects of additive, dominance and interaction components on the generation means. But as we shall see later (Section 21) they have a particular significance with regard to the specification of heterosis.

By inspection of the expectations of the generation means for pairs of linked, interacting genes (Table 27) it is possible to find relationships between them similar to, but of course considerably more complex than, those used as the basis for testing the adequacy of additive-dominance models. Van der Veen (1959) has described three such relationships, two of which (X and Y in Table 29) involve only commonly grown types of family and the third (Z in Table 29) involving a more unusual family obtained by crossing the first backcross generations B_1 and B_2 to their F_1 parents (B_1F_1 and B_2F_1). These relationships which sum to zero in the absence of trigenic or higher order interactions can therefore be used as unambiguous tests for the latter.

TABLE 28

The effect of linkage on the estimates of the parameters of the digenic interaction model assuming no linkage

Estimate	No Linkage		Linkage	
	Association	Dispersion	Coupling	Repulsion
$\frac{1}{2}(\bar{P}_1 - \bar{P}_2)$	$d_a + d_b$	$d_a - d_b$	$d_a + d_b$	$d_a - d_b$
$6\bar{B}_1 + 6\bar{B}_2 - 8\bar{F}_2 - \bar{F}_1 - 1\frac{1}{2}\bar{P}_1 - 1\frac{1}{2}\bar{P}_2$	$h_a + h_b$	$h_a + h_b$	$h_a + h_b - (1-2p)i_{ab}$ $+ (1-2p)(1+4p)l_{ab}$	$h_a + h_b + (1-2p)i_{ab}$ $+ (1-2p)(1+4p)l_{ab}$
$2\bar{B}_1 + 2\bar{B}_2 - 4\bar{F}_2$	i_{ab}	$-i_{ab}$	$2pi_{ab} + 2p(1-2p)l_{ab}$	$-2pi_{ab} + 2p(1-2p)l_{ab}$
$2\bar{B}_1 - \bar{P}_1 - 2\bar{B}_2 + \bar{P}_2$	$j_{ab} + j_{ba}$	$j_{ab} - j_{ba}$	$2pj_{ab} + 2pj_{ba}$	$2pj_{ab} - 2pj_{ba}$
$\bar{P}_1 + \bar{P}_2 + 2\bar{F}_1 + 4\bar{F}_2 - 4\bar{B}_1 - 4B_2$	l_{ab}	l_{ab}	$4p^2 l_{ab}$	$4p^2 l_{ab}$
$\frac{1}{2}\bar{P}_1 + \frac{1}{2}\bar{P}_2 + 4\bar{F}_2 - 2\bar{B}_1 - 2\bar{B}_2$	m	m	$m + (1-2p)i_{ab} - 2p(1-2p)l_{ab}$	$m - (1-2p)i_{ab} - 2p(1-2p)l_{ab}$

TABLE 29
Scaling tests for detecting trigenic and higher order interactions

Test	Expectation in the absence of trigenic and higher order interactions
X	$\frac{1}{2}(\bar{P}_1 - \bar{P}_2) - (\bar{B}_{11} + \bar{B}_{12}) + (\bar{B}_{21} + \bar{B}_{22}) = 0$
Y	$\bar{F}_1 - \frac{1}{2}(\bar{P}_1 + \bar{P}_2) + (\bar{B}_{11} - \bar{B}_{12}) - (\bar{B}_{21} - \bar{B}_{22}) = 0$
Z	$(\bar{B}_1 - \bar{B}_2) - 2(\bar{B}_1\bar{F}_1 - \bar{B}_2\bar{F}_1) = 0$

Two of these tests, X and Y, can be carried out on the data of Hill (1966) on flowering-time in *Nicotiana rustica* where, as we have seen, a digenic interaction model is inadequate. The results of the tests for each of the three years in which the experiment was grown are summarized in Table 30. X is significantly different from zero ($P \leqslant 0.05$) in all three years, but its magnitude and sign differ from one year to another. Y is significant only in 1961 and differs significantly in magnitude and sign from year to year. There is, therefore, evidence of trigenic and higher order interactions for flowering-time but the magnitude and direction of these interactions is not consistent over years.

TABLE 30
The results of the X and Y scaling tests on flowering-time in *N. rustica* (Hill, 1966)

Test	Year		
	1960	1961	1962
X	$4{\cdot}325 \pm 0{\cdot}946$	$3{\cdot}300 \pm 0{\cdot}814$	$-1{\cdot}350 \pm 0{\cdot}661$
Y	$1{\cdot}245 \pm 1{\cdot}089$	$-2{\cdot}210 \pm 0{\cdot}988$	$-0{\cdot}440 \pm 0{\cdot}794$

The next step is to fit a trigenic interaction model to these data. On doing this Hill found a satisfactory fit for 1961 but not for 1960 or 1962. In 1960 and 1962, therefore, there appears to be either linkage of interacting genes or interactions involving four or more genes at a time.

The contributions of trigenic interactions to the means of the 21 generations in Table 27b are given in Table 31. By comparing these expectations with those in Table 27b it can be seen that F_2, $F_2 \times F_1$ and $F_{2\,\text{bip}}$ have the same means in the presence of trigenic interactions but not in the presence of linked digenic interactions. In the absence of linkage these three generations have identical gene and genotype frequencies but in its presence they have different genotype frequencies because the opportunities for recombination increase in the sequence F_2, $F_2 \times F_1$ and $F_{2\,\text{bip}}$. This also applies to the B_1 and $F_2 \times P_1$ and B_2 and $F_2 \times P_2$ generations. We can, therefore, specifically test for the presence of linked interacting genes by comparing the observed means of the appropriate

TABLE 31
Expected generation means in the presence of digenic and trigenic non-allelic interactions

Generation	$[m]$	$[d]$	$[h]$	$[i_{ab/}]$	$[j_{a/b}]$	$[l_{/ab}]$	$[i_{abc/}]$	$[j_{ab/c}]$	$[j_{a/bc}]$	$[l_{/abc}]$
P_1	1	1		1			1			
P_2	1	-1		1			-1			
F_1	1		1			1				1
F_2	1		$\frac{1}{2}$			$\frac{1}{4}$				$\frac{1}{8}$
B_1	1	$\frac{1}{2}$	$\frac{1}{2}$	$\frac{1}{4}$	$\frac{1}{4}$	$\frac{1}{4}$	$\frac{1}{8}$	$\frac{1}{8}$	$\frac{1}{8}$	$\frac{1}{8}$
B_2	1	$-\frac{1}{2}$	$\frac{1}{2}$	$\frac{1}{4}$	$-\frac{1}{4}$	$\frac{1}{4}$	$-\frac{1}{8}$	$\frac{1}{8}$	$-\frac{1}{8}$	$\frac{1}{8}$
F_3	1		$\frac{1}{4}$			$\frac{1}{16}$				$\frac{1}{64}$
$F_2 \times P_1(L_1)$	1	$\frac{1}{2}$	$\frac{1}{2}$	$\frac{1}{4}$	$\frac{1}{4}$	$\frac{1}{4}$	$\frac{1}{8}$	$\frac{1}{8}$	$\frac{1}{8}$	$\frac{1}{8}$
$F_2 \times P_2(L_2)$	1	$-\frac{1}{2}$	$\frac{1}{2}$	$\frac{1}{4}$	$-\frac{1}{4}$	$\frac{1}{4}$	$-\frac{1}{8}$	$\frac{1}{8}$	$-\frac{1}{8}$	$\frac{1}{8}$
$F_2 \times F_1(L_3)$	1		$\frac{1}{2}$			$\frac{1}{4}$				$\frac{1}{8}$
$F_{2\,bip}$	1		$\frac{1}{2}$			$\frac{1}{4}$				$\frac{1}{8}$
B_{11}	1	$\frac{3}{4}$	$\frac{1}{4}$	$\frac{9}{16}$	$\frac{3}{16}$	$\frac{1}{16}$	$\frac{27}{64}$	$\frac{9}{64}$	$\frac{3}{64}$	$\frac{1}{64}$
B_{12}	1	$-\frac{1}{4}$	$\frac{3}{4}$	$\frac{1}{16}$	$-\frac{3}{16}$	$\frac{9}{16}$	$-\frac{1}{64}$	$\frac{3}{64}$	$-\frac{9}{64}$	$\frac{27}{64}$
B_{22}	1	$-\frac{3}{4}$	$\frac{1}{4}$	$\frac{9}{16}$	$-\frac{3}{16}$	$\frac{1}{16}$	$-\frac{27}{64}$	$\frac{9}{64}$	$-\frac{3}{64}$	$\frac{1}{64}$
B_{21}	1	$\frac{1}{4}$	$\frac{3}{4}$	$\frac{1}{16}$	$\frac{3}{16}$	$\frac{9}{16}$	$\frac{1}{64}$	$\frac{3}{64}$	$\frac{9}{64}$	$\frac{27}{64}$
$B_1 \times F_1$	1	$\frac{1}{4}$	$\frac{1}{2}$	$\frac{1}{16}$	$\frac{1}{8}$	$\frac{1}{4}$	$\frac{1}{64}$	$\frac{1}{32}$	$\frac{1}{16}$	$\frac{1}{8}$
$B_2 \times F_1$	1	$-\frac{1}{4}$	$\frac{1}{2}$	$\frac{1}{16}$	$-\frac{1}{8}$	$\frac{1}{4}$	$-\frac{1}{64}$	$\frac{1}{32}$	$-\frac{1}{16}$	$\frac{1}{8}$
$B_{1\,bip}$	1	$\frac{1}{2}$	$\frac{3}{8}$	$\frac{1}{4}$	$\frac{3}{16}$	$\frac{9}{64}$	$\frac{1}{8}$	$\frac{3}{32}$	$\frac{9}{128}$	$\frac{27}{512}$
$B_{2\,bip}$	1	$-\frac{1}{2}$	$\frac{3}{8}$	$\frac{1}{4}$	$-\frac{3}{16}$	$\frac{9}{64}$	$-\frac{1}{8}$	$\frac{3}{32}$	$-\frac{9}{128}$	$\frac{27}{512}$
B_{1S}	1	$\frac{1}{2}$	$\frac{1}{4}$	$\frac{1}{4}$	$\frac{1}{8}$	$\frac{1}{16}$	$\frac{1}{8}$	$\frac{1}{16}$	$\frac{1}{32}$	$\frac{1}{64}$
B_{2S}	1	$-\frac{1}{2}$	$\frac{1}{4}$	$\frac{1}{4}$	$-\frac{1}{8}$	$\frac{1}{16}$	$-\frac{1}{8}$	$\frac{1}{16}$	$-\frac{1}{32}$	$\frac{1}{64}$

pairs of generations (Jinks, 1978). By combining these comparisons with those in Table 29 we can detect trigenic or higher order interactions and linked digenic interactions unambiguously even when both are present.

The experiment of Hill (1966) did not include the generations necessary for testing for the presence of linked interacting genes. Jinks and Perkins (1969) have, however, described an experiment which is an extension of that of Hill to include all 21 generations. Their means and amounts of information (the reciprocal of the variances of the mean) for the character final height are given in Table 32.

Of the three tests (X, Y and Z of Table 29) for trigenic and higher order interactions only Y approaches significance for final height ($P = 0.04 - 0.05$). On the other hand, while the tests for linked interacting genes show no significant differences between the means of B_1 and $F_2 \times P_1$ and of B_2 and $F_2 \times P_2$ they show highly significant differences among the means of F_2, $F_2 \times F_1$ and $F_{2\,bip}$ ($P < 0.001$). There is, therefore, stronger evidence for linkage than for higher order interactions.

Jinks and Perkins (1969) have tested this conclusion by fitting models to all 21 generations in Table 32. The χ^2s testing the goodness of fit of the weighted least squares estimates of the parameters of an additive-dominance and a digenic interaction model show that both are inadequate (Table 33). Hence,

TABLE 32
Generation means averaged over two blocks and their corresponding
weights for the character plant height in *Nicotiana rustica*

Generation	Mean	Number of plants	Weights
P_1	153·57	160	0·7972
P_2	126·80	100	0·9327
F_1	155·70	140	1·3961
F_2	145·39	300	1·3590
B_1	156·29	160	0·8751
B_2	142·14	160	1·0367
F_3	146·18	400	1·1540
$F_2 \times P_1 (L_1)$	157·73	400	2·1154
$F_2 \times P_2 (L_2)$	142·49	400	2·1025
$F_2 \times F_1 (L_3)$	149·68	400	1·6383
F_{2bip}	151·74	400	1·3672
B_{11}	157·48	140	0·7532
B_{12}	147·52	140	0·6834
B_{22}	135·13	140	0·8470
B_{21}	155·35	140	0·6641
$B_1 \times F_1$	151·56	280	1·1320
$B_2 \times F_1$	144·58	280	0·9773
B_{1bip}	155·70	210	1·1288
B_{2bip}	143·99	210	0·7627
B_{1S}	152·17	280	1·2300
B_{2S}	141·78	280	0·9249

TABLE 33
χ^2 tests of the goodness of fit of models of increasing complexity for final height in
Nicotiana rustica

Model	No. of parameters	χ^2	Degrees of freedom	P
Additive-dominance	3	68·27	18	<0·001
Digenic interactions	6	60·77	15	<0·001
Trigenic interactions	10	38·38	11	<0·001
Linked digenics	11	17·80	10	0·10–0·05

while epistasis is present it is more complex than digenic interaction. However, a model which allows for both digenic and trigenic interaction among unlinked genes is also inadequate. On the other hand, digenic interactions between linked pairs of genes give a satisfactory description of the differences among the generation means. We can conclude, therefore, that linkage rather than higher order interactions is responsible for the failure of the digenic interaction model.

19. GENOTYPE × ENVIRONMENT INTERACTION: SPECIFICATION AND DETECTION

Although we have had occasion to comment on the similarities and differences between the results of growing the same cross in different years or in different localities, all the analyses so far undertaken have avoided as far as possible the additional complication of deliberately imposed environmental differences by being based on families which have been grown simultaneously in the same locality and in the same year. On the other hand, some environmental differences arise inevitably even within a single experiment grown in the same locality in the same season due to heterogeneity of the environment over which the individuals must be distributed. This leads to differences between individuals belonging to the same non-segregating generation grown in the same experiment. And it is, of course, the source of the error variances of the generation means.

The specification of the environmental contribution to the phenotype depends on the experimental design and this, in turn, determines the specification of the genotype × environment interactions. Following Mather and Jones (1958) we will begin by considering the simplest case, that of two true breeding lines differing by a single gene A–a grown in two environments X and Y (Table 34). Three parameters are necessary to describe completely the differences among the four phenotypes. One is d. A similar parameter, e_1, may be used to represent the differences between the two environments, being the amount by which the mean phenotypes in the two environments differ from the overall mean (m). The third parameter, g_{d1}, measures the interaction of the genetical and environmental components d and e_1. The relative signs of the three parameters are so defined that they are orthogonal to one another when equal numbers of individuals are observed in each of the four situations.

The first and most obvious development from this simple case is to consider two true breeding lines differing by an arbitrary number of genes. This requires only that $[d]$ is substituted for d in Table 34. A practical application of this model is provided by an experiment in which true breeding varieties, 1 and 5 of *Nicotiana rustica,* and generations derived from an initial cross between them

TABLE 34
The four phenotypes of two genotypes in two environments expressed as deviations from the overall mean, m

Environments	Genotypes		Mean
	AA	aa	
X	$d + e_1 + g_{d1}$	$-d + e_1 - g_{d1}$	e_1
Y	$d - e_1 - g_{d1}$	$-d - e_1 + g_{d1}$	$-e_1$
Mean	d	$-d$	0

were grown in each of two locations in 1965 (Bucio Alanis, 1966). In the design used plants belonging to the same family were individually randomized within each locality. The mean final heights of varieties 1 and 5 in the two localities and their standard errors based on the differences between individuals belonging to the same variety randomized within the same locality are given in Table 35.

TABLE 35
The final heights of varieties 1 and 5 of *Nicotiana rustica* grown in two locations

Variety	Location	Mean and standard error	Weight	m	$[d]$	e_1	g_{d1}
1	1	$153\cdot57 \pm 1\cdot499$	$0\cdot4446$	1	1	1	1
1	2	$124\cdot97 \pm 2\cdot807$	$0\cdot1269$	1	1	-1	-1
5	1	$126\cdot82 \pm 1\cdot112$	$0\cdot8006$	1	-1	1	-1
5	2	$104\cdot04 \pm 2\cdot497$	$0\cdot1602$	1	-1	-1	1

The analysis may proceed along a number of alternative lines. For example, we can estimate each of the four parameters (including m) and their standard errors and test the significance of the deviation of each parameter from zero. Then

$$\hat{m} = \tfrac{1}{4}(153\cdot57 + 124\cdot97 + 126\cdot82 + 104\cdot04) = 127\cdot35$$

$$[\hat{d}] = \tfrac{1}{4}(153\cdot57 + 124\cdot97 - 126\cdot82 - 104\cdot04) = 11\cdot92$$

$$\hat{e}_1 = \tfrac{1}{4}(153\cdot57 - 124\cdot97 + 126\cdot82 - 104\cdot04) = 12\cdot85$$

$$\hat{g}_{d1} = \tfrac{1}{4}(153\cdot57 - 124\cdot97 - 126\cdot82 + 104\cdot04) = 1\cdot45$$

The variance of the estimates is given by

$$\tfrac{1}{16}(1\cdot499^2 + 2\cdot807^2 + 1\cdot112^2 + 2\cdot497^2) = 1\cdot099\,86$$

and the standard error becomes $\pm 1\cdot049$ from which it can be seen that there is no evidence of genotype × environment interaction although there are significant genetic and environmental effects. The formulae for the estimation of g_{d1} and its standard error constitute, of course, a scaling test, which tells us whether a model which assumes that genotype × environmental interactions are absent is adequate and as such it is strictly comparably with the scaling tests for detecting non-allelic interactions. Indeed we can approach the whole analysis from the viewpoint of a joint scaling test by fitting a model consisting of m, $[d]$ and e_1 only, which assumes that there are no interactions. This allows us to obtain weighted least squares estimates of the three parameters leaving one degree of freedom for testing the adequacy of the non-interactive model. The outcome of such an analysis on the *N. rustica* data is summarized in the second column of Table 36. This confirms the significance of $[d]$ and e_1 and the adequacy of the model.

TABLE 36
Estimates of the genetic and environmental components of the generation means for final height in *Nicotiana rustica* (Bucio Alanis, 1966) and tests of the adequacy of non-interactive models

Components	Parents only	Parents and F_1s
m	127.45 ± 1.069	127.43 ± 1.052
$[d]$	12.67 ± 0.904	12.67 ± 0.907
$[h]$		14.43 ± 2.593
e_1	12.67 ± 1.067	12.78 ± 1.003
χ^2	$1.84(1\ df)$	$1.80(2\ df)$

A further elaboration of the model is to include heterozygotes and their interactions with the environment. For the single gene case this requires in addition to h and e_1, which we have already defined, g_{h1} a parameter corresponding with g_{d1} and representing the interaction of h with e_1.

Environments	Genotype Aa
X	$h + e_1 + g_{h1}$
Y	$h - e_1 - g_{h1}$

Again, the model can be extended to any number of heterozygous loci by substituting $[h]$ for h. We cannot estimate the new parameters from data consisting solely of F_1 generations grown in the two environments. We can do so, however, by combining observations on both parents and their F_1s grown simultaneously in two environments. The *N. rustica* varieties 1 and 5 and the F_1 between them will serve as an illustration. These provide us with six means from which to estimate the six parameters of a full interaction model. We can, however, test the need for including the two parameters for genotype × environmental interactions by fitting a model consisting of m, $[d]$, $[h]$ and e_1 by weighted least squares, and testing its goodness of fit using a χ^2 for two degrees of freedom (last column Table 36). The outcome of the analysis confirms that based on parents alone in showing that a model which assumes no interactions is perfectly adequate.

If we return to the earlier analysis of the tomato cross Johannisfeuer × Danmark which was grown in 1938 and 1939 (Tables 13, 15 and 23) we find that the estimates of $[d]$ differ significantly in the two seasons. If we carry out the analysis just described on the parents and F_1s grown in the two years for the characters, number of locules per fruit and mean plant spread, we find that a model which assumes no genotype × environment interaction is inadequate ($\chi^2_{(2)} = 28.94$ and 25.73, respectively). We must, therefore, fit the interaction model. The estimates of the six parameters for each character are listed in Table 37. These confirm the need for the inclusion of the interaction parameters in that the estimate of g_{d1} is highly significant for both characters.

TABLE 37

Estimates of the genetic, environmental and genotype × environment interaction components of the generation means in tomato

Component	Character	
	Number of locules	Mean spread
m	7.45 ± 0.038	99.47 ± 1.309
$[d]$	1.62 ± 0.038	12.42 ± 1.309
$[h]$	1.68 ± 0.075	25.54 ± 2.677
e_1	0.15 ± 0.038	26.55 ± 1.309
g_{d1}	0.20 ± 0.038	-6.15 ± 1.309
g_{h1}	0.12 ± 0.075	3.76 ± 2.677

1939, the year in which the characters had the greater mean expression, was defined as $+e_1$ and 1938 as $-e_1$ in the model.

By contrast the estimates of g_{h1} are smaller and not significantly different from zero although $[h]$ is as great or greater than $[d]$ in magnitude. Hence, both absolutely and relatively, the homozygotes show a greater interaction with the environment than do the heterozygotes. This is a frequent finding to which we shall have occasion to refer again.

So far we have considered only two environments but the model is readily extendable to any number of environments since no matter how numerous they may be, their differences are expressible by a series of orthogonal comparisons equal in number to the degrees of freedom among them (Mather and Jones, 1958). With two environments there is one such comparison represented in our simple case by e_1. With three environments there are two comparisons which can be represented by parameters donated by e_1 and e_2 and so on. With many environments the average phenotypes in the different environments can be combined in a variety of ways to give appropriate sets of orthogonal comparisons but unless there is a relation of special interest to us among the environments we may choose whatever set of comparisons happens to be most convenient for representing the environmental differences. This is illustrated for three different situations involving four environments in Table 38. All represent common situations in experimental breeding.

TABLE 38

Appropriate orthogonal comparisons for the differences between four environments. The first set are arbitrary and are represented as four years, the second set are factorial and are represented as two sites in two years and the third set are hierarchical and are represented as two blocks within each of two sites

Environments – arbitrary				Environments – factorial					Environments – hierarchical				
Years	e_1	e_2	e_3	Years	Sites	e_1	e_2	e_3	Sites	Blocks	e_1	e_2	e_3
1	1	1	1	1	A	1	1	1	A	I	1	1	
2	1	−1	−1	1	B	1	−1	−1	A	II	1	−1	
3	−1	1	−1	2	A	−1	1	−1	B	III	−1		1
4	−1	−1	1	2	B	−1	−1	1	B	IV	−1		−1

In the first case the four environments are four years. In the absence of any meteorological or other basis for grouping the years, the es must be assigned in a purely arbitrary way. The set of comparisons given in Table 38 is just one of many alternatives that would be equally satisfactory. It follows from this, of course, that the estimates of the three es have no particular meaning in terms of the environmental factors contributing differences between years. Indeed under these circumstances it is questionable whether or not anything is gained from subdividing the differences between environments by the use of orthogonal components. It may well be more useful to define four parameters e_1, e_2, e_3 and e_4 where $e_1 + e_2 + e_3 + e_4 = 0$ and e_1 is the sum of all environmental effects in environment 1, e_2 the same for environment 2, etc.

In the second case the four environments are the factorial combinations of two years and two sites. In this case the orthogonal set of comparisons is fixed by the design and each has a fixed meaning. Thus e_1 represents the differences between years, e_2 the difference between sites and e_3 the interaction between years and sites. In the third case the design again determines the comparisons, the four environments consisting of two sites each subdivided into two blocks. The es are then fixed as e_1 representing the difference between sites and e_2 and e_3 the difference between blocks in site A and site B, respectively. More complex designs can be accommodated by extending the appropriate example in Table 38 or by combining two or more of them.

As the number of es is increased to accommodate more environments so the number of gs will grow too. With three environments there will be four gs corresponding to the interaction of additive [d] and dominance [h] effects with e_1 and e_2, namely, g_{d1} and g_{d2} and g_{h1} and g_{h2}, respectively. With four environments there will be six gs and so on. Since there are a number of ways of assigning the es there are a corresponding number of ways of assigning the gs. But the g comparisons must always follow from the particular partition used for the es, each taking a coefficient and sign that is the product of the coefficients and signs of the corresponding [d] or [h] and e. Where, because of its arbitrary nature, there is no point in partitioning the differences between the average phenotype in each environment between an orthogonal set of es, there is equally no point in partitioning the genotype × environment interactions among any orthogonal set of gs. The approach to be used under these circumstances can be illustrated by an experiment with N. rustica in which varieties 1 and 5 and the F_1 between them were grown along with other generations derived from them over 16 years (Bucio Alanis, 1966).

Let us begin by considering the expected performance of the two varieties in any one of the environments, for example in environment j, where j may be 1 to 16.

$$\bar{P}_{1j} = m + [d] + e_j + g_{dj}$$

and
$$\bar{P}_{2j} = m - [d] + e_j - g_{dj},$$

where P_1 is variety 5 and P_2 is variety 1.

The mean of these two values is

$$\tfrac{1}{2}(\bar{P}_{1j} + \bar{P}_{2j}) = m + e_j$$

and half the difference between them is

$$\tfrac{1}{2}(\bar{P}_{1j} - \bar{P}_{2j}) = [d] + g_{dj}.$$

Now m is the mid-parent value averaged over all environments and $[d]$ is half the mean difference between the parents averaged over all environments. Equally the sum of all es over environments is zero and the sum of all gs over environments is zero. Hence the e for any one environment is the mid-parent value in that environment minus the average mid-parent value over all environments. Similarly, the g_d for any one environment is half the difference between the parents in that environment minus the average value of this difference over all environments.

Analysis of the F_1 generation grown in the same 16 environments follows identical lines. In any one environment, for example in environment j,

$$\bar{F}_{1j} = m + [h] + e_j + g_{hj}.$$

The deviation of this F_1 from the mid-parent value in the same environment is

$$\bar{F}_{1j} - \tfrac{1}{2}(\bar{P}_{1j} + \bar{P}_{2j}) = [h] + g_{hj}.$$

The average value of this deviation over all environments is equal to $[h]$. Equally the sum of all $g_h s$ over environments is zero. Hence, the difference between this deviation for any one environment and its average value over all environments estimates g_h for that environment. The estimates of m, $[d]$ and $[h]$ for the 16 environments are

$$\hat{m} = 112{\cdot}27$$
$$[\hat{d}] = 7{\cdot}34$$
$$[\hat{h}] = 13{\cdot}94$$

and the estimates of the es, $g_d s$ and $g_h s$ are listed in Table 39. These estimates could be subdivided among any arbitrary set of 15 orthogonal es, $g_d s$ and $g_h s$ if useful purpose would be served by doing so. The 16 environments do, in fact, include unavoidable changes in location as well as deliberate changes in plant husbandry but these are confounded with one another and with the naturally occurring differences between years. No useful purpose is, therefore, served by attempting a partition.

It is particularly noticeable that in general the es are greater than the $g_d s$ which are in turn greater than the $g_h s$ and that all three are correlated in magnitude and sign over the 16 environments. It appears that once again the heterozygotes interact less with the environment than the homozygotes. Furthermore, the magnitudes of these interactions are determined by the environment over the range used. Following Bucio Alanis (1966) we find that

TABLE 39
Estimates of the environmental and genotype × environment interaction components of the generation means for final height in N. rustica grown in 16 years

Year (j)	e_j	g_{dj}	g_{hj}
1	0·08	5·03	0·05
2	1·70	6·04	1·30
3	−0·96	3·68	−0·38
4	−12·83	−2·21	−3·30
5	−6·68	−3·86	−3·53
6	−16·10	−8·56	−1·78
7	−12·34	−4·29	−8·28
8	−2·31	−3·18	2·54
9	−4·75	−6·91	−1·75
10	−25·02	−11·13	−6·12
11	4·42	−3·56	2·46
12	−8·97	2·64	1·47
13	26·72	7·92	8·69
14	21·08	5·84	5·38
15	13·11	3·15	1·17
16	22·84	10·03	2·08

the regressions of the gs on the corresponding es are significant and linear (Fig. 8), the regression coefficients being $b_d = 0·352 \pm 0·066$ and $b_h = 0·236 \pm 0·043$, respectively. That is, $g_{dj} = b_d e_j$ and $g_{hj} = b_h e_j$. In this particular set of results, therefore, we can rewrite the expectations of the generation means in any environment, for example, environment j, as

$$\bar{P}_{1j} = m + [d] + (1 + b_d)e_j$$
$$\bar{P}_{2j} = m - [d] + (1 - b_d)e_j$$
$$\bar{F}_{1j} = m + [h] + (1 + b_h)e_j.$$

On substitution of the estimates these become

$$\bar{P}_{1j} = 119·61 + 1·35\, e_j$$
$$\bar{P}_{2j} = 104·93 + 0·65\, e_j$$
$$\bar{F}_{1j} = 126·21 + 1·24\, e_j.$$

Put in this form we can see explicitly that:

(i) The expected means are the sum of two components, mean performance and environmental sensitivity. Mean performance which is the expected mean in the average environment ($e_j = 0$) is determined solely by the values of m, $[d]$ and $[h]$ which are independent of the environment. Environmental sensitivity which is the expected rate of change of the mean with change in the additive environmental value is determined solely by the value of $1 + b_d$, $1 - b_d$ and $1 + b_h$ which are the coefficients of the linear regressions of the means of P_{1j}, P_{2j} and F_{1j} on e_j, respectively.

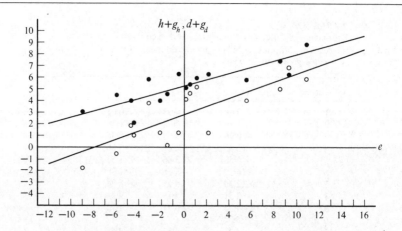

Figure 8 The regressions of g_d (open circles) and g_h (filled circles) measuring the genotype × environment interaction, on e measuring the mean effect of the environment in an experiment with *Nicotiana rustica*. (Reproduced by permission, with minor changes, from Bucio Alanis and Hill, 1966.) Note the coordinates are graduated in inches whereas cm are used in the text.

Not surprising, therefore, these three equations can be derived directly by regressing the observed $\bar{P}_{1j}$, $\bar{P}_{2j}$ and $\bar{F}_{1j}$ on to $\frac{1}{2}(\bar{P}_{1j} + \bar{P}_{2j})$ or e_j.

(ii) The estimates of $[h]$ and b_h are both positive, the F_1 resembling the higher scoring parent, P_1 in mean performance and in environmental sensitivity. Increasing alleles are therefore dominant more often than recessive for both aspects of the phenotype and P_1 has more of the dominant increasing alleles than P_2.

(iii) The difference between the two parents is a linear function of the environment and it is larger in good environments $(+es)$ than in poor environments $(-es)$. In the poorest environments, however, the difference first becomes zero and then increases again but with P_2 now the larger. The e_j value at which this cross over occurs can be estimated as

$$\bar{P}_{1j} = \bar{P}_{2j};\ m + [d] + (1 + b_d)e_j = m - [d] + (1 - b_d)e_j$$

whereupon $\quad 2[d] = -2b_d e_j$

and $\quad e_j = -\dfrac{[d]}{b_d}.$

On substitution this becomes $-\dfrac{7\cdot34}{0\cdot35} = -20\cdot97$, or $8\cdot2$ inches (Fig. 8).

When, therefore, the environment deteriorates below an e_j value of $-20\cdot97$, $\bar{P}_{2j} > \bar{P}_{1j}$. Reference to Table 39 shows that such an environment occurred in year 10 $(e_j = -25\cdot02)$ and this resulted in $\bar{P}_{2j}$ exceeding $\bar{P}_{1j}$ by $3\cdot79$ cm.

(iv) Because $\bar{F}_{ij} > \bar{P}_{1j}$ there is heterosis in most environments but because $b_{\mathrm{d}} > b_{\mathrm{h}}$ it is smaller for good environments than for poor environments (see Section 21).

20. GENOTYPE × ENVIRONMENT INTERACTION: SEGREGATING GENERATIONS

The same parameters and procedures can be used to specify and analyse genotype × environment interactions in segregating generations. The specification for an F_2 segregating in respect of a single gene difference A–a grown in two environments X and Y is given in Table 40.

TABLE 40
The six phenoypes expected from the three genotypes in an F_2 segregating for a single gene difference in two environments

Genotype Frequencies	AA $\frac{1}{4}$	Aa $\frac{1}{2}$	aa $\frac{1}{4}$	Mean
Environments				
X	$d + e_1 + g_{\mathrm{d}1}$	$h + e_1 + g_{\mathrm{h}1}$	$-d + e_1 - g_{\mathrm{d}1}$	$e_1 + \frac{1}{2}h + \frac{1}{2}g_{\mathrm{h}1}$
Y	$d - e_1 - g_{\mathrm{d}1}$	$h - e_1 - g_{\mathrm{h}1}$	$-d - e_1 + g_{\mathrm{d}1}$	$-e_1 + \frac{1}{2}h - \frac{1}{2}g_{\mathrm{h}1}$
Mean	d	h	$-d$	$\frac{1}{2}h$

These expectations for a single gene difference in two environments can be extended to any number of genes, in any number of environments by the appropriate procedure described for parental and F_1 generations. We can also derive comparable expressions for any generation in which the proportions of the homozygotes and heterozygotes are known in respect of any gene difference. There is, however, no need to go through these derivations since they follow automatically once the specification of the genetic and environmental contributions to a generation mean is known. We merely have to add to them the appropriate gs defined in the way already described. For example, let us consider the backcross to the larger parent (B_1) in respect of any number of genes, grown in two environments. The genetic part of the expectation is known to be

$$\bar{B}_1 = m + \tfrac{1}{2}[d] + \tfrac{1}{2}[h].$$

The specification of the environment is $+e_1$ and $-e_1$. We need, therefore a $g_{\mathrm{d}1}$ and a $g_{\mathrm{h}1}$ with coefficients that are the products of those of the corresponding e_1s and those of the $[d]$ and $[h]$ components, respectively. The complete specification is then

$$\bar{B}_{1\cdot1} = m + \tfrac{1}{2}[d] + \tfrac{1}{2}[h] + e_1 + \tfrac{1}{2}g_{\mathrm{d}1} + \tfrac{1}{2}g_{\mathrm{h}1}$$

and

$$\bar{B}_{1\cdot2} = m + \tfrac{1}{2}[d] + \tfrac{1}{2}[h] - e_1 - \tfrac{1}{2}g_{\mathrm{d}1} - \tfrac{1}{2}g_{\mathrm{h}1}$$

for the two environments.

If we assume that an additive dominance model is adequate for specifying the genetical component of the generation means we can combine parents, F_1s, F_2s and backcrosses grown in two or more environments to obtain weighted least squares estimates of the genetical, environmental and genotype × environmental interaction parameters and at the same time test the goodness of fit of the model in the now familiar way (Bucio Alanis *et al.*, 1969).

If the procedure used for non-segregating generations is followed, the first step in the analysis is to determine whether or not a model which assumes no genotype × environment interactions is adequate. If the χ^2 for testing the goodness of fit of such a model is not significant a model consisting of m, $[d]$, $[h]$ and the appropriate number of es is satisfactory. If, however, this χ^2 is significant one or more of the following explanations is possible.

(i) The specification of the genetical contribution is inadequate, that is, non-allelic interaction is present.
(ii) The specification of the environmental contribution is inadequate.
(iii) Genotype × environmental interaction is present.

To conclude that genotype × environmental interaction is present it is first necessary to exclude the first two explanations. Since the genetical parameters in the model are defined and estimated as the average effects over all environments, the presence of non-allelic interaction can be detected by fitting an m, $[d]$ and $[h]$ model to the generation means averaged over all environments. If this model is satisfactory the presence of non-allelic interactions is not the cause of the failure of the first model. If, however, this model is unsatisfactory it is necessary to include parameters for non-allelic interactions in the specification of the generation means.

Finding an m, $[d]$ and $[h]$ model satisfactory when fitted to the family mean averaged over all environments does not exclude the possibility that non-allelic interactions are present in some of the individual environments but in such a form that their effects balance out when averaged over all environments. If this were the case it might be necessary to include the interactions of the non-allelic interactions with the environment in the model even though overall the non-allelic interactions themselves were undetectable. To test whether or not this is a possible cause of the failure of the first m, $[d]$, $[h]$ and e model it is necessary to test the goodness of fit of an m, $[d]$, $[h]$ model to the family means in each individual environment in turn. If this model is satisfactory not only for the means averaged over all environments, but also for each individual environment, it can be concluded that it is not necessary to include non-allelic interactions in the model in any form.

Incomplete specification of the environment cannot be the cause of the inadequacy of the first m, $[d]$, $[h]$ and e model providing that they have been correctly allotted a set of parameters which is one less than the number of environments. It could be the cause if the number of es had been reduced below this number by unjustified assumptions about the relationships between the environments. This situation can be readily remedied by reverting to a complete specification.

Once we have satisfied ourselves that the genetical and environmental specifications are adequate, any failure of the m, $[d]$, $[h]$ and e model can be ascribed to the presence of genotype × environment interactions. The full model can then be fitted still leaving sufficient degrees of freedom to confirm that there is no further cause of failure of the model.

The procedures just described may be illustrated by the analysis of data already referred to, namely final height in the generations derived from crossing varieties 1 and 5 of *N. rustica* and mean plant spread in the tomato cross Danmark × Johannisfeuer. For each of the crosses the F_2 and backcross generations grown in two environments are available in addition to the non-segregating generations previously analysed (Tables 35 to 37). The full data available are summarized in Table 41.

A preliminary analysis showed that an additive-dominance model is satisfactory for each set of generation means in each environment as well as for the generation means averaged over environments. We can take it, therefore, that non-allelic interactions and their interactions with the environment are absent. To test for the presence of genotype × environment interactions an m, $[d]$, $[h]$ and e_1 model was fitted to the generation means for both environments. This model was satisfactory for the *N. rustica* experiment ($\chi^2_{(8)} = 7.74$) thus confirming the analyses based on non-segregating generations (Tables 35 and 36), and estimates of the four parameters of this model are given in Table 41. These show the expected close agreement with the earlier estimates based on the parental and F_1 generations and a reduction in the standard errors of the estimates because of the increased amount of information.

The non-interactive model proved to be unsatisfactory for the tomato experiment, the $\chi^2_{(8)} = 44.95$ testing the goodness of fit being highly significant. Again this confirms the earlier analysis of non-segregating generations (Table 37). To complete the analysis of this experiment we must, therefore, fit the full interaction model given in Table 41. This is satisfactory ($\chi^2_{(6)} = 4.80$), as indeed it must be if the conclusions based on the earlier analysis of these data were correct. The weighted least squares estimates of the six parameters in the full model are given in Table 41. Apart from their lower standard errors, which are expected, these estimates are almost identical with those obtained from parental and F_1 generation means (Table 37). These results show once again that the interactions are small relative to the genetic and environmental effects. Again the interaction with heterozygous genotypes is smaller than that with homozygotes, though in this case not significantly so.

In the case of the *Nicotiana rustica* experiment carried out by Bucio Alanis *et al.* (1969) and mentioned earlier, an additive-dominance genetic model was adequate for the P_1 (variety 5), P_2 (variety 1), F_1, F_2, B_1, B_2 and the genotype × environmental interactions of the P_1, P_2 and F_1s were linear functions of the

TABLE 41
The analysis of the generation means for final height in *Nicotiana rustica* and plant spread in tomato grown in two environments

Generation	Environment	N. rustica Mean	N. rustica Weight	Tomato Mean	Tomato Weight	m	d	h	e_1	g_{d1}	g_{h1}
P_1	1	153·57	0·4446	144·59	0·2261	1	1		1	1	
P_1	2	124·97	0·1269	79·20	0·0812	1	1		-1	-1	
P_2	1	126·82	0·8006	107·45	0·1422	1	-1		1	-1	
P_2	2	104·04	0·1602	66·65	0·2744	1	-1		-1	1	
F_1	1	155·70	0·7166	155·32	0·3787	1		1	1		1
F_1	2	128·88	0·1255	94·70	0·0521	1		1	-1		-1
F_2	1	146·43	0·6835	139·65	0·1450	1		$\frac{1}{2}$	1		$\frac{1}{2}$
F_2	2	123·34	0·1792	84·25	0·1622	1		$\frac{1}{2}$	-1		$-\frac{1}{2}$
B_1	1	156·31	0·4496	153·49	0·2510	1	$\frac{1}{2}$	$\frac{1}{2}$	1	$\frac{1}{2}$	$\frac{1}{2}$
B_1	2	124·36	0·6229	88·15	0·1409	1	$\frac{1}{2}$	$\frac{1}{2}$	-1	$-\frac{1}{2}$	$-\frac{1}{2}$
B_2	1	142·16	0·5191	130·46	0·2663	1	$-\frac{1}{2}$	$\frac{1}{2}$	1	$-\frac{1}{2}$	$\frac{1}{2}$
B_2	2	113·69	0·1506	77·05	0·1988	1	$-\frac{1}{2}$	$\frac{1}{2}$	-1	$\frac{1}{2}$	$-\frac{1}{2}$

Component		
m	127·15 ± 0·866	99·75 ± 1·165
$[d]$	12·85 ± 0·810	13·55 ± 1·119
$[h]$	15·21 ± 0·410	24·69 ± 2·253
e_1	13·23 ± 0·615	26·43 ± 1·165
g_{d1}		6·13 ± 1·119
g_{h1}		4·85 ± 2·253
χ^2	7·74 (8 df)	4·80 (6 df)

additive-environmental components. Since, as we have seen, the expected F_2 and first backcross generation means in environment j are given by the expressions

$$\bar{F}_{2j} = m + \tfrac{1}{2}[h] + e_j + \tfrac{1}{2}g_{hj}$$
$$\bar{B}_{1j} = m + \tfrac{1}{2}[d] + \tfrac{1}{2}[h] + e_j + \tfrac{1}{2}g_{dj} + \tfrac{1}{2}g_{hj}$$

and
$$\bar{B}_{2j} = m - \tfrac{1}{2}[d] + \tfrac{1}{2}[h] + e_j - \tfrac{1}{2}g_{dj} + \tfrac{1}{2}g_{hj},$$

whereupon

$$\bar{F}_{2j} = m + \tfrac{1}{2}[h] + (1 + \tfrac{1}{2}b_h)e_j$$
$$\bar{B}_{1j} = m + \tfrac{1}{2}[d] + \tfrac{1}{2}[h] + (1 + \tfrac{1}{2}b_d + \tfrac{1}{2}b_h)e_j$$

and
$$\bar{B}_{2j} = m - \tfrac{1}{2}[d] + \tfrac{1}{2}[h] + (1 - \tfrac{1}{2}b_d + \tfrac{1}{2}b_h)e_j.$$

If we now insert the estimates of m, $[d]$, $[h]$, b_d and b_h obtained from our earlier analysis of the parents and F_1 (pages 108 and 109) these become:

$$\bar{F}_{2j} = 119\cdot25 + 1\cdot12e_j$$
$$\bar{B}_{1j} = 122\cdot91 + 1\cdot29e_j$$
$$\bar{B}_{2j} = 115\cdot57 + 0\cdot94e_j.$$

To test these expectations we can regress the F_2, B_1 and B_2 means in each of the sixteen environments against the additive environmental values e_j in Table 38 or alternatively against $\tfrac{1}{2}(\bar{P}_{1j} + \bar{P}_{2j}) = m + e_j$. The three resulting regression equations are

$$\bar{F}_{2j} = 121\cdot41 + 1\cdot10e_j$$
$$\bar{B}_{1j} = 122\cdot50 + 1\cdot24e_j$$
$$\bar{B}_{2j} = 116\cdot89 + 0\cdot93e_j$$

which do not differ significantly in either their mean performances or environmental sensitivities from those predicted from the analysis of parents and F_1.

Since the model involving m, $[d]$, $[h]$, e, b_d and b_h is clearly adequate for the data from all the generations we can, of course, combine these results to obtain weighted least squares estimates of the parameters (Jinks and Perkins, 1970), instead of using estimates from the parental and F_1 generations to predict the F_2 and first backcrosses.

An investigation of the sensitivity of mean number of sternopleural chaetae in the Samarkand (P_1) and Wellington (P_2) inbred lines of *Drosophila melanogaster* and their F_1 and F_2 to differences among six environments created by all combinations of two temperatures, 18 and 25°C and three types of culture container has produced similar results (Mather and Caligari, 1974; Mather and Jinks, 1977). The responses of these four types of family can once again be completely described by linear regression equations obtained by regressing the mean of each genotype in each of the six environments on their additive environmental values e_j. The resulting equations

$$\bar{P}_{1j} = 20 \cdot 56 - 0 \cdot 226 e_j$$
$$\bar{P}_{2j} = 18 \cdot 79 + 2 \cdot 226 e_j$$
$$\bar{F}_{1j} = 19 \cdot 46 + 1 \cdot 836 e_j$$
$$\bar{F}_{2j} = 19 \cdot 45 + 1 \cdot 603 e_j$$

can also be obtained by regressing, for example, the deviation of the mean of each genotype from the mid-parent value in each environment against the mid-parent value in that environment $(m + e_j)$ or by any of the other methods illustrated by the N. rustica data. It might be noted that as the coefficient of e_j is negative in the expression for $\bar{P}_{1j}$, while in the expression for $\bar{P}_{2j}$ it is greater than 2, $b_d > 1$ and $\bar{P}_{1j}$ and $\bar{P}_{2j}$ are not merely changing differently as e changes but must, in fact, be changing in opposite directions.

Assuming an additive dominance model for both mean performance and environmental sensitivity we can, as described earlier, predict the regression equation for the F_2 from the P_1, P_2 and F_1. We can do so either by first estimating m, $[d]$, $[h]$, b_d and b_h from the regression equation of $\bar{P}_{1j}$, $\bar{P}_{2j}$ and $\bar{F}_{1j}$ and then substituting them into the formula

$$\bar{F}_{2j} = m + \tfrac{1}{2}[h] + (1 + \tfrac{1}{2}b_h)e_j$$

or directly by using the standard scaling test formula $\bar{F}_{2j} = \tfrac{1}{4}\bar{P}_{1j} + \tfrac{1}{4}\bar{P}_{2j} + \tfrac{1}{2}\bar{F}_{1j}$. By either method the expected $\bar{F}_{1j} = 19 \cdot 57 + 1 \cdot 418 e_j$ which is a good fit with observation for mean performance and while lower than observed for environmental sensitivity it is not significantly so.

The plot of deviation of $\bar{P}_{1j}$, $\bar{P}_{2j}$, $\bar{F}_{1j}$ and $\bar{F}_{2j}$ from the mid-parent value in each of the six environments against the additive environmental values is shown in Fig. 9. This figure brings out very clearly, as do also the regression coefficients, the close similarity of F_1 and F_2 to Wellington (P_2) in their patterns of sensitivity to environmental change: in fact the Wellington pattern must be determined by alleles which are dominant to their counterparts in Samarkand (P_1). Because of the considerable differences in environmental sensitivity of Wellington and Samarkand the chaetae numbers of P_1 and P_2 change relative to one another and to those of F_1 and F_2 with environment. In environments 1, 2 and 3 (the three at 18 °C) the F_1 and F_2 are very close to halfway between P_1 and P_2 in chaetae number, i.e. they show little or no net directional dominance, while in environments 4, 5 and 6 (at 25°C) they are much closer to P_2, i.e. they show a large negative directional dominance. The dominance relations of the genes controlling chaetae number thus depend on the environment in which they are measured.

All the specifications, models and analyses so far considered have been confined to two inbred lines and the generations which can be derived from them, when grown in two or more environments. The approach used can be extended to any number of lines and multiple crosses among them using the joint regression analysis described by Yates and Cochran (1938) and put into a

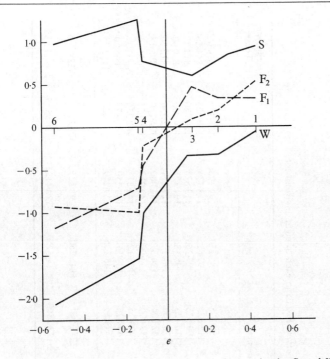

Figure 9 The sensitivity diagram for sternopleural chaetae in the S and W lines of *Drosophila*, together with their F_1 and F_2. F_1 and F_2 follow the response pattern of W more than that of S, thus indicating the dominance of the relevant genes in W. (From Mather and Jinks, 1977.)

biometrical genetical context by Perkins and Jinks (1968*a*, *b*). We shall illustrate their analysis with data from an experiment in which ten inbred lines of *Nicotiana rustica* were grown in each of eight environments. The lines were a stratified sample drawn from a random sample of 82 inbred lines derived from the cross of varieties 1 and 5. Eight completely randomized replicates of each line were grown in each environment. The means over replicates for plant height are given in Table 42.

In the general case the mean, Y_{ij}, of the $r = 8$ replicates of the *i*th genotype in the *j*th environment is expected to be the sum of four components,

$$Y_{ij} = \mu + d_i + e_j + g_{ij}$$

with *i* taking values from 1 to *s*, the number of genotypes and *j* taking values from 1 to *t*, the number of environments.

μ, the overall mean, is estimated as

$$Y../st = \overset{s}{\underset{i=1}{S}} \overset{t}{\underset{j=1}{S}} Y_{ij}/st \text{ which for these data}$$

$$= \frac{9388 \cdot 875}{80} = 117 \cdot 361.$$

TABLE 42

Mean plant heights (over eight replicate individuals) of ten inbred lines of *Nicotiana rustica* in each of eight environments

Genotypes	Environments								Genotype Totals
	1	2	3	4	5	6	7	8	
1	87·000	75·750	77·250	70·000	86·875	81·250	99·125	101·750	679·000
2	106·500	96·375	98·875	93·750	112·625	99·625	113·125	120·625	841·500
3	106·875	101·625	94·375	90·000	100·625	87·250	113·875	120·250	814·875
4	109·375	96·250	108·375	85·625	110·125	95·125	121·125	124·000	850·000
5	114·750	106·750	107·375	94·000	128·500	111·125	127·250	122·750	912·500
6	130·625	118·125	118·500	115·250	132·000	124·375	131·000	125·250	995·125
7	132·250	112·625	121·500	101·500	148·000	133·000	143·250	139·625	1031·750
8	127·750	120·000	128·000	117·375	149·875	135·125	144·875	142·375	1065·375
9	130·750	116·625	126·000	111·625	144·625	130·250	148·250	136·875	1045·000
10	149·875	135·000	137·375	119·250	159·250	148·125	149·250	155·625	1153·750
Environment Totals	1195·750	1079·125	1117·625	998·375	1272·500	1145·250	1291·125	1289·125	9388·875
Means $\mu + e_j$	119·5750	107·9125	111·7625	99·8375	127·2500	114·5250	129·1125	128·9125	$\mu = 117·360$

d_i is the genetical deviation of the ith genotype and is estimated as

$$(Y_i./t) - \mu = \left(\overset{t}{\underset{j=1}{S}} Y_{ij} \middle/ t \right) - \mu.$$

The values of $Y_i.$ for the ten genotypes (the row totals of Table 42) and the corresponding estimates of $\mu + d_i$ are given in Table 44. In the special case of highly inbred lines these genetical deviations are, of course, attributable to additive gene action.

e_j is, as before, the additive environmental deviation of the jth environment and is estimated as

$$(Y._j/s) - \mu = \left(\overset{t}{\underset{j=1}{S}} Y_{ij} \middle/ s \right) - \mu.$$

The values of $Y._j$ for the eight environments are the column totals of Table 42, which also includes the corresponding estimates of $\mu + e_j$.

Finally, g_{ij} the genotype × environment interaction of the ith genotype and the jth environment is estimated as

$$Y_{ij} - \mu - d_i - e_j$$

there being $st = 80$ such interaction components.

We can, of course, subject the data to a standard two way analysis of variance which will test the need for including each of the components and in particular the genotype × interaction component in the model. In the general case the three items of the analysis Genotypes, Environments and their interaction will have $(s-1)$, $(t-1)$ and $(s-1)(t-1)$ degrees of freedom. The degrees of freedom, sums of squares and mean squares for the final height data are given in Table 43. To complete the analysis we must include our estimate of the error variation which for these data comes from the $r = 8$ replications of each genotype in each environment. These yield $st = 80$ sums of squares each for $r - 1 = 7$ degrees of freedom, which on summing gives an overall replicates sum of squares for $st(r - 1) = 560$ degrees of freedom. The loss of

TABLE 43
Joint regression analysis of genotype × environment interactions of ten inbred lines in eight environments

Item	SS	df	MS	P
Genotypes	23276·6463	9	2586·2940	< 0·001
Environments	8099·6233	7	1157·0890	< 0·001
G × E	1886·6054	63	29·9461	< 0·001
(a) Heterogeneity of regression	415·3922	9	46·1547	0·01–0·001
(b) Remainder	1471·2132	54	27·2447	0·01–0·001
Replicate error	9314·8406	559	16·6634	

one replicate in this experiment reduces it to 559. Since the other three items in the analysis are based on the means of 8 replicates this error sum of squares must be divided by 8 to make it appropriate for an analysis of means.

All three items in this analysis are highly significant. A model which includes genotype × environment interactions is, therefore, required. The 80 values of g_{ij} are not, however, particularly informative so we may go on to ask whether the eight values for each genotype are a linear function of the additive environmental values, i.e.

$$g_{ij} = b_i e_j,$$

and, if so, whether these linear functions differ among the ten genotypes. We shall, therefore, test the adequacy of the model

$$Y_{ij} = \mu + d_i + (1 + b_i)e_j$$

by a joint regression analysis in which the sum of squares for genotype × environment interactions are partitioned into linear and non-linear portions.

We must begin by carrying out a linear regression analysis of the t values of g_{ij} on the corresponding e_j values for each genotype separately. There will, therefore, be $(t-1)$ degrees of freedom for the variation in g_{ij} of which 1 will be for the linear regression sum of squares and $t-2$ for the remainder. With s genotypes this implies a total of $s(t-1)$ degrees of freedom whereas the total for the genotype × environment interactions is only $(s-1)(t-1)$. Clearly the s separate regressions are not independent because of restraints which are readily indentified as

$$\underset{i=1}{\overset{s}{S}} g_{ij} = 0, \quad \underset{j=1}{\overset{t}{S}} g_{ij} = 0, \text{ and } \underset{j=1}{\overset{t}{S}} e_j = 0 \text{ leading to } \underset{i=1}{\overset{s}{S}} b_i = 0.$$

For each of the s genotypes we estimate b_i in the usual way as

$$\left. \underset{j=1}{\overset{t}{S}} g_{ij} e_j \middle/ \underset{j=1}{\overset{t}{S}} e_j^2 \right.$$

and the corresponding regression sum of squares as

$$\left(\underset{j=1}{\overset{t}{S}} g_{ij} e_j \right)^2 \middle/ \underset{j=1}{\overset{t}{S}} e_j^2 = b_i^2 \underset{j=1}{\overset{t}{S}} e_j^2.$$

On summing over all s regression sums of squares we have a total sum of squares for s degrees of freedom of

$$\underset{i=1}{\overset{s}{S}} \left(b_i^2 \underset{j=1}{\overset{t}{S}} e_j^2 \right).$$

Since $\underset{j=1}{\overset{t}{S}} e_j^2$ is the same for all s regressions $\underset{i=1}{\overset{s}{S}} \underset{j=1}{\overset{t}{S}} e_j^2$ is equal to

$$s \underset{j=1}{\overset{t}{S}} e_j^2.$$

In the standard joint regression analysis this is partitioned into a joint regression sum of squares for 1 degree of freedom and a heterogeneity of regression sum of squares for $s - 1$ degrees of freedom. Because of the restraint

$$\sum_{i=1}^{s} Sb_i = 0$$

the joint regression sum of squares is zero and the heterogeneity sum of squares is the total sum of squares for regression for $s - 1$ degrees of freedom.

Similarly in each of the s separate regression analyses there will be a remainder sum of squares which will be the sum of squares for the genotype × environment interactions minus the regression sum of squares, that is

$$\sum_{j=1}^{t} Sg_{ij}^2 - b_i^2 \sum_{j=1}^{t} Se_j^2.$$

On summing over all s remainder sums of squares we have a total remainder sum of squares of

$$\sum_{i=1}^{s} \left[\sum_{j=1}^{t} S g_{ij}^2 \right] - \sum_{i=1}^{s} \left[b_i^2 \sum_{j=1}^{t} Se_j^2 \right]$$

which because of the restraints has $(s-1)(t-1) - (s-1) = (s-1)(t-2)$ degrees of freedom.

The final form of analysis is illustrated in Table 43. In practice, however, the analysis rarely proceeds through the estimation of either g_{ij} or e_j although the final outcome is the same. The regression analyses for the individual genotypes are carried out by regressing the t values of Y_{ij} for the ith genotype against the corresponding t column means $Y_{.j}/s = \mu + e_j$. Since μ and d_i are constants this is effectively regressing $e_j + g_{ij}$ on to e_j. As a result the regression sum of squares has the expectation

$$(1+b_i)^2 \sum_{j=1}^{t} Se_j^2$$

corresponding with a regression coefficient of $(1 + b_i)$. Clearly both are inflated by the inclusion of the regression of e_j on e_j. The remainder sum of squares is not, however, affected since the total sum of squares is inflated by exactly the same amount as the regression sum of squares.

On summing the regression sums of squares over the s genotypes the total sum of squares becomes

$$\sum_{i=1}^{s} \left[b_i^2 \sum_{j=1}^{t} Se_j^2 \right] + s \sum_{j=1}^{t} Se_j^2 \text{ because the term}$$

$$\sum_{i=1}^{s} \left[2b_i \sum_{j=1}^{t} Se_j^2 \right] = 0 \text{ since } \sum_{i=1}^{s} Sb_i = 0 \text{ as noted earlier.}$$

If we now partition the total regression sum of squares into a joint regression and heterogeneity of regression sum of squares the former is simply

$$\underset{j=1}{\overset{t}{s}}\ Se_j^2.$$

This sum of squares, which corresponds with a joint regression coefficient of 1, is, of course, the environment sum of squares for $(t-1)$ degrees of freedom while the heterogeneity sum of squares is identical with that obtained previously, namely

$$\underset{i=1}{\overset{s}{S}} \left[b_i^2 \ \underset{j=1}{\overset{t}{Se_j^2}} \right].$$

The stages in the analysis of the plant height data following this more usual procedure are summarized in Table 44 and the final form of the analysis is given in Table 43.

TABLE 44
Regression analyses of ten inbred lines in eight environments

Genotypes	Mean performance $\mu + d_i$	Environmental sensitivity $1 + b_i$	SP xy (7 df)	Regn SS (1 df)	Rem SS (6 df)
1.	84·8750	0·9713	786·7219	764·1483	108·0080
2.	105·1875	0·8431	682·8648	575·7112	61·9138
3.	101·8594	0·7974	645·9047	515·0766	404·3434
4.	106·2500	1·1120	900·6422	1001·4742	213·3071
5.	114·0625	1·0692	866·0492	926·0199	41·6051
6.	124·3906	0·5297	429·0486	227·2732	71·4280
7.	128·96875	1·4230	1152·5402	1640·0133	144·1351
8.	133·1719	1·0145	821·6990	833·6058	150·0485
9.	130·6250	1·1238	910·2219	1022·8918	101·3582
10.	144·21875	1·1160	903·9308	1008·8012	175·0660
Totals	1173·6093	10·0000	8099·6233	8515·0155	1471·2132
Mean	117·3609	1·0000			

Both the heterogeneity of regression and remainder items are significant when tested against the replicate error. Some of the genotype × environment interactions are, therefore, a linear function of the additive environmental value, and the linear regression coefficients (b_i) thus differ significantly between genotypes. There are, however, residual interactions which are not accounted for by the linear model. Whether the latter is true of all genotypes or of some only, can be examined by returning to the remainder mean squares of the individual regression analyses (Table 43). Because the replicate error mean squares are heterogeneous over genotypes (Bartlett test of homogeneity, $P < 0.001$) which means that the genotypes are showing interactions with the microenvironment variation within the eight environments, the remainder mean squares must be tested against their own replicate errors for each genotype separately. These tests show that the remainder mean squares are significant for genotypes 3 and 4 only. We may, therefore, inquire into the nature of their non-linearity. By comparing the improvement in goodness of fit of a number of different models Jinks and Pooni (1980) showed that two intersecting straight lines gave the best representation of the relationship between g_{ij} and e_j for genotypes 3 and 4. A wider survey of 148 different genotypes in up to 15 environments showed that two intersecting straight lines is the most satisfactory alternative model when a single linear regression is inadequate (Pooni and Jinks, 1980). This is the result that Mather (1975) predicted would arise when genes at different loci control sensitivity to different factors which are varying in the environment.

Let us now return to Table 44 where we have estimates of the mean performances over environments $(\mu + d_i)$ and linear sensitivities to the environments $(1 + b_i)$ of the ten genotypes of the sample of inbred lines derived from the cross of varieties 1 and 5. The corresponding estimates for these two parental genotypes when raised in the same range of environments are

	mean performance	environmental sensitivity
Variety 5	$m + [d] = 118\cdot33$	$(1 + b_d) = 1\cdot15$
Variety 1	$m - [d] = 103\cdot08$	$(1 - b_d) = 0\cdot85$

More than half of the ten inbred lines fall outside the parental range for mean performance and environmental sensitivity. The extreme inbreds (1 and 10 of Table 44) differ in mean performance by 59.34 cm which is almost four times the 15.25 cm difference between parental varieties 1 and 5. Similarly, the extreme inbreds (6 and 7 of Table 44) differ by $0\cdot89$ in their environmental sensitivities which is almost three times the $0\cdot30$ difference between parental varieties. Clearly mean performance and environmental sensitivity are each controlled by two or more loci the increasing and decreasing alleles at which must have been dispersed $(r_d < 1)$ in the parental varieties. If we assume that in inbred lines 1 and 10, being the extremes of a stratified sample for mean performance, these alleles are associated $(r_d = 1)$, it follows that half the difference between them is an estimate of $\overset{k}{\underset{i=1}{S}} d_i$ and since half the difference between varieties 1 and 5 is an estimate of $[d] = r_d \overset{k}{\underset{i=1}{S}} d_i$ we can estimate r_d which measures the degree of association (page 66). Since

$$\overset{k}{\underset{i=1}{S}} d_i = 29\cdot67 \text{ and } [d] = 7\cdot63, \hat{r}_d = 0\cdot26.$$

With equal additive genetic effects at the k loci

$$r_d = \frac{k - 2k'}{k} \quad \text{(page 66)}$$

from which it follows that $k'/k = (1 - 0.26)/2 = 0.37$, i.e. the parent with the higher mean performance, variety 5, has 37% of the decreasing and 63% of the increasing additive genetic effects while the other parent, variety 1 has the reverse.

We could carry out a similar calculation for environmental sensitivity with a similar result although it is somewhat less justified because the stratified sample was chosen on the basis of mean performance and not environmental sensitivity. It will not, therefore, include the extreme inbreds for the latter unless, of course, mean performance and environmental sensitivity are completely correlated, i.e. they are pleiotropic manifestations of the same genes – a point we can now examine.

In the parental varieties, higher mean performance and higher environmental sensitivity go together, as they often do. If the genes conferring high mean performance also confer high environmental sensitivity they should be positively correlated among the ten inbred lines. In fact, the correlation while positive is low, $r = 0.29$, and non-significant. The genes controlling mean performance and environmental sensitivity are, therefore, largely independent. Inspection of the estimates for the ten inbreds shows that we would have no difficulty in producing inbred lines from this cross which combine above average mean performance with above or below average environmental sensitivity and the same for below average mean performance. This conclusion has been verified in the random sample of 82 inbred lines from which the stratified sample was drawn (Perkins and Jinks, 1973) and in samples of recombinant inbred lines and selections derived from a cross of varieties 2 and 12 (Jinks et al., 1977).

A more direct demonstration of separable genetical control of mean performance and environmental sensitivity is provided by Caligari and Mather's (1975; and see also Mather and Caligari, 1976) investigation of the Wellington and Samarkand inbred lines of D. melanogaster referred to earlier (page 115). They created the eight substitution lines (Section 2) which comprise all the possible true breeding combinations of the three major chromosomes (X, II and III) from Wellington and Samarkand. The powerful technique of chromosome assay could then be used to investigate the response of number of sternopleural chaetae and total yield of offspring to change in temperature and type of culture container.

Chaeta number changed more with temperature than with types of container whereas the reverse was true of yield of offspring. In respect of chaeta number the genes chiefly responsible for response to environmental change were located on a different chromosome (II) from those chiefly responsible for differences in mean performance (III) and there were indications of a similar situation in respect of yield of offspring. Environmental sensitivity and mean performance are, therefore, separately adjustable by selection.

The genotypes and environments in the Nicotiana and Drosophila data are fixed effects and have been analysed as such. The use of the environmental means $(\mu + e_j)$ obtained from the sums of the s genotype means within each environment as the independent variable in the regression analyses, therefore, poses no statistical or interpretative problems. When s, the number of genotypes, is large the environmental means are for all practical purposes independent of the individual genotype means within each environment (Y_{ij}) which are regressed on to them. There are also many simple devices for making them completely independent, for example, by regressing each genotype on to an environmental mean based on the remaining $(s-1)$ genotypes (Mather and Caligari, 1974) and by using independent replicates of the same genotypes or even other genotypes raised in each environment for the sole purpose of providing estimates of the environmental values (Perkins and Jinks, 1973). Where, as a result of adopting one of these devices, independent estimates of the environmental values are available, the restraints (page 120) are removed and the analysis may be applied to genotypes and environments which are random effects (Perkins and Jinks, 1973).

21. HETEROSIS

Heterosis may be defined as the amount by which the mean of an F_1 family exceeds its better parent. Since the F_1 and parental generation means may be specified in terms of genetic parameters, the expected magnitude of heterosis can be similarly specified. Before doing so, however, we must define what we mean by the better parent and by the F_1 exceeding it. We have so far adopted the convention that P_1 corresponds to the parent with the greater mean value and P_2 to the parent with the smaller mean value, but for the present purpose, either P_1 or P_2 may be the better parent, according to the character under consideration. For example, heterosis for a character such as yield usually implies that the F_1 has a greater yield than its greater yielding parent $(\overline{F}_1 > \overline{P}_1)$. On the other hand, heterosis for characters such as earliness, or time taken to reach a particular stage in development, usually implies that the F_1 has a lower value than its lower parent $(\overline{F}_1 < \overline{P}_2)$, although the reverse will be the case if the same characters are measured in terms of rates of development. A general specification of heterosis must, therefore, be able to accommodate heterosis both in the positive $(\overline{F}_1 > \overline{P}_1)$ and in the negative $(\overline{F}_1 < \overline{P}_2)$ directions.

If heterosis is measured on a scale on which an additive-dominance model is adequate, then for positive heterosis, its expected magnitude is given by

$$\text{Heterosis} = \overline{F}_1 - \overline{P}_1 = [h] - [d]$$

and for heterosis to occur $[h]$ must be positive and greater than $[d]$. For negative heterosis the comparable expectation is given by

$$\text{Heterosis} = \overline{F}_1 - \overline{P}_2 = [h] - (-[d])$$

and heterosis will occur only when $[h]$ is negative and greater than $[d]$.

For $\pm[h]$ to be greater than $[d]$ one or both of two conditions must be satisfied.

(i) Sh must be greater than Sd, i.e. there must be superdominance or over-dominance $(h/d > 1)$ at some or all loci.

(ii) r_d must be less than one and approaching zero; $Sh \leqslant Sd$ at all loci then being sufficient to produce heterosis, i.e. there must be dispersion of completely or incompletely dominant genes $(Sh/Sd \leqslant 1)$.

In addition, heterosis is more likely to occur if the signs of all the hs are the same, i.e. dominance is unidirectional, with the consequence that $[h]$ is not reduced by the internal cancellation of hs of different signs.

The distinction between these two conditions, either of which can give rise to heterosis, has very great practical and economic importance. Unfortunately, neither the degree of dominance (h/d) nor the degree of association (r_d) can be estimated from generation means. Therefore, we cannot distinguish between these two causes of heterosis without recourse to second degree statistics which describe the variation within generations (see Chapter 6).

Jayasekara and Jinks (1976) and Pooni and Jinks (1981) have, however, provided direct demonstrations that positive heterosis for plant height in *Nicotiana rustica* is due to dispersion and not overdominance. The parental and F_1 means for the most heterotic of all the crosses examined, that between varieties 2 and 12, are given in Table 45.

TABLE 45
Plant height in two crosses in *Nicotiana rustica* grown in 1975 D10 and D17 being extreme selections from the cross of V2 and V12

V_2	$86·56 \pm 2·13$	D_{10}	$75·90 \pm 2·05$
V_{12}	$118·43 \pm 2·12$	D_{17}	$160·33 \pm 1·30$
F_1	$140·27 \pm 1·93$	F_1	$138·38 \pm 1·84$
Heterosis	$21·84 \pm 2·86$	Heterosis	–

From the F_2 of this cross 60 true breeding lines were obtained by single seed descent. The tallest and shortest of these lines (D_{17} and D_{10} of Table 45) differed considerably more in height than the parental varieties V_2 and V_{12}. Since D_{17} and D_{10} can only differ for the same alleles at the same loci as V_2 and V_{12}, the alleles of like effect must be largely associated in D_{17} and D_{10} $(r_d = 1)$ and dispersed in V_2 and V_{12} $(r_d < 1)$. The three tallest inbreds, which include D_{17}, are also as tall as, or taller than the heterotic F_1 of the cross of varieties V_2 and V_{12} (Table 45). Furthermore, on crossing D_{17} and D_{10}, which is equivalent to crossing V_2 and V_{12} except that the alleles of like effect are now associated rather than dispersed, the F_1 is, as expected since it is heterozygous for the same alleles at the same loci, as tall as the original F_1 cross of V_2 and V_{12}, and it does not, of course, show heterosis over its better parent D_{17}. The heterosis displayed by the original F_1 over its better parent V_{12} could, therefore, only have been because of dispersion.

If heterosis is measured on a scale on which an additive-dominance model is inadequate its specification becomes more complex. If we make allowance for interactions between pairs of genes then for positive heterosis,

$$\text{Heterosis} = \bar{F}_1 - \bar{P}_1 = ([h] + [l]) - ([d] + [i])$$

and for negative heterosis,

$$\text{Heterosis} = \bar{F}_1 - \bar{P}_2 = ([h] + [l]) - (-[d] + [i]).$$

For positive heterosis to occur $([h] + [l])$ must be positive and greater in value than $([d] + [i])$. However, either $[h]$ or $[l]$, but not both, may be negative, providing that the balance is positive and greater than $([d] + [i])$. Since the parent involved is P_1, $[d]$ must have a positive sign but $[i]$ may be positive or negative within the same restrictions. The same arguments apply to negative heterosis except that $([h] + [l])$ must on balance be negative and greater than $(-[d] + [i])$ although either $[h]$ or $[l]$ may be positive and $[i]$ positive or negative within this restriction.

In the presence of digenic interactions there are clearly many ways in which heterosis could arise. Nevertheless, it is more likely to arise, and to arise with a greater magnitude, when one or more of the following conditions are satisfied.

(i) $[h]$ and $[l]$ have the same sign, i.e. interaction is predominantly of a complementary kind.

(ii) The genes are dispersed so that their contribution to r_d is very small or zero and hence the contribution of $[d]$ is negligible.

(iii) There are as many or more dispersed as associated pairs of interacting genes so that their contribution to r_i is very small or negative thus making the contribution of $[i]$ negligible or the opposite sign to Si. For the classical interactions the latter would make the contribution of $[i]$ and $[l]$ to heterosis the same sign (see Section 16).

Since linkage, even of interacting pairs of genes, does not affect the specification of the parental and F_1 means, the specification of heterosis is independent of linkage. But as we have seen (Table 27) linkage biases the estimates of three of the four components of heterosis. So if linkage is present it will distort the relative magnitudes of these components and affect the interpretation of the cause of heterosis. The direction of this bias can be demonstrated by a simple numerical example.

Let us consider a classical complementary interaction between two genes A–a, and B–b dispersed in the parents so that the cross is AAbb × aaBB with $d_a = d_b = h_a = h_b = i_{ab} = j_{ab} = j_{ba} = l_{ab} = 0.25$. The means of the parental and F_1 generations are then $\bar{P}_1 = m - 0.25$, $\bar{P}_2 = m - 0.25$ and $\bar{F}_1 = m + 0.75$. Heterosis $= \bar{F}_1 - \bar{P}_1$ (or $\bar{P}_2$ in this case) $= ([h] + (l)) - ([d] + [i]) = 1.0$, making the contribution of the components to heterosis; $[d] = 0$, $[h] = 0.5$, $[i] = 0.25$ and $[l] = 0.25$.

If we now assume that the dispersed genes are linked in repulsion the magnitude of the heterosis will remain constant at a value of 1·0. The estimates of the components of heterosis made from the parents, F_1, F_2, B_1 and B_2 generation means, however, will change with the frequency of recombination (p) in the way shown in Fig. 10. Thus while the true values of $[h]$, $[i]$ and $[l]$ remain the same the apparent value of $[h]$ rises while the values of $-[i]$ and $[l]$ fall by an equivalent amount as the value of p falls. So linkage will lead to an overestimate of the contribution of dominance and an underestimate of the contribution of interaction to the magnitude of heterosis. In the limiting case of complete linkage ($p = 0$) dominance appears to be solely responsible for the heterosis. This is to be expected since if no recombination occurs between the two genes they are indistinguishable from a single gene with the summed effects of the individual genes. The non-allelic interaction then becomes dominance. These conclusions can be generalized to linkage in the coupling phase and to any number of linked pairs of genes. In general, therefore, linkage will always lead to an underestimate of the importance of interactions as a cause of heterosis.

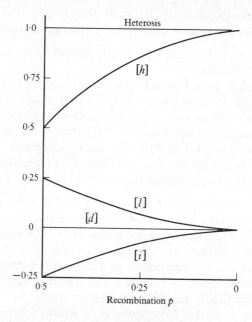

Figure 10 The apparent contributions of $[d]$, $[h]$, $[i]$ and $[l]$ to heterosis arising from complementary genes in the presence of repulsion linkage.

Trigenic and higher order interactions, unlike linkage, affect the specification of heterosis. For trigenic interactions it takes the form

$$\text{Positive heterosis} = \bar{F}_1 - \bar{P}_1 = ([h] + [l_{ab}] + [l_{abc}]) - ([d] + [i_{ab}] + [i_{abc}])$$

where $[l_{ab}]$ and $[i_{ab}]$ are the interactions between pairs of genes and $[l_{abc}]$ and $[i_{abc}]$ the corresponding terms for trigenic interactions for heterozygous combinations and homozygous combinations, respectively. The expression to cover even higher order interactions follows quite obviously from this.

To investigate the cause of heterosis in a particular cross we must first determine the appropriate model for its specification. That is, we must determine, by applying the appropriate scaling tests, whether or not an additive-dominance model is adequate, and if not, whether an adequate model requires the inclusion of digenic interactions, higher order interactions or linkage of interacting genes. When an adequate model is established the component parameters can be estimated from the family means by one or other of the procedures described in earlier sections.

With one exception none of the crosses examined so far in this chapter show heterosis, the reason being that either $[h] \leqslant [d]$ for the crosses where an additive-dominance model applies or $[h]$ and $[l]$ are the same order of magnitude but of opposite sign, with the result that their contributions balance out for the crosses where a digenic interaction model is required (Tables 15 and 23). Nevertheless, in characters other than those considered so far, the same crosses show heterosis. In the tomato crosses plant height and mean plant spread show positive heterosis on a scale which is adequate for fitting an additive-dominance model. The analysis of three examples from crosses grown in 1938 is presented in Table 46.

TABLE 46
The observed and expected magnitude of heterosis and its components in tomato crosses grown in 1938

Cross	Danmark × Red Currant		Johannisfeuer × Red Currant
Character	Plant height	Plant spread	Plant spread
Heterosis			
Observed $\bar{F}_1 - \bar{P}_1$	$9\cdot60 \pm 2\cdot09$	$30\cdot90 \pm 6\cdot67$	$38\cdot40 \pm 2\cdot56$
Expected $[h] - [d]$	$8\cdot53$	$22\cdot33$	$38\cdot62$
$[d]$	$3\cdot40 \pm 0\cdot43$	$21\cdot64 \pm 2\cdot13$	$7\cdot48 \pm 1\cdot15$
$[h]$	$11\cdot93 \pm 1\cdot01$	$43\cdot97 \pm 4\cdot52$	$46\cdot10 \pm 2\cdot32$
$\chi^2_{(3)}$	$5\cdot13$	$5\cdot89$	$1\cdot27$

The observed and expected heterosis show the satisfactory agreement that is expected when the model used is adequate although the agreement is clearly better for the cross Johannisfeuer × Red Currant which gives the smallest $\chi^2_{(3)}$ for testing the goodness of fit of the additive-dominance model. $[h] > [d]$ is obviously the cause of heterosis but in general we cannot distinguish between overdominance and dispersion of dominant genes as explanations of this situation without the aid of procedures which will be discussed in later chapters.

In a survey of the occurrence and basis of heterosis for final height in the 28 crosses that are possible between eight inbred strains of *Nicotiana rustica*, grown in three successive years, it was found that no cross for which an additive-dominance model was adequate showed heterosis in all three seasons. The analysis of heterosis for one such cross (4 × 5) for the one season in which it showed heterosis is summarized in Table 47. Crosses which consistently showed heterosis failed to fit an additive-dominance model particularly in the seasons when the heterosis was most marked. The analysis of a typical example, cross 4 × 6, which showed heterosis in all three seasons, is presented in Table 47 for the season in which it showed the greatest heterosis.

TABLE 47
The observed and expected heterosis and its components for plant height in two crosses in *Nicotiana rustica* grown in 1953

Cross	4 × 5	4 × 6
Heterosis		
Observed $\bar{F}_1 - \bar{P}_1$	21.1 ± 4.3	51.0 ± 4.8
Expected	21.1	51.0*
[d]	24.1 ± 9.1	20.8 ± 2.3
[h]	45.2 ± 17.0	−43.2 ± 48.8
[i]	—	−50.3 ± 19.0
[l]	—	64.8 ± 30.7

* Perfect fit solution.

For cross 4 × 6 it is the combined effects of [l] and [i] both of which make positive contributions to the heterosis which are responsible for its high value. Unfortunately, [h] is not significantly different from zero, hence we cannot classify the predominant type of interaction involved.

Because the most heterotic F_1s are crosses between parents in which the genes are dispersed, their [d], [j] and [i] components will be less than Sd, Sj and Si, respectively, because r_d, r_j and r_i, will be less than one and the sign of [i] may even be the reverse of that of Si (Section 16). In the most heterotic *N. rustica* cross, that between varieties 2 and 12, we can examine the estimates of these components undisturbed by dispersion by estimating them from the cross between D_{10} and D_{17}, the extreme true breeding selections from this cross in which the genes are largely associated. This has been done by Pooni and Jinks (1980) with the results given in Table 48.

All of the estimates are highly significant. The opposing signs of [h] and [l] indicate duplicate interactions between dominant increasing alleles. In contrast the same signs for [d], [h], [i] and [j] suggest complementary interactions. Thus the interactions between heterozygous loci are negative while those involving one or more homozygous loci are positive. This has no parallel among the classical non-allelic interactions that were equated to the six

TABLE 48
The components for plant height in a cross between extreme selections in *Nicotiana rustica* grown in 1975

Cross	$D_{10} \times D_{17}$
Component	
m	$98{\cdot}73 \pm 5.72$
$[d]$	$42{\cdot}22 \pm 1.21$
$[h]$	70.44 ± 14.59
$[i]$	$19{\cdot}38 \pm 5.59$
$[j]$	$21{\cdot}13 \pm 4.49$
$[l]$	-30.79 ± 9.66

components of the digenic interactions model on page 84. The consequences, however, are clear. In spite of the 2×12 cross being the most heterotic, the tallest pure breeding line whose expected mean is

$$m + \underset{k}{Sd} + \underset{\frac{1}{2}k(k-1)}{Si}$$

will be taller than the F_1 whose mean is

$$m + \underset{k}{Sh} + \underset{\frac{1}{2}k(k-1)}{Sl} \, ,$$

an expectation which has already beeen confirmed (Table 45 and Section 60).

22. HETEROSIS AND GENOTYPE × ENVIRONMENT INTERACTION

If the relative performance of parents and F_1s are compared over a number of environments the expectations for the magnitude of heterosis must allow for the possibility of genotype × environmental interactions (Bucio Alanis *et al.*, 1969). Measured on a scale on which an additive dominance model is adequate, then for positive heterosis,

Heterosis in environment $j = \bar{F}_{1j} - \bar{P}_{1j} = [h] - [d] + g_{hj} - g_{dj}$

and for negative heterosis,

Heterosis in environment $j = \bar{F}_{1j} - \bar{P}_{2j} = [h] - (-[d]) + g_{hj} + g_{dj}.$

The corresponding equations in the presence of non-allelic interactions may be readily derived by introducing g_{lj} and g_{ij} with the appropriate signs and coefficients. For example, positive heterosis in the presence of digenic interactions is expected to be

$$\bar{F}_{1j} - \bar{P}_{1j} = ([h] + [l]) - ([d] + [i]) + (g_{hj} + g_{lj}) - (g_{dj} + g_{ij})$$

Where, as is frequently observed,

$$g_{dj} = b_d e_j$$

and

$$g_{hj} = b_h e_j$$

these expectations may be rewritten in the form

$$\bar{F}_{1j} - \bar{P}_{1j} = ([h] - [d]) + (b_h - b_d)e_j, \quad \text{etc.}$$

The effect of environmental changes on the magnitude of heterosis can be predicted from the relative magnitudes of b_d and b_h and to a lesser extent from the relative magnitudes of g_{dj} and g_{hj}. Thus if $b_d = b_h$ or $g_{dj} = g_{hj}$ in all environments, then heterosis is constant in magnitude over environments. If, on the other hand, $b_d > b_h$ then the magnitude of the heterosis decreases as the environment improves at a rate equal to $(b_h - b_d)e_j$; whereas, if $b_d < b_h$ the magnitude of heterosis increases as the environment improves at a rate equal to $(b_h - b_d)e_j$. Hence, in the former case heterosis is greatest in the worst environments while in the latter case it is greatest in the best environments.

To take a specific case: the two inbred lines of *N. rustica* varieties 1 and 5 and their F_1 have been grown in sixteen environments from 1946 to 1964. Bucio Alanis *et al.* (1969) showed that their final heights in these environments could be represented adequately by the formulae

$$\bar{P}_{2j} = m - [d] + (1 - b_d)e_j$$
$$\bar{P}_{1j} = m + [d] + (1 + b_d)e_j$$

and

$$\bar{F}_{1j} = m + [h] + (1 + b_h)e_j$$

where

$$\hat{m} = 112 \cdot 27$$
$$[\hat{d}] = 7 \cdot 34$$
$$[\hat{h}] = 13 \cdot 94$$
$$\hat{b}_d = 0 \cdot 35$$

and

$$\hat{b}_h = 0 \cdot 24$$

all the estimates of the parameters being highly significant.

The expected heterosis, which is positive, is therefore given by the expression

$$13 \cdot 94 - 7 \cdot 34 + (0 \cdot 24 - 0 \cdot 35)e_j = 6 \cdot 60 - 0 \cdot 11 e_j.$$

Hence, the magnitude of the heterosis is expected to fall off linearly as the environment improves – an expectation which was found to be in complete agreement with the heterosis observed in each of the sixteen environments.

23. THE LIMITATIONS OF FIRST DEGREE STATISTICS

In this chapter we have been solely concerned with the specification, analysis and interpretation of first degree statistics such as family and generation means. The procedures described have many attractions not least of which are their relative simplicity and statistical reliability, attractions which we shall be in a better position to appreciate after we have considered the corresponding procedures for handling higher order statistics in later chapters. We must not, however, be misled by these attractions into questioning the need for the greater complexities and in some ways the statistically less satisfactory analyses based on higher degree statistics. As we have already had occasion to note, such analyses are essential for the determination of the degree of dominance as opposed to the potence ratio and for discriminating between dispersed dominants and overdominance as causes of heterosis, and as we shall see in later chapters, they are essential for detecting and estimating the effects of linkage between non-interacting genes.

On the other hand, the detection, estimation and interpretation of non-allelic interactions has progressed much further at the level of first degree statistics because their effects are less confounded with one another at this level and the kinds of experiments required for their analysis are both smaller and simpler to carry out. Nevertheless, if we wish to estimate the contribution of non-allelic interactions unconfounded by the effects of gene distribution in the original parental lines, the greater complexity of an analysis of second degree statistics and the more complicated experiments that are required must be faced.

Indeed, there is no reason why the relative contributions of the various genetic, environmental and interactive sources of variation should be the same for the differences between the means of families and generations and for the differences among individuals in the same family or generation. Thus there can be no question that the approaches based on first, second and higher order statistics are complementary, rather than alternative. A complete understanding generally requires that the advantages of the analyses at all levels are exploited.

6 Components of variation

24. VARIATION IN F_2 AND BACKCROSSES

The last two chapters have shown the range of information that can be gained from comparisons among family means and the use to which these first degree statistics can be put in the genetical analysis of continuous variation. We must now turn to consider the constitution of the second degree statistics, the variances and covariances that can be calculated from the families raised in genetical experiments, and the information that they yield. In doing so it is convenient to proceed as we did with means and look first at the simple case in which dominance is taken into account but non-allelic interaction and genotype × environment interaction are neglected. The consequences of these interactions will be taken up in the next chapter, as will the effects of linkage.

The variation of the measurements of true breeding parental lines and their F_1 must be exclusively non-heritable, apart from the effects of mutation which, though detectable by appropriate experiments, are so small as to be safely neglectable (Durrant and Mather, 1954; Paxman, 1957). The variances of these measurements consequently afford estimates of the non-heritable contribution to the variances of the measurements in families of later generations, such as F_2, where because of segregation heritable components of variation will also be present. In the absence of differential fertilization and differential viability the constitution of F_2 in respect of the segregating pair of alleles A–a will be $\frac{1}{4}AA; \frac{1}{2}Aa; \frac{1}{4}aa$, and these genes will contribute increments of $d_a; h_a;$ and $-d_a$ to the measurements of individuals in the three classes, respectively. Thus, as indeed we have already seen, the contribution of A–a to the F_2 mean, expressed as a deviation from the mid-parent, will be $\frac{1}{2}h_a$; and taking all genes into account the contribution to the F_2 mean will be $\frac{1}{2}S(h_a) = \frac{1}{2}[h]$, i.e., of course, half that to the F_1 mean.

The contribution of A–a to the sum of squares of deviations from the mid-parent will clearly be $\frac{1}{4}d_a^2 + \frac{1}{2}h_a^2 + \frac{1}{4}(-d_a)^2 = \frac{1}{2}d_a^2 + \frac{1}{2}h_a^2$. The contribution to the sum of squares of deviations from the F_2 mean will thus become $\frac{1}{2}d_a^2 + \frac{1}{2}h_a^2 - (\frac{1}{2}h_a)^2 = \frac{1}{2}d_a^2 + \frac{1}{4}h_a^2$, the term correcting for the departure of the mean from the mid-parent being the square of the mean itself since we are using the proportionate frequencies of the three genotypes, which sum to unity. For the

135

same reason the contribution to the sum of squares is also the contribution to the mean square measuring heritable variance.

Provided that non-allelic genes neither interact nor are linked, the total heritable variance given by k genes in F_2 will be the sum of the k individual contributions, namely

$$\tfrac{1}{2}S(d_a^2) + \tfrac{1}{4}S(h_a^2) = \tfrac{1}{2}D + \tfrac{1}{4}H$$

where $D = S(d_a^2)$ and $H = S(h_a^2)$. The heritable variation of F_2 is, therefore, divisible into two portions: the D or additive portion which because it depends on d_a etc., measuring the differences between homozygotes, must in principle always be fixable by the selection of true breeding strains; and the H or dominance portion which because it depends on h_a etc., measuring the behaviour of alleles in heterozygotes, cannot be fixed in the selection of true breeding strains and may therefore be described as unfixable. The i type non-allelic interactions are also fixable since they can be utilized in selecting true breeding strains; whereas the j and l types of interactions, which both characterize heterozygotes or partially heterozygous genotypes, are unfixable.

Returning to the variance of F_2, it must also, of course, contain a non-heritable component which in the absence of genotype × environment interaction can be denoted by a separate term, E. We may, therefore, write the variance of F_2 as

$$V_{1F2} = \tfrac{1}{2}D + \tfrac{1}{4}H + E.$$

The reason for the use of V_{1F2} rather than the simpler V_{F2} to denote this variance will appear later.

In backcrossing the F_1 to either of its true breeding parents half the progeny will be homozygous and half heterozygous in respect of each gene pair by which the parents differ. The contribution of A–a to the mean magnitudes of the two backcrosses, as measured from the mid-parent, will therefore be

$$B_1 : \tfrac{1}{2}(d_a + h_a) \quad \text{and} \quad B_2 : \tfrac{1}{2}(h_a - d_a)$$

where B_1 is the backcross to the AA parent, and B_2 that to the aa parent.

The contributions to the variance of the two backcrosses can easily be shown to be

$$B_1 : \tfrac{1}{4}(d_a - h_a)^2 \quad \text{and} \quad B_2 : \tfrac{1}{4}(d_a + h_a)^2.$$

The contributions of d and h to these variances are not separable as they stand, but if the two variances are summed, A–a contributes $\tfrac{1}{2}(d_a^2 + h_a^2)$ to the total, the contributions of d and h now being separable.

Then, with genes independent in expression and transmission the heritable components of the two variances and their sum will be

$$V_{B1} = \tfrac{1}{4}S(d_a - h_a)^2 \quad V_{B2} = \tfrac{1}{4}S(d_a + h_a)^2$$
$$V_{B1} + V_{B2} = \tfrac{1}{2}S(d^2) + \tfrac{1}{2}S(h^2) = \tfrac{1}{2}D + \tfrac{1}{2}H.$$

This last compares with $\frac{1}{2}D + \frac{1}{4}H$ for the heritable component of V_{1F2} and hence the excess of the summed heritable variances of the backcrosses over that of F_2 must be a measure of $\frac{1}{4}H$. It will indicate the presence of dominance and this irrespective of the direction of dominance at each locus because $H = S(h^2)$. V_{B1} and V_{B2} will, however, both include the non-heritable component E. The full expression for their sum will thus be $V_{B1} + V_{B2} = \frac{1}{2}D + \frac{1}{2}H + 2E$ and if $E > 0$, $V_{B1} + V_{B2}$ will exceed $V_{1F2} = \frac{1}{2}D + \frac{1}{4}H + E$ even though dominance be absent.

The difference between the variances of the two backcrosses is $S(dh)$, the sign of h here being taken into account. The difference must be zero wherever either the parents are equal in expression of the character or the F_1 equals the mid-parent, for in these cases the d and h items are balanced respectively in the sense that the sum of the dh products with positive sign will be equalled by the sum of those with negative sign. Where, however, $S(dh)$ does not equal zero it supplies additional evidence of dominance. Furthermore it supplements the evidence from the relation of F_1 to mid-parent in showing which parents carry the preponderance of dominant alleles, since as we have seen the backcross to this parent has the lower variance.

This method has been used by Powers (1942) for showing the dominance of the genes governing fruit weight (expressed on a log scale) in the tomato cross Danmark × Red Current to which reference has already been made. The parental means were 1·672 and $-0·053$ giving a mid-parent of 0·809. The F_1 mean was 0·707, suggesting preponderant dominance of the genes from the smaller parent, Red Currant. The heritable portions of the variances $(_H V)$ Powers estimated as $_H V_{BRC} = 0·013\,53$, $_H V_{BD} = 0·019\,07$ and $_H V_{1F2} = 0·31\,92$. $_H V_{BRC} + _H V_{BD}$ exceeds $_H V_{1F2}$ by 0·000 68, so giving $H = 0·002\,72$. It thus follows that $D = 0·062\,48$ and $S(dh) = -0·005\,54$. Dominance seems clear though not marked. Furthermore, $S(dh)$ indicates that genes from the smaller fruited parent are preponderantly dominant, as the comparison of F_1 mean with mid-parent had already suggested.

Dominance appears to be more marked in the case of corolla length (measured as 1000 times the sum of log tube length and log lobe length) in the species cross *Nicotiana Langsdorfii* × *N. Sanderae* (Smith, 1937). The parental means were $\bar{S} = 1292$ and $\bar{L} = 37$ giving a mid-parent of 664·5 as compared with an F_1 mean of 742. The average variance of the reciprocal backcrosses to *N. Sanderae* was 85·5 and to *N. Langsdorfii* 98·5. $S(dh)$ is thus in the direction agreeing with the excess of F_1 over mid-parent, the genes from *N. Sanderae* showing the preponderant dominance. The non-heritable item included in V_{BL}, V_{BS} and V_{1F2} may be estimated from the average of parental and F_1 variances, as 42. Deduction of this leaves 43·5 and 56·5 as the heritable components of the two average backcross variances and 88·5 as that of the average F_2 variance. Then $\frac{1}{2}D + \frac{1}{2}H = 43·5 + 56·5 = 100$ and $\frac{1}{2}D + \frac{1}{4}H = 88·5$ so giving $H = 4 \times 11·5 = 46$ and $D = 2(100 - 23) = 154$.

If we care to assume that h and d are constant in magnitude (though not necessarily in sign) for all the k gene pairs, segregating in this cross,

$$\sqrt{\left(\frac{H}{D}\right)} = \sqrt{\left(\frac{46}{154}\right)} = 0.55 = \frac{h}{d},$$

provides, therefore, an estimate of the degree of dominance. If h and d are not constant in magnitude $\sqrt{(H/D)}$ provides an estimate of the average dominance of the genes.

Now $S(dh)$ will equal $\sqrt{(DH)}$ if h/d, the measure of dominance, is constant in both magnitude and sign for all the gene-pairs involved. Variation in h/d results in $S(dh)$ falling short of $\sqrt{DH}$ and the greater the variation the greater the shortfall. In the present case

$$S(dh) = V_{BL} - V_{BS} = 13$$
$$\sqrt{(DH)} = \sqrt{(46 \times 154)} = 84.$$

Thus the ratio h/d must vary from one gene pair to another: indeed with such a large difference between $S(dh)$ and $\sqrt{(DH)}$ there is good reason to believe that even though the genes from *N. Sanderae* show a preponderance of dominance, some of them must be recessive to their alleles from *N. Langsdorfii*.

25. GENERATIONS DERIVED FROM F₂ AND BACKCROSSES

In the F_3 generation, derived in Mendel's fashion by selfing the F_2 individuals, all the families from AA and aa F_2 individuals will be wholly AA and aa, respectively, while those from heterozygous F_2 individuals will repeat the F_2 segregation. The means of families from AA, Aa and aa parents will therefore be d_a, $\frac{1}{2}h_a$ and $-d_a$ in respect of this gene pair. The overall mean will thus depart by $\frac{1}{4}h_a$ from the mid-parent and the contribution of A–a to the variance of F_3 means will be

$$\tfrac{1}{4}d_a^2 + \tfrac{1}{2}(\tfrac{1}{2}h_a)^2 + \tfrac{1}{4}(-d_a)^2 - (\tfrac{1}{4}h_a)^2 = \tfrac{1}{2}d_a^2 + \tfrac{1}{16}h_a^2$$

given that the F_3 families are large enough for the sampling variation of their means to be negligible. Since the situation will be similar for other genes, the contribution to the overall mean of the F_3 will be $\frac{1}{4}[h]$ and the total heritable variance of F_3 means will be

$$\tfrac{1}{2}S(d^2) + \tfrac{1}{16}S(h^2) = \tfrac{1}{2}D + \tfrac{1}{16}H.$$

It can be shown similarly that the covariance of F_3 mean with its F_2 parent's measurement will be $\frac{1}{2}S(d^2) + \frac{1}{8}S(h^2) = \frac{1}{2}D + \frac{1}{8}H$ and the mean variance of F_3 families will be $\frac{1}{4}S(d^2) + \frac{1}{8}S(h^2) = \frac{1}{2}D + \frac{1}{8}H$. These results are given by Fisher *et al.* (1932), but before leaving them two points must be made. In the first place, in addition to their basic genetical variance the means of the F_3 families will be subject to a sampling variance equal to $1/m$ the mean variance of the F_3 families, where m is the number of individuals in the family (or the harmonic mean of the numbers if they differ among the families). Furthermore while the covariance will be free from non-heritable effects (aside, of course, from cases such as that of mammals where the parents exert an effect on the offspring beyond that of merely providing their genes), both the mean variance of the F$_3$ families and the variance of their means will contain a non-heritable component. These will not, however, be alike. They will be denoted by E_1, the non-heritable variance of individuals, and E_2, the non-heritable variance of family means. In general E_2 will be less than E_1 because each mean is based on m individuals; but it will only be $(1/m)E_1$ where the differences in environment between individuals in different families are no greater than those to which members of the same family are subject. This situation may be secured by appropriate design of the experiments from which the observations are obtained; but in general, where members of the same family are raised in environments more alike than those of different families – as where families of plants are raised each in its own plot, or families of animals each in its own cage or culture vessel – we must expect E_2 to exceed $(1/m)E_1$ and in extreme cases it could even exceed E_1 itself. It is useful, therefore, to recognize a different subdivision of the non-heritable variation into two components, E_w and E_b where E_w is the variation within families and E_b the additional variation between families. Then $E_1 = E_w$ and $E_2 = E_b + (1/m)E_w$. This item $(1/m)E_w$ is, of course, included in the sampling variance, found as $1/m$ times the mean variance within F$_3$ families, which the variance of F$_3$ means will include.

So, denoting the variance of F$_3$ means by V_{1F3} and the mean variance of F$_3$ families by V_{2F3}, we find

$$V_{1F3} = \tfrac{1}{2}D + \tfrac{1}{16}H + E_b + \frac{1}{m}V_{2F3}$$

$$V_{2F3} = \tfrac{1}{4}D + \tfrac{1}{8}H + E_w$$

and the covariance

$$W_{1F23} = \tfrac{1}{2}D + \tfrac{1}{8}H.$$

The variance of F$_3$ means resembles the variance of F$_2$ in that its heritable portion reflects the genetical differences produced by segregation at gametogenesis in the F$_1$. These are then both first rank variances and as such take the subscript 1. The mean variance of F$_3$ families, on the other hand, reflects the effects of segregation at gametogenesis in the F$_2$ individuals and so is of the second rank, denoted by 2 in the subscript (see Mather and Vines, 1952). The covariance is of course of the first rank. Rank is, as we shall see in Chapter 7, of special importance in relation to linkage.

Proceeding to F$_4$, three variances are distinguishable. The first rank variance V_{1F4} is between the means of groups of families which can each be traced via an F$_3$ family to a single F$_2$ individual. The second rank variance, V_{2F4}, is between family means within groups averaged over groups. The third rank variance, V_{3F4}, is the variance within families, averaged over all the F_4 families. There are correspondingly, two covariances: W_{1F34} between F_3 family means and F_4 group means; and W_{2F34} between F_3 individuals and F_4 family means, calculated within groups but averaged over groups. Assuming that E_b is no greater between families in different groups than between families of the same group and allowing for the appropriate sampling variation of group and family means, with m' families per group and m individuals per family we find

$$V_{1F4} = \tfrac{1}{2}D + \tfrac{1}{64}H + \frac{1}{m'}V_{2F4}$$

$$V_{2F4} = \tfrac{1}{4}D + \tfrac{1}{32}H + E_b + \frac{1}{m}V_{3F4}$$

$$V_{3F4} = \tfrac{1}{8}D + \tfrac{1}{16}H + E_w$$
$$W_{1F34} = \tfrac{1}{2}D + \tfrac{1}{32}H$$
$$W_{2F34} = \tfrac{1}{4}D + \tfrac{1}{16}H.$$

The selfing series can be extended to later generations, F_5 yielding four variances and three covariances and so on. Leaving aside the non-heritable components and the items representing sampling variances, the general formula for the heritable variance of rank j in generation F_n (where, of course, $j < n$) is given by

$$V_{jFn} = (\tfrac{1}{2})^j D + (\tfrac{1}{2})^{2n-j-1}H.$$

The general formula for the covariance of rank j between generations F_{n-1} and F_n is

$$W_{jFn(n-1)} = (\tfrac{1}{2})^j D + (\tfrac{1}{2})^{2n-j-2}H.$$

These results are collected together in Table 49.

TABLE 49
Components of variation in F_2, backcrosses and their derivatives (for further explanation, see in the text)

Statistic	D	H	E_w	E_b	Sampling variation
V_{1F2}	$\tfrac{1}{2}$	$\tfrac{1}{4}$	1		
V_{1F3}	$\tfrac{1}{2}$	$\tfrac{1}{16}$		1	$\dfrac{1}{m}V_{2F3}$
V_{2F3}	$\tfrac{1}{4}$	$\tfrac{1}{8}$	1		
W_{1F23}	$\tfrac{1}{2}$	$\tfrac{1}{8}$			
V_{1F4}	$\tfrac{1}{2}$	$\tfrac{1}{64}$			$\dfrac{1}{m}V_{2F4}$
V_{2F4}	$\tfrac{1}{4}$	$\tfrac{1}{32}$		1	$\dfrac{1}{m}V_{3F4}$
V_{3F4}	$\tfrac{1}{8}$	$\tfrac{1}{16}$	1		
W_{1F34}	$\tfrac{1}{2}$	$\tfrac{1}{32}$			

W_{2F34}	$\frac{1}{4}$	$\frac{1}{16}$		
V_{jFn}	$(\frac{1}{2})^j$	$(\frac{1}{2})^{2n-j-1}$	as appropriate	
$W_{jFn(n-1)}$	$(\frac{1}{2})^j$	$(\frac{1}{2})^{2n-j-2}$		
V_{1S3}	$\frac{1}{4}$	$\frac{1}{16}$	1	$\frac{1}{m}V_{2S3}$
V_{2S3}	$\frac{1}{4}$	$\frac{3}{16}$	1	
W_{1S23}	$\frac{1}{4}$			
V_{1S4}	$\frac{1}{4}$	$\frac{3}{128}$		$\frac{1}{m'}V_{2S4}$
V_{2S4}	$\frac{1}{8}$	$\frac{5}{128}$	1	$\frac{1}{m}V_{3S4}$
V_{3S4}	$\frac{1}{4}$	$\frac{11}{64}$	1	
W_{1S34}	$\frac{1}{4}$	$\frac{1}{32}$		
W_{2S34}	$\frac{1}{8}$	$\frac{1}{32}$		
V_{1M3}	$\frac{1}{8}$		1	$\frac{1}{m}V_{2M3}$
V_{2M3}	$\frac{3}{8}$	$\frac{1}{4}$	1	
W_{1M23}	$\frac{1}{4}$			
$V_{1F2P}(= V_{1F2P1} + V_{1F2P2})$	$\frac{1}{4}$	$\frac{1}{4}$	2	$\frac{1}{m}V_{2F2P}$
V_{2F2P}	$\frac{1}{4}$	$\frac{1}{4}$	2	
W_{1F2P}	$\frac{1}{2}$	$\frac{1}{2}$	2	
$V_{B1} + V_{B2}$	$\frac{1}{2}$	$\frac{1}{2}$	2	
$V_{1B11} + V_{1B12}$ ⎫ $V_{1B21} + V_{1B22}$ ⎬	$\frac{1}{8}$	$\frac{1}{8}$	2	$\frac{1}{m}V_2{}^*$
$V_{2B11} + V_{2B12}$ ⎫ $V_{2B21} + V_{2B22}$ ⎭	$\frac{1}{4}$	$\frac{1}{4}$	2	
$W_{1B1/B11} + W_{1B2/B22}$	$\frac{1}{4}$	$\frac{1}{4}$		
$W_{1B1/B12} = W_{1B2/B21}$	$\frac{1}{8}$	$-\frac{1}{8}$		
$V_{1B1S} + V_{1B2S}$	$\frac{1}{2}$	$\frac{1}{8}$	2	$\frac{1}{m}V_2{}^*$
$V_{2B1S} = V_{2B2S}$	$\frac{1}{4}$	$\frac{1}{8}$	1	
$W_{1B1/B1S} + W_{1B2/B2S}$	$\frac{1}{2}$	$\frac{1}{4}$		
$W_{B11/B12} = W_{B21/B22}$	$\frac{1}{16}$	$-\frac{1}{16}$		
$W_{B11/B1S} + W_{B22/B2S}$	$\frac{1}{4}$	$\frac{1}{8}$		
$W_{B12/B1S} + W_{B21/B2S}$	$\frac{1}{4}$	$-\frac{1}{8}$		

* V_2 means the corresponding sum of rank 2 variances.

In addition to results from true F$_3$s, Fisher *et al.* considered also those from families obtained by intercrossing pairs of individuals taken at random from F$_2$. These biparental families, BIPs as Mather (1949a) termed them, are not uncommonly referred to as F$_3$s, especially by animal geneticists. Allowing for sampling variation and non-heritable components, they yield however

$$V_{1S3} = \tfrac{1}{4}D + \tfrac{1}{16}H + E_b + \frac{1}{m}V_{2S3}$$
$$V_{2S3} = \tfrac{1}{4}D + \tfrac{3}{16}H + E_w$$
$$W_{1S23} = \tfrac{1}{4}D.$$

The S in the subscript indicates sib-mating, and continuing the sib-mating we find for the fourth generation

$$V_{1S4} = \tfrac{1}{4}D + \tfrac{3}{128}H + \frac{1}{m'}V_{2S4}$$

$$V_{2S4} = \tfrac{1}{8}D + \tfrac{5}{128}H + E_b + \frac{1}{m}V_{3S4}$$

$$V_{3S4} = \tfrac{1}{4}D + \tfrac{11}{64}H + E_w$$

$$W_{1S34} = \tfrac{1}{4}D + \tfrac{1}{32}H$$

$$W_{2S34} = \tfrac{1}{8}D + \tfrac{1}{32}H$$

where V_{1S4} is the variance of means of groups tracing through an S_3 family to a single pair of F_2 individuals, V_{2S4} is the variance of family means within groups averaged over groups, and V_{3S4} is the average variance within families, with W_{1S34} and W_{2S34} defined correspondingly. Expressions for the variances and covariances for generations up to S_{10} are given by Hayman (1953).

Finally, Fisher *et al.*, who were discussing data from plants, considered families of the third generation obtained by exposing a single F_2 individual as female to pollination by the pollen of the F_2 as a whole (maternal progenies). This gives

$$V_{1M3} = \tfrac{1}{8}D + E_b + \frac{1}{m}V_{2M3} \quad \text{and} \quad V_{2M3} = \tfrac{3}{8}D + \tfrac{1}{4}H + E_w$$

to which may be added the covariance of family mean with F_2 mother

$$W_{1M23} = \tfrac{1}{4}D.$$

It is hard to imagine an experimental situation in which a single F_2 plant would be exposed to a truly random sample of the pollen from the F_2 as a whole. The essential character of this population of pollen is, however, that half its grains carry A and half a. Now this is, of course, characteristic of the gametes produced by F_1. If, therefore, individuals from F_2 are separately crossed to individuals from F_1 whether as males or females, the resulting families will display the characteristics of Fisher *et al.*'s maternal progenies. Such families are easy to produce and have in fact been used experimentally by Opsahl (1956) who independently found the same expressions for the constitution of the two variances and the covariance.

Opsahl also considered the properties of the groups of families obtained by crossing F_2 individuals to the two true-breeding parental lines P_1 and P_2. He used the notation F_2P_1 for the group of families obtained from the cross to P_1, and F_2P_2 for that from the cross to P_2, and he found

$$V_{1F2P} = V_{1F2P1} + V_{1F2P2} = \tfrac{1}{4}D + \tfrac{1}{4}H + 2E_b + \frac{1}{m}V_{2F2P}$$

$$W_{1F2P} = W_{1F2P1} + W_{1F2P2} = \tfrac{1}{2}D$$

$$V_{2F2P} = V_{2F2P1} + V_{2F2P2} = \tfrac{1}{4}D + \tfrac{1}{4}H + 2E_w,$$

where V_1 is, of course, the variance of means and V_2 the mean variance, the term $(1/m)V_{2F2P}$ being included in V_1 although Opsahl did not in fact consider it.

All the groups and kinds of families we have considered so far in the section are derived from the F$_2$. Groups of families may be derived similarly from the first backcrosses, B$_1$ and B$_2$, by further backcrossing, selfing and other systems of mating. Continuing with the notation used in earlier chapters, B$_{11}$ is the backcross of B$_1$ individuals (themselves from the backcross F$_1$ × P$_1$) to P$_1$, B$_{12}$ the backcross of these same individuals to P$_2$ and B$_{1S}$ the families obtained by selfing B$_1$ individuals. The contributions of A–a to the mean variance of the B$_{11}$ and B$_{21}$ progenies are each $\frac{1}{8}(d_a - h_a)^2$ and of B$_{12}$ and B$_{22}$ are $\frac{1}{8}(d_a + h_a)^2$ and we therefore find

$$V_{2B11} + V_{2B12} = V_{2B21} + V_{2B22} = \tfrac{1}{4}D + \tfrac{1}{4}H + 2E_w.$$

The corresponding first rank variances, found as the variances of the means of B$_{11}$ etc. give

$$V_{1B11} + V_{1B12} = V_{1B21} + V_{1B22} = \tfrac{1}{8}D + \tfrac{1}{8}H + 2E_b + \frac{1}{m}V_2$$

where V_2 is the corresponding second rank variance. And just as $V_{B1} - V_{B2}$ yields an estimate of $S(dh)$

$$V_{1B11} - V_{1B12} = V_{1B21} - V_{1B22} = \tfrac{1}{4}S(dh)$$

and

$$V_{2B11} - V_{2B12} = V_{2B21} - V_{2B22} = \tfrac{1}{2}S(dh).$$

Turning to the covariances of the means of the second backcross families with the expression of the character in their first backcross parents

$$W_{1B1/B11} + W_{1B2/B22} = \tfrac{1}{4}D + \tfrac{1}{4}H$$

while

$$W_{1B1/B12} = W_{1B2/B21} = \tfrac{1}{8}D - \tfrac{1}{8}H.$$

Then the sum of the covariances from the two recurrent backcrosses (i.e. where the second backcross is to the same parent as the first) yields a direct estimate of $\frac{1}{4}D + \frac{1}{4}H$, while the sum of the covariances from the two non-recurrent backcrosses gives a direct estimate of $\frac{1}{4}D - \frac{1}{4}H$. Prospectively, therefore, these covariances have a special value for the separation of the fixable and unfixable components of the genetic variance, though for a variety of reasons little use has been made of them for this purpose.

The above results for the second backcross generations were obtained by Mather (1949a). They have been added to by Hill (1966) whose results show that

$$V_{1B1S} + V_{1B2S} = \tfrac{1}{2}D + \tfrac{1}{8}H + 2E_b + \frac{1}{m}V_2$$

$$V_{2B1S} = V_{2B2S} = \tfrac{1}{4}D + \tfrac{1}{8}H + E_w$$

to which we may add

$$W_{1B1/B1S} + W_{1B2/B2S} = \tfrac{1}{2}D + \tfrac{1}{4}H.$$

144 · COMPONENTS OF VARIATION

Hill also gives the covariances of the mean of the families derived from the same first backcross individuals by recurrent backcrosses, non-recurrent backcrosses and selfing. The covariances of the two types of second backcross are of interest because they yield estimates of the differences between D and H thus

$$W_{B11/B12} = W_{B21/B22} = \tfrac{1}{16}D - \tfrac{1}{16}H.$$

The covariances of the second backcrosses and the selfs give

$$W_{B11/B1S} + W_{B22/B2S} = \tfrac{1}{4}D + \tfrac{1}{8}H$$
$$W_{B12/B1S} + W_{B21/B2S} = \tfrac{1}{4}D - \tfrac{1}{8}H.$$

These results for the generations derived from the first backcrosses are collected together in Table 49.

26. THE BALANCE SHEET OF VARIATION

As a result of crossing, segregation and recombination genetic variability may be redistributed among the various states in which it can exist (Mather, 1943, 1973), but in the absence of mutation, random change and certain forms of selection, its total quantity must remain unchanged. One aspect of this conservation of variability is revealed by the heritable variances that we have been discussing in the last section.

The phenotypic differences between homozygotes contribute to the D portion of the variation detectable among the individuals of a family or generation. Not all the fixable variability displays itself in this way, however: some lies hidden in the homozygous potential state where non-allelic genes tend to balance out one another's effects. In the absence of linkage and interaction, however, D bears a constant relation to the total fixable variation within the descendants of a cross between two true breeding lines. The consequences of linkage and interaction will be considered later.

The contribution of heterozygotes to the phenotypic variation is displayed in two ways. It may appear as differences among the individuals of a family or generation and is then measured by the H portion of the variation. Like D, H is constant within the descendants of a cross in the absence of linkage and interaction. The contribution of heterozygotes may, however, also appear as the departure from the mid-parent of the generation mean, which depends on $S(h) = [h]$. In contrast to D and H which are quadratic, $[h]$ is a linear quantity, and the departure of the mean from the mid-parent must be squared if it is to be made comparable to the D and H parts of the variation. The free variation, expressed in the phenotypes of a generation, may thus be written as $xD + yH + z[h]^2$ and in the absence of complication from non-allelic interaction and linkage the conservation of variation requires that x, y and z must sum to unity in each generation (Mather, 1949b).

In an F_1, $x = y = 0$ and $z = 1$; while in the F_2 derived from it $x = \frac{1}{2}$, $y = \frac{1}{4}$ and since the mean is $\frac{1}{2}[h]$ the contribution of this mean to the free variation is $(\frac{1}{2}[h])^2 = \frac{1}{4}[h]^2$ giving $z = \frac{1}{4}$. Thus $x + y + z = \frac{1}{2} + \frac{1}{4} + \frac{1}{4} = 1$. Proceeding to F_3, there are two variances to be taken into account: the heritable portion of $V_{1F3} = \frac{1}{2}D + \frac{1}{16}H$ and that of $V_{2F3} = \frac{1}{4}D + \frac{1}{8}H$, giving jointly $\frac{3}{4}D + \frac{3}{16}H$. It will be observed that this joint coefficient of D reflects the fact that three-quarters of the genes are homozygous in this generation just as the term $\frac{1}{2}D$ in V_{1F2} reflects the homozygosity of half the genes in F_2. The overall mean of F_3 is $\frac{1}{4}[h]$ and its contribution to the free variation is thus $\frac{1}{16}[h]^2$ giving $z = \frac{1}{16}$. Thus once again $x + y + z = \frac{3}{4} + \frac{3}{16} + \frac{1}{16} = 1$. And the same applies to F_4 (see Table 50).

Taking biparental progenies of the third generation, the sum of the heritable portions of V_{1S3} and V_{2S3} is $\frac{1}{2}D + \frac{1}{4}H$, the square of the overall mean being $\frac{1}{4}[h]^2$. Thus again $x + y + z = \frac{1}{2} + \frac{1}{4} + \frac{1}{4} = 1$ and from Table 50 the same is seen to be true of the fourth generation derived by sib-mating. The table also shows it to be true of the third generation where this consists of maternal progenies.

Turning back to the selfing series, the proportion of loci that are heterozygous is halved in each generation from F_1 onwards. Hence in F_n a proportion $(\frac{1}{2})^{n-1}$ of the loci will be heterozygous and $1 - (\frac{1}{2})^{n-1}$ homozygous. The coefficient of D in the overall heritable variation of F_2 will thus be $1 - (\frac{1}{2})^{n-1}$. The mean of the generation will be $(\frac{1}{2})^{n-1}[h]$ and its contribution to the variation thus $(\frac{1}{2})^{2(n-1)}[h]^2$, leaving $(\frac{1}{2})^{n-1} - (\frac{1}{2})^{2(n-1)} = (\frac{1}{2})^{n-1}$ $[1 - (\frac{1}{2})^{n-1}]$ as the coefficient of H. When n becomes large, the generation will, of course, consist of true breeding lines, its mean will be the mid-parent and its heritable variation will be D. The coefficients of D, H and $[h]^2$ in the nth generation derived by sib-mating can be found similarly by using the Fibonacci series instead of $(\frac{1}{2})^{n-1}$ but the general expressions are clumsy to write and so will not be set out here.

The same principle of conservation of variability applies to backcrosses, provided all the backcross families in a given generation are taken together. Thus taking the first backcross generation $V_B = \frac{1}{2}(V_{B1} + V_{B2}) = \frac{1}{4}D + \frac{1}{4}H$. The means of the two backcrosses (measured from m, the mid-parental value) are $\bar{B}_1 = \frac{1}{2}([d] + [h])$ and $\bar{B}_2 = \frac{1}{2}([h] - [d])$, the overall mean of the generation thus being $\frac{1}{2}(\bar{B}_1 + \bar{B}_2) = \frac{1}{2}[h]$. The heritable variation of the backcross means is, therefore,

$$V_B = \frac{1}{2}[\frac{1}{2}([d] + [h])]^2 + \frac{1}{2}[\frac{1}{2}([h] - [d])]^2 - (\frac{1}{2}[h])^2 = \frac{1}{4}[d]^2$$

and the departure of the overall generation mean from the mid-parent accounts for $(\frac{1}{2}[h])^2 = \frac{1}{4}[h]^2$ of the variability. A component of variability, $[d]^2$, which we have not previously encountered, thus appears, and we should note that in the parental generation, again measuring from m, $V_P = \bar{P}_1^2 = \bar{P}_2^2$ $= [d]^2$. In the backcross generation the coefficients of D, H, $[d]^2$ and $[h]^2$ sum to unity in accordance with the principle of conservation.

The same is true of the second backcross generation when we bring us all the

TABLE 50
The balance sheet of variability

Generation		D	H	Coefficient of $[d]^2$	$[h]^2$
Parents		0	0	$1^2 = 1$	0
F_1		0	0	0	$1^2 = 1$
F_2		$\frac{1}{2}$	$\frac{1}{4}$	0	$(\frac{1}{2})^2 = \frac{1}{4}$
F_3	V_{1F3}	$\frac{1}{2}$	$\frac{1}{16}$		
	V_{2F3}	$\frac{1}{4}$	$\frac{1}{8}$		
	Total	$\frac{3}{4}$	$\frac{3}{16}$	0	$(\frac{1}{4})^2 = \frac{1}{16}$
F_4	V_{1F4}	$\frac{1}{2}$	$\frac{1}{64}$		
	V_{2F4}	$\frac{1}{4}$	$\frac{1}{32}$		
	V_{3F4}	$\frac{1}{8}$	$\frac{1}{16}$		
	Total	$\frac{7}{8}$	$\frac{7}{64}$	0	$(\frac{1}{8})^2 = \frac{1}{64}$
S_3 (BIP)	V_{1S3}	$\frac{1}{4}$	$\frac{1}{16}$		
	V_{2S3}	$\frac{1}{4}$	$\frac{3}{16}$		
	Total	$\frac{1}{2}$	$\frac{1}{4}$	0	$(\frac{1}{2})^2 = \frac{1}{4}$
S_4	V_{1S4}	$\frac{1}{4}$	$\frac{3}{128}$		
	V_{2S4}	$\frac{1}{8}$	$\frac{5}{128}$		
	V_{3S4}	$\frac{1}{4}$	$\frac{11}{64}$		
	Total	$\frac{5}{8}$	$\frac{15}{64}$	0	$(\frac{3}{8})^2 = \frac{9}{64}$
Maternal	V_{1M3}	$\frac{1}{8}$	0		
	V_{2M3}	$\frac{3}{8}$	$\frac{1}{4}$		
	Total	$\frac{1}{2}$	$\frac{1}{4}$	0	$(\frac{1}{2})^2 = \frac{1}{4}$
1st Backcross (B_1 and B_2)	$V_{\bar{B}}$	0	0	$\frac{1}{4}$	
	$\bar{V}_B$	$\frac{1}{4}$	$\frac{1}{4}$	0	
	Total	$\frac{1}{4}$	$\frac{1}{4}$	$\frac{1}{4}$	$(\frac{1}{2})^2 = \frac{1}{4}$
2nd Backcross (B_{11}, B_{12}, B_{21} and B_{22})	$V_{\bar{B}}$	0	0	$\frac{5}{16}$	$\frac{1}{16}$
	$\bar{V}_B$	$\frac{3}{16}$	$\frac{3}{16}$	0	0
	Total	$\frac{3}{16}$	$\frac{3}{16}$	$\frac{5}{16}$	$\frac{1}{16} + (\frac{1}{2})^2 = \frac{5}{16}$

four second backcross families, B_{11}, B_{12}, B_{21} and B_{22}, again as will be seen from Table 50. The principle of conservation will apply whenever the frequencies of allelic genes are equal over the range of families brought into the assessment, but will not apply where the equality of allele frequencies has been lost, since the total variability has thereby been changed. The effect of allele frequencies in the totality of the variability is discussed by Mather (1973) and need not detain us further here.

The coefficients of D, H, $[d]^2$ and $[h]^2$ sum to unity because all the variability must be accounted for, but each of these items has its own special relation to the expression of variability among the phenotypes. The D and $[d]^2$ depend on the genetically additive effects of the genes, while H and $[h]^2$ depend on their dominance effects. The additive effects express themselves in different ways in D and $[d]^2$ as do the dominance effects in H and $[h]^2$. $[d]^2$ is affected by association and dispersion whereas D is not, and dominance in opposite directions tends to balance out $[h]^2$ but not H. Furthermore $[d]^2$ does not equal D nor $[h]^2$ equal H except in the trivial case where only one gene difference is involved. Even where all the genes are in association $[d]^2$ exceeds D, and where all dominance is in the same direction $[h]^2 > H$, in both cases by a factor which depends on the number of gene pairs that are involved as well as on the extent to which the ds and hs vary from one gene pair to another. We shall return to these relations when we come to consider the estimation of the number of segregating gene pairs in Chapter 11.

27. PARTITIONING THE VARIATION

Since D, H and E appear in these second degree statistics with coefficients which vary in both absolute and relative values, a suitable group of variances and covariances can be used to provide estimates of the components of variation. In suitable cases estimates of D and H can be obtained directly from an analysis of variance. Three experimental designs yielding data which can be analysed in this way have been described by Comstock and Robinson (1952). These types of experiment are commonly known as North Carolina designs 1, 2 and 3, respectively, and of them number 3 illustrates the case of the analysis of variance technique at its most powerful. The basic material of this type of experiment consists of pairs of families, each pair being obtained by backcrossing an individual from F_2 to both true breeding parental lines.

The contribution of a single gene difference to the means and variances of the pairs of backcross families (L_{1i} and L_{2i} for the backcrosses of the ith F_2 individual to P_1 and P_2, respectively) are set out in Table 51. Assuming no linkage or interaction between the genes, for any number of genes and allowing for the environmental components of variation, these expectations become

$$\text{Variance of } \tfrac{1}{2}(\bar{L}_{1i} + \bar{L}_{2i}) = \tfrac{1}{8}D + \tfrac{1}{2}E_b$$
$$\text{Variance of } \tfrac{1}{2}(\bar{L}_{1i} - \bar{L}_{2i}) = \tfrac{1}{8}H + \tfrac{1}{2}E_b$$
$$\text{Mean variance of all families} = \tfrac{1}{8}D + \tfrac{1}{8}H + E_w.$$

If we now consider the analysis of variance of actual data in which n F_2 individuals ($i = 1$ to n) are each backcrossed to P_1 and P_2 and m individuals of

TABLE 51
Backcrosses of F_2 to both parents

F_2 genotype	AA	Aa	aa	Mean
Frequency	$\frac{1}{4}$	$\frac{1}{2}$	$\frac{1}{4}$	
Mean of backcross to $P_1(\bar{L}_1)$	d	$\frac{1}{2}(d+h)$	h	$\frac{1}{2}(d+h)$
Mean of backcross to $P_2(\bar{L}_2)$	h	$\frac{1}{2}(h-d)$	$-d$	$\frac{1}{2}(h-d)$
Sum of means $(\bar{L}_1+\bar{L}_2)$	$d+h$	h	$h-d$	h
Difference of means $(\bar{L}_1-\bar{L}_2)$	$d-h$	d	$d+h$	d
Heritable variance of backcross to $P_1(L_1)$	0	$\frac{1}{4}(d-h)^2$	0	$\frac{1}{8}(d-h)^2$
Heritable variance of backcross to $P_2(L_2)$	0	$\frac{1}{4}(d+h)^2$	0	$\frac{1}{8}(d+h)^2$

Variance of $\frac{1}{2}$ sums	$\frac{1}{8}d^2$
Variance of $\frac{1}{2}$ differences	$\frac{1}{8}h^2$
Average heritable variance of all families	$\frac{1}{8}d^2 + \frac{1}{8}h^2$

each progeny family are grown in each of r replicates the expected mean squares are

Item	df	Expected MS
Sums (L_1+L_2)	$n-1$	$\sigma_w^2 + 2m\sigma_{rs}^2 + 2mr\sigma_s^2$
Differences (L_1-L_2)	$n-1$	$\sigma_w^2 + 2m\sigma_{rd}^2 + 2mr\sigma_d^2$
Sums × replicates	$(r-1)(n-1)$	$\sigma_w^2 + 2m\sigma_{rs}^2$
Differences × replicates	$(r-1)(n-1)$	$\sigma_w^2 + 2m\sigma_{rd}^2$
Within families	$2nr(m-1)$	σ_w^2

where σ_w^2 is the mean variance of all families $= \frac{1}{8}D + \frac{1}{8}H + E_w$; σ_{rs}^2 is the environmental component of variation over replicates for sums; σ_{rd}^2 is the environmental component of variation over replicates for differences, both of which are estimates of $\frac{1}{2}E_b$; $\sigma_d^2 = \frac{1}{8}H$ and $\sigma_s^2 = \frac{1}{8}D$.

The appropriate tests of significance for D and H follow from this and we see that they are tested independently and with equal sensitivity. Although we have indicated that σ_{rs}^2 and σ_{rd}^2 measure $\frac{1}{2}E_b$, the precise meaning of this environmental component of variation will depend on the experimental design. Other items, for example, block differences, overall differences, etc. (see Table 52) may enter the analysis but these have no expectation on the biometrical model.

Let us apply these findings to the analysis of the data obtained by Dr Bucio Alanis in a study of plant height in *Nicotiana rustica* carried out in the Department of Genetics at the University of Birmingham. His observations covered the pairs of families obtained by backcrossing 40 plants ($n = 40$) from the F_2 of the cross between inbred varieties 1 and 5 to each of these two parental lines. The $40 \times 2 = 80$ families were raised as completely randomized plants, five plants of each family ($m = 5$) being grown in each of

TABLE 52
Analysis of variance from North Carolina type 3 experiment with *Nicotiana rustica* for the character final height

Item	SS	df	MS	χ^2	P
Blocks	129·065	1	129·065		
Sums	24 698·145	39	633·285	152·82	<0·001
Overall differences	45 930·425	1	45 930·425		
Differences	8 863·755	39	227·275	54·85	0·05
Blocks × sums	7 158·115	39	183·541	44·29	NS
Blocks × differences	7 214·275	40	180·355	44·64	NS
Within families	102 621·955	635	161·610		

two replicate blocks ($r = 2$). The data for the analysis are thus 160 values each being the mean of five plants and the corresponding variances of these sets of five plants.

The appropriate analysis of variance of these 160 means is set out in Table 52. There are 159 degrees of freedom of which 1 is for the overall difference between the blocks. The 40 sums of means of corresponding pairs of families ($\bar{L}_{11} + \bar{L}_{21}, \bar{L}_{12} + \bar{L}_{22}, \ldots, \bar{L}_{140} + \bar{L}_{240}$) yield a sum of squares of sums for 39 degrees of freedom ($n - 1$) and the 40 differences of these pairs of means ($\bar{L}_{11} - \bar{L}_{21}, \bar{L}_{12} - \bar{L}_{22}, \ldots, \bar{L}_{140} - \bar{L}_{240}$) a sum of squares of differences also for 39 degrees of freedom ($n - 1$). A further degree of freedom can be assigned to the sum of these 40 differences within the pairs but it is of no interest to us as it merely reflects the overall differences between the backcrosses (the corresponding sum of the 40 sums of the pairs is, of course, the overall sum of the 160 means). We are left with 79 degrees of freedom for the block interactions of the differences among the 40 sums and 40 differences each for 39 degrees of freedom, i.e. $(r - 1)(n - 1)$, and the grand overall difference for 1 degree of freedom. The last two items could be formally separated but are, in fact, pooled in the analysis of variance since they do not differ significantly from one another. To calculate the sums of squares for the sums and differences items and their interactions over the replicate blocks we have summed or differenced pairs of family means ($\bar{L}_{1i}$ and $\bar{L}_{2i}$, etc.) and then summed or differenced these over the two replicates, we must therefore use a divisor of 4.

Where, as in the present experiment, the individuals are completely randomized the average value of the within family variance provides an additional estimate of the error variance. However, to include it in the analysis of variance just described we must either multiply all the sums of squares by 5 or divide the mean variance within families by 5. This is necessary because the mean variance within families is based on differences among the scores of individuals whereas all other items in the analysis have been based on the mean scores of five such individuals. In Table 52 all items except the mean variance within families have been multiplied by a factor of $m = 5$. The mean variance within families has 640 degrees of freedom, i.e. $2nr(m-1)$, but the loss of 5 plants has reduced it to 635.

The significance of the block interactions can be tested by comparison with the within-family variance. Because it is based on 635 degrees of freedom, the sums of squares of blocks × sums and blocks × differences can be divided by the within-family mean square and treated as χ^2s for 39 and 40 degrees of freedom, respectively. Neither interaction is significant, as might be expected, since the complete randomization should lead to E_b being equal to zero. The interactions could, therefore, be pooled with the within-family variance but because of the large number of degrees of freedom of the latter it makes no difference to the outcome of the analysis. In the absence of block interactions the within-family variance is the appropriate error variance for testing sums and differences. The sums item is highly significant but the differences item is borderline ($P = 0.05$). Indeed if the differences item is tested against its own block interaction or even against the pooled block items it is not significant ($P = 0.20 - 0.10$). While, therefore, there can be no doubt that D contributes to the variation of the F_2 sample there is not such clear evidence for an H component of the variation.

There are, of course, other items in the analysis of variance whose significance can be tested but these are of marginal interest as they have no expectation in terms of the D, H and E_w model.

The direct tests of significance of D and H are the first important results obtained from an experiment of North Carolina type 3. The analysis of variance will, however, also yield estimates of D and H. Since in these data there are no significant block interactions the expected and observed mean squares in this particular case reduce to

$$\hat{\sigma}_w^2 + 20\hat{\sigma}_s^2 = 633.285$$
$$\hat{\sigma}_w^2 + 20\hat{\sigma}_d^2 = 227.275$$
$$\hat{\sigma}_w^2 \qquad\quad = 161.610$$

Hence,

$$\hat{\sigma}_s^2 = \tfrac{1}{8}\hat{D} = 23.584$$
$$\hat{\sigma}_d^2 = \tfrac{1}{8}\hat{H} = \ 3.283$$
$$\hat{\sigma}_w^2 = \tfrac{1}{8}\hat{D} + \tfrac{1}{8}\hat{H} + \hat{E}_w = 161.610$$

From which it follows that

$$\hat{D} = 188.67 \qquad \sqrt{\frac{\hat{H}}{\hat{D}}} = 0.37$$

$$\hat{H} = 26.27$$

and
$$\hat{E}_w = 134.74 \ (= \hat{\sigma}_w^2 - \hat{\sigma}_s^2 - \hat{\sigma}_d^2).$$

We can, of course, estimate E_w directly from the within-family variances of the two inbred parents and of their F_1 which were included in the experiment.

These yield a joint estimate of $\hat{E}_w = 129\cdot81$ which is in remarkably good agreement with the estimate obtained from the analysis of variance.

We can go further and estimate the expected variance of the F_2 parental population from which the sample of 40 plants was obtained for the design 3 experiment. In terms of D, H and E, we have seen that $V_{1F2} = \frac{1}{2}D + \frac{1}{4}H + E$. Substituting the values obtained for these three parameters, $V_{1F2} = 235\cdot64$. The estimate of V_{1F2} from a sample of F_2 plants included in the design 3 experiment equalled $224\cdot84$ and $203\cdot74$ in the two replicates, respectively – which is again in good agreement with the expectation.

Although the good agreement between the observed and expected values of E_w and V_{1F2} suggest that the D, H and E_w model is adequate and hence that the assumptions underlying this model hold for these data, the N.C.3 design itself provides no tests of these assumptions. Kearsey and Jinks (1968) proposed an extension of this design which provides a test of one of the assumptions, namely, the absence of non-allelic interactions. The extended design is a triple test cross in which each F_2 individual is crossed not only to the inbred parents from which the F_2 was derived but also to their F_1. The contributions of a single gene difference to the means and variances of the $F_2 \times F_1$ crosses are given in Table 53.

TABLE 53
Crosses of F_2 to the F_1

F_2 genotype	AA	Aa	aa	
Frequency	$\frac{1}{4}$	$\frac{1}{2}$	$\frac{1}{4}$	Mean
Mean of cross to F_1	$\frac{1}{2}d + \frac{1}{2}h$	$\frac{1}{2}h$	$-\frac{1}{2}d + \frac{1}{2}h$	$\frac{1}{2}h$
Heritable variance of cross to F_1	$\frac{1}{4}(d-h)^2$	$\frac{1}{2}d^2 + \frac{1}{4}h^2$	$\frac{1}{4}(d+h)^2$	$\frac{3}{8}d^2 + \frac{1}{4}h^2$
Variance of means	$\frac{1}{8}d^2$			
Average heritable variance of all families	$\frac{3}{8}d^2 + \frac{1}{4}h^2$			

Since the gametic output of an F_1 is equal to a mixture of equal parts of that of its two inbred parents, for any number of independent gene pairs, it is not surprising that for the ith set of crosses

$$\bar{L}_{1i} + L_{2i} - 2\bar{L}_{3i} = 0$$

where $\bar{L}_{3i}$ is the mean of the family obtained by crossing the ith individual in the F_2 sample to the F_1. This relationship, therefore, provides a test for non-independence of the genes in action, i.e. for the presence of non-allelic interactions; it will hold within the limits of the sampling errors of the family means if the additive-dominance model is adequate. We may test the relationship as a t for each of the $i = 1$ to n sets of families in a manner comparable to the scaling tests described in Chapter 4. Alternatively we may use an analysis of variance to test all n sets of families simultaneously.

We will illustrate the latter approach by reference once more to the experiment of Bucio Alanis. In addition to being crossed to the two inbred lines each of the 40 F_2 plants was also crossed to the F_1 and all progenies were grown together in the completely randomized and replicated block design. We have therefore 40 values of $\bar{L}_{1i} + \bar{L}_{2i} - 2\bar{L}_{3i}$ for $i = 1$ to 40 for each of the two replicates. After summing over replicates we can obtain the sum of the 40 squared deviations of $\bar{L}_{1i} + \bar{L}_{2i} - 2\bar{L}_{3i}$ from zero divided by 6×2 for 40 degrees of freedom (n). The error for testing the significance of this item is then the sum of the 40 squared deviations from zero after differencing over replicates divided by 6×2, again for 40 degrees of freedom $n(r-1)$. Because the experiment was completely randomized a further estimate of the error variance is provided by the mean variance within the $\bar{L}_1$, $\bar{L}_2$ and $\bar{L}_3$ types of families. This is expected to have 960 degrees of freedom ($3nr(m-1)$) but the loss of 9 plants reduces it to 951 degrees of freedom for these data. It is, of course, necessary to multiply the other items by 5 to bring them into line with this error based on comparisons among individuals. Because the variances within L_1, L_2 and L_3 families have different expectations (Table 51 and 53) they may well be significantly heterogeneous. When they are, they must be weighted in the proportions of the square of the coefficients of $\bar{L}_{1i}$, $\bar{L}_{2i}$ and $\bar{L}_{3i}$, i.e. $1:1:4$ to obtain an estimate of the mean variance within families appropriate for the analysis of variance. This analysis is given in Table 54. It is quite clear that there is no evidence of non-allelic interaction and hence the additive-dominance model investigated earlier is adequate.

In the absence of non-allelic interactions the $F_2 \times F_1$ families can obviously be used to provide additional information about the additive component of variation (D). Of the alternative ways of using this information that proposed by Jinks and Perkins (1970) is the most useful because it retains the analysis of variance approach and hence the orthogonality of the estimates of D and $\hat{H}$. For each set of families $\bar{L}_{1i}$, $\bar{L}_{2i}$ and $\bar{L}_{3i}$ we can make the following orthogonal comparisons.

	$\bar{L}_{1i}$	$\bar{L}_{2i}$	$\bar{L}_{3i}$
Comparison 1	1	1	1
2	1	-1	
3	1	1	-2

The first of these replaces the variance of $\bar{L}_{1i} + \bar{L}_{2i}$ as the estimate of D the only modification being that the coefficient of σ_s^2 in the expected mean square is changed from $2mr$ to $3mr$. Comparison 2 is, of course, the unchanged comparison for estimating H and comparison 3 is the one just described for testing for the presence of non-allelic interactions.

In North Carolina design 1 and 2 experiments, F_2 individuals are also used. In design 1 certain of them are used as male parents and each mated to a

TABLE 54
Test for the presence of non-allelic interactions in a triple test cross experiment
with *Nicotiana rustica* for the character final height

Item	SS	df	MS	χ^2	P
Non-allelic interactions	8 369·79	40	209·24	45·96	NS
Replicates	8 748·92	40	218·72	48·05	NS
Within families	173 184·94	951	182·11		

different group of females. In design 2 every male is mated to every female. In the analysis of design 1, the mean square between the male parents yields a direct test of significance and an estimate of D. The mean square among females within males also contains a component dependent upon H and so can be made to yield an estimate of H by comparison with the mean square between males. In design 2 the mean squares between males and females yield similar estimates of D uncontaminated by H and the mean square for interaction between male and female parents provides a test of significance and an estimate of H uncontaminated by D (Comstock and Robinson, 1952, and see Section 38).

This approach through the analysis of variance suffers, however, from two limitations. In the first place, only certain types of family yield data which afford an immediate test of significance and estimates of one or more components of variation, and in any case the number of mean squares which appear in the analysis of variance and from which the estimates are derived, is so restricted that little can be done towards testing the validity of the assumption of no linkage and no interaction. Secondly, the technique of analysis of variance offers no means of combining several different generations into a single analysis and so multiplying the number of statistics that can be used in estimating D, H and E as we have defined them, in testing their adequacy for representing the variation or in estimating any further components that might be required by a more complex structure of the variation.

A more general approach, which can be extended to include any number of statistics from any number of generations, is through a form of multivariate analysis. A simple form of this devised by Mather (1949a) and developed further by Mather and Vines (1952) may be illustrated by its application to the data on sternopleural chaetae in *Drosophila* recorded and analysed by Cooke *et al.* (1962) and Cooke and Mather (1962). This experiment began with reciprocal crosses between two long inbred lines, Samarkand (S) and the Birmingham line of Oregon (B). F_2s were raised from the resulting F_1s. Pair matings were made to the number of 346 from the cross B × S and 364 from S × B. These families are, of course, BIPs.

The experiment was spread over time and the BIPs were raised in groups of 30 to 40. Alongside each group of BIPs a few (1 to 4) F_2 cultures were also raised as were a single culture of each parent and of each reciprocal F_1. Parents, F_1s and F_2s were raised at other times also. The sternopleural chaetae were counted on 10 females and 10 males from each culture, parent, F_1, F_2 and BIP. The conditions were held sufficiently constant for the culture means not to vary significantly with occasion among the F_2s, which could be tested by an analysis of variance, and there was no trend with time obvious to inspection in parents and F_1s. The distinction between occasions has therefore been omitted from the present consideration of the results.

The numbers of cultures and the mean numbers of sternopleural chaetae in both males and females are shown for each generation in Table 55. Scaling tests were carried out on these means, males and females being taken separately, using Cavalli's (1952) technique as described in Chapter 4. These gave a clear indication of the presence of non-allelic interaction with $\chi^2_{(2)}$ = 25·3 for males and $\chi^2_{(2)}$ = 48·6 for females. No attempt was, however, made to scale out these interactions since, as Cooke and Mather showed, they were not large by comparison with the main effects and in any case were not of a kind which could easily be removed in this way.

TABLE 55
Mean numbers of sternopleural chaetae in two inbred lines of *Drosophila* and their descendants (Cooke *et al.*, 1962; Cooke and Mather, 1962)

	Number of cultures	Mean	
		Males	Females
Parents			
B	29	22·47	22·93
S	29	18·20	18·52
F_1			
B × S	29	19·58	20·46
S × B	29	20·11	20·31
F_2			
B × S	33	19·71	20·12
S × B	37	19·82	20·46
BIP			
B × S	346	19·48	20·06
S × B	364	19·84	20·36

We will follow Cooke *et al.* in treating both the sexes and the descendants of the two reciprocal crosses separately so as to give a quadruplicated experiment. The statistics available for estimating the components of variation are shown separately for the four parts of the experiment as well as from the four parts combined, in Table 56. In Mather's simple form of multivariate analysis no allowance is made for the inflation of a rank 1 variance by sampling variation

TABLE 56

The statistics available for estimating the components of variation of sternopleural chaetae number in *Drosophila* (Cooke *et al.*, 1962). The number of degrees of freedom on which each variance or covariance is based is shown in brackets below it

Statistic	BS males	BS females	SB males	SB females	Overall average
V_{1F2}	4·8660	3·1475	4·4303	4·1654	4·1606
$(\frac{1}{2}D + \frac{1}{4}H + E_1)$	(297)	(297)	(333)	(333)	
V_{1S3}	1·0673	1·1921	1·2002	1·1960	1·1648
$(\frac{1}{4}D + \frac{1}{16}H + E_2)$	(345)	(345)	(363)	(363)	
W_{1S23}	0·5797	1·0544	0·5376	0·7318	0·7236
$(\frac{1}{4}D)$	(345)	(345)	(363)	(363)	
V_{2S3}	3·2767	2·8733	3·3076	3·1315	3·1491
$(\frac{1}{4}D + \frac{3}{16}H + E_1)$	(3114)	(3114)	(3276)	(3276)	
E_1	2·6854	2·4906	2·2630	1·8873	2·3316
	(522)	(522)	(522)	(522)	
E_2	0·4881	0·4268	0·3308	0·3502	0·3990
	(56)	(56)	(56)	(56)	

stemming from the corresponding rank 2 variance. It is, therefore, convenient to treat the non-heritable components of variation within and between families as separate parameters, E_1 and E_2.

The direct estimates of E_1 and E_2 are derived from the variances respectively within cultures and between culture means of parents and F_1. The B parent showed somewhat higher variances than the S and the F_1s. E_1 and E_2 were therefore estimated as the sum of one-quarter the variance of each parental line and one-half the variance of the appropriate F_1.

Before we turn to the further analysis of the data set out in Table 56, it should be noted that they show certain inconsistencies. In particular, by comparison with the rest, the females from B × S show a very high value for W_{1S23} together with low values for V_{1F2} and V_{2S3}. These are not merely inconsistent with the other three parts of the experiment; they are inconsistent with one another, for W_{1S23} whose expectation is $\frac{1}{4}D$ falls short of V_{1S3}, with expectation $\frac{1}{4}D + \frac{1}{16}H + E_2$, by only 0·14, which is less than one-third of the direct estimate of E_2. Again, it is only 1·83 less than V_{1F2} whose expectation includes E_1 of which the direct estimate is 2·49. While observing these inconsistencies, Cooke *et al.* could give no reason for them. As we shall see they are, not surprisingly, reflected in the estimates of the components of variation.

Table 56 shows that we have, from each part of the experiment, six statistics from which to estimate four components of variation. A least squares technique of estimation is therefore used. While recognizing that the various statistics, V_{1F2} etc. are not known with equal precision and so should be weighted if a rigorous analysis is desired, in the interests of simplicity Mather (1949a) used no weights, so effectively treating the statistics as if they were all known with the same precision. A consequent advantage of this simple unweighted analysis, in which the element of sampling variation in the rank 1 variance is also neglected, is that the matrices developed for the analysis of any one experiment can be used for all others of the same design, as we shall see.

The six basic equations for the males of B × S, one for each line of Table 56 are:

$$V_{1F2} = \tfrac{1}{2}D + \tfrac{1}{4}H \;\; + E_1 = 4\cdot 8660$$
$$V_{1S3} = \tfrac{1}{4}D + \tfrac{1}{16}H + E_2 = 1\cdot 0673$$
$$W_{1S23} = \tfrac{1}{4}D \qquad\qquad = 0\cdot 5797$$
$$V_{2S3} = \tfrac{1}{4}D + \tfrac{3}{16}H + E_1 = 3\cdot 2767$$
$$E_1 = 2\cdot 6854$$
$$E_2 = 0\cdot 4881.$$

These are combined by the technique Mather described to give four normal equations yielding least squares estimates for the four components of variation in a way similar to that used in Section 14 but without weights, as unweighted least squares is being used in the present analysis. Each of the six equations is multiplied through by the coefficient of D which it contains and the six are then summed. Where D does not appear the equation is omitted. Then we have

$$\tfrac{1}{4}D + \tfrac{1}{8}H \;\; + \tfrac{1}{2}E_1 \qquad\quad = 2\cdot 433\,000$$
$$\tfrac{1}{16}D + \tfrac{1}{64}H \qquad\quad + \tfrac{1}{4}E_2 = 0\cdot 266\,825$$
$$\tfrac{1}{16}D \qquad\qquad\qquad\quad = 0\cdot 144\,925$$
$$\tfrac{1}{16}D + \tfrac{3}{64}H + \tfrac{1}{4}E_1 \qquad\quad = 0\cdot 819\,175$$

$$\text{Total } \tfrac{7}{16}D + \tfrac{3}{16}H + \tfrac{3}{4}E_1 + \tfrac{1}{4}E_2 = 3\cdot 663\,925$$

The three further equations are then found similarly but using the coefficients of H, E_1 and E_2 as multipliers in turn. The solution of the four equations obtained in this way gives the desired estimates of the components of variation in the BS males. Similar estimates can be found by applying the same method to the data from the other parts of the experiment.

The left sides of the equation are, however, concerned only with D, H, E_1 and E_2; they will be the same for all four parts of the experiment and indeed for all experiments of this kind. The right sides of the equation, on the other hand, are derived from the values observed for V_{1F2} etc. and the equations as they stand can refer only to the data in question. The four parts of the experiment, and other similar experiments, may nevertheless be analysed using the same solution if the right sides are replaced successively by 1, 0, 0, 0; 0, 1, 0, 0 and so on, to give four sets of four equations in the way Fisher (1946) sets out for multiple regression analyses. In doing this, the equations become

(1) $0.437\,500D + 0.187\,500H + 0.750\,000E_1 + 0.250\,000E_2 = 1, 0, 0, 0$
(2) $0.187\,500D + 0.101\,563H + 0.437\,500E_1 + 0.062\,500E_2 = 0, 1, 0, 0$
(3) $0.750\,000D + 0.437\,500H + 3.000\,000E_1 \qquad\qquad = 0, 0, 1, 0$
(4) $0.250\,000D + 0.062\,500\,H \qquad\qquad + 2.000\,000E_2 = 0, 0, 0, 1$

In solving these four sets of equations we obtain a matrix of multipliers of which c_{DD} is the value of D in the first of the four sets, and c_{DH}, c_{D1} and c_{D2} the value of D in the second, third and fourth sets. Similarly c_{HD}, c_{HH}, c_{H1} and c_{H2} are the values of H in the four sets and so on. This matrix turns out to be as shown in Table 57. It will be observed that as $c_{DH} = c_{HD}$ etc. the matrix is symmetrical about one diagonal.

TABLE 57
The c-matrix for unweighted analysis of the *Drosophila* experiment

		D		H		E_1		E_2
From equation set								
1	c_{DD}	12.394 366	c_{HD}	−24.338 028	c_{1D}	0.450 704	c_{2D}	−0.788 732
2	c_{DH}	−24.338 028	c_{HH}	75.718 310	c_{1H}	−4.957 746	c_{2H}	0.676 056
3	c_{D1}	0.450 704	c_{H1}	−4.957 746	c_{11}	0.943 662	c_{21}	0.098 592
4	c_{D2}	−0.788 732	c_{H2}	0.676 056	c_{12}	0.098 592	c_{22}	0.577 465

Now in the BS males the sum of the F_2 variance, variance of F_3 means and so on, each multiplied by the corresponding coefficient of D, is 3.6639. This we may denote as $S(Dy)$. Similarly

$$S(Hy) = \quad 1.8976$$
$$S(E_1 y) = 10.8281$$
$$S(E_2 y) = \quad 1.5554$$

all rounded off to four decimal places. The unweighted least squares estimate of D in these BS males is then found from the D column of Table 57 as

$$\hat{D} = c_{DD}S(Dy) + c_{DH}S(Hy) + c_{D1}S(E_1 y) + c_{D2}S(E_2 y)$$
$$= (12.394\,37 \times 3.6639) - (24.338\,03 \times 1.8976)$$
$$\quad + (0.450\,70 \times 10.8281) - (0.788\,73 \times 1.5554)$$
$$= 2.8813.$$

$\hat{H}$ is found from the H column as

$$\hat{H} = c_{HD}S(Dy) + c_{HH}S(Hy) + c_{H1}S(E_1 y) + c_{H2}S(E_2 y)$$
$$= 1.8795.$$

Similarly $\hat{E}_1 = 2\cdot6149$ and $\hat{E}_2 = 0\cdot3588$. (The value obtained here for E_1 does not agree with that given by Cooke *et al.* which thus appears to be in error.)

The estimates of the components of variation can be obtained for the other parts of the experiment, and also for the experiment as a whole using the overall average figures of Table 56, by use of the same c-matrix as given in Table 57. The values of $S(Dy)$ etc. are, however, found among the figures appropriate to the part of the experiment in question. Thus taking the BS females as an example

$$S(Dy) = \tfrac{1}{2}(3\cdot1475) + \tfrac{1}{4}(1\cdot1921) + \tfrac{1}{4}(1\cdot0544) + \tfrac{1}{4}(2\cdot8733) = 2\cdot8537$$

$$S(Hy) = \tfrac{1}{4}(3\cdot1475) + \tfrac{1}{16}(1\cdot1921) + \tfrac{3}{16}(2\cdot8733) \qquad = 1\cdot4001$$

$$S(E_1y) = 3\cdot1475 + 2\cdot8733 + 2\cdot4906 \qquad\qquad = 8\cdot5114$$

$$S(E_2y) = 1\cdot1921 + 0\cdot4268 \qquad\qquad\qquad = 1\cdot6189$$

from which is found for this part of the experiment

$$\hat{D} = \quad (12\cdot394\,37 \times 2\cdot8537) - (24\cdot338\,03 \times 1\cdot4001) \ldots = \quad 3\cdot8533$$

$$\hat{H} = \ -(24\cdot338\,03 \times 2\cdot8537) + (75\cdot718\,31 \times 1\cdot4001) \ldots = \ -4\cdot5431$$

$$\hat{E}_1 = \quad 2\cdot5363$$

$$\hat{E}_2 = \quad 0\cdot4698.$$

The estimates of D, H, E_1 and E_2 so obtained from the four parts of the experiment and from the overall averages are set out in Table 58. One anomaly is at once apparent. All four quantities are quadratic and cannot therefore be negative yet the value yielded by the BS females for H is negative. It will be recalled that internal inconsistencies have already been noted in the results from this part of the experiment. We shall return later to the significance to be attached to this negative value for H.

TABLE 58
The components of variation in the *Drosophila* experiment

	BS males	BS females	SB males	SB females		Overall	Empirical standard error
D	2·8813	3·8533	2·5123	2·9550	±0·8818	3·0480 ± 0·4409	±0·5694
H	1·8795	−4·5431	3·3469	3·0118	±2·1795	0·9550 ± 1·0898	±3·6983
E_1	2·6149	2·5363	2·2175	1·8834	±0·2433	2·3125 ± 0·1217	±0·3341
E_2	0·3588	0·4698	0·3469	0·3096	±0·1903	0·3711 ± 0·0902	±0·0689

The estimates of D, H, E_1 and E_2 can be used to reconstruct values which they lead us to expect for V_{1F2}, V_{1S3}, W_{1S23}, V_{2S3}, E_1 and E_2 in the various parts of the experiment. Thus for the BS males

$$V_{1F2} = \tfrac{1}{2}D + \tfrac{1}{4}H + E_1$$

$$= \tfrac{1}{2}(2\cdot8813) + \tfrac{1}{4}(1\cdot8795) + 2\cdot6149$$

$$= 4\cdot5254.$$

TABLE 59
Values expected for the statistics and the deviations of those observed (Table 56) from these expectations

	BS males	BS females	SB males	SB females	Overall
Expectation					
V_{1F2}	4·525 4	3·327 1	4·310 4	4·113 9	4·075 3
V_{1S3}	1·196 6	1·149 2	1·184 2	1·236 6	1·192 8
W_{1S23}	0·720 3	0·963 3	0·628 1	0·738 8	0·762 0
V_{2S3}	3·687 6	2·647 8	3·473 1	3·186 9	3·253 6
E_1	2·614 9	2·536 3	2·217 5	1·883 4	2·312 5
E_2	0·358 8	0·469 8	0·346 9	0·309 6	0·371 1
Deviation					
V_{1F2}	0·340 6	−0·179 6	0·119 9	0·051 5	0·085 3
V_{1S3}	−0·129 3	0·042 9	0·016 0	−0·040 6	−0·028 0
W_{1S23}	−0·140 6	0·091 1	−0·090 5	−0·007 0	−0·038 4
V_{2S3}	−0·410 9	0·225 5	−0·165 5	−0·055 4	−0·104 5
E_1	0·070 5	−0·004 57	0·045 5	0·003 9	0·019 1
E_2	0·129 3	−0·043 0	−0·016 1	0·040 6	0·027 9
Sum of squares	0·343 02	0·097 27	0·052 54	0·009 08	0·021 60

These expectations are set out in Table 59 which also shows the amounts by which the values actually observed for these statistics (see Table 56) deviate from the expectations. There are six such deviations, one from each statistic, in each part of the experiment, but the sum of the squares of the six corresponds to only two degrees of freedom since the four quantities D, H, E_1 and E_2, have been estimated from the six statistics. The sum of squares is, of course, a measure of the adequacy of the four quantities to account for the variation observed in that part of the experiment. It also provides the basis for finding the standard errors of the estimates of the four quantities. Each sum of squares is, however, based on only two degrees of freedom and so is itself subject to a large sampling variation. This difficulty is overcome, in some measure, by adding the sum of squares from the four parts of the experiment to give a combined sum of squares of

$$0·343\ 02 + 0·097\ 37 + 0·052\ 54 + 0·009\ 08 = 0·501\ 91$$

for 8 degrees of freedom. Of these 8 df, 2 may be picked out to test the adequacy of the overall estimates of D, H, E_1 and E_2 to account for the variation as measured by overall average values of V_{1F2} etc. The sum of squares of deviations of these statistics as observed from their expectation based on the overall estimates of D etc. is 0·021 60; but as it stands this is not comparable with the other sum of squares as the values of V_{1F2} etc. are the averages of the four V_{1F2}s etc., one from each of the four parts of the experiment.* The sum of squares from the overall comparison must therefore be multiplied by 4, to give 0·086 40, to make it comparable with the other sums of squares. The remaining 6 df represent heterogeneity among the four parts of the experiment in the estimates of D etc., which they yield. They correspond to the remainder sum of the squares found as $0·501\,91 - 0·086\,40 = 0·415\,51$. This analysis of variance is set out in Table 60. The mean square for overall deviation is slightly, though of course not significantly, lower than that for heterogeneity among the parts of the experiment. The overall fit is thus no poorer than would be expected from the differences among the parts of the experiments: there is, therefore, no evidence that the four components of variation, D, H, E_1 and E_2, are inadequate to account for the variation in number of sternopleural chaetae observed in this experiment.

TABLE 60
Analysis of variance of Drosophila experiment

Item	SS	df	MS
Overall deviation	0·086 40	2	0·043 20
Heterogeneity	0·415 51	6	0·069 25
Total	0·501 91	8	0·062 74
Breakdown of heterogeneity			
BS females versus the rest	0·270 70	2	0·135 35
Among the rest	0·144 81	4	0·036 20

In view of the suspicion that the data from the BS females are internally inconsistent, the analysis may be taken a little further. Omitting the BS females and using only the other three parts of the experiment, new overall estimates of the four quantities can be found. These yield a further analysis of variance with 2 df for overall deviations, as before, but with only 4 df for heterogeneity, since with only three parts of the experiment there are but 6 df in all. Using unweighted averages of the statistics from BS males, SB males and SB females to give the overall averages, itself not strictly accurate but adequate for our present purpose (see footnote below), we find the sum of squares for heterogeneity among these three parts of the experiment to be 0·144 81.

This sum of squares must be included in the sum of squares for heterogeneity among all four parts of the experiment as set out in the analysis of variance of Table 60. Deducting 0·144 81 from the heterogeneity sum of

* The treatment as given is not fully exact, since corresponding statistics in the four parts of the experiment are not always based on the same number of degrees of freedom, and the overall averages are indeed weighted averages. The error introduced into the analysis of variance by assuming equal precision and weighting of the four parts of the experiment will, however, be negligible in the present case.

squares of Table 60 leaves $0.41551 - 0.14481 = 0.27070$ for heterogeneity between the BS females on the one hand and the rest of the experiment on the other. This will correspond to 2 df giving a mean square of $0.27070 \div 2 = 0.1354$. The mean square for heterogeneity among the other three parts of the experiment is $0.14481 \div 4 = 0.0362$. Thus the BS females do stand out as more different from the rest of the experiment than the other three parts are from one another, once more suggesting something suspicious about these females. Again, however, it can be no more than suspicion, for the variance ratio is only $0.1354/0.0362 = 3.7$, whereas for mean squares based on 2 and 4 df a variance ratio of 6.9 is needed before the probability is down to 0.05. Since the difference is not significant there is no real justification for separating off the BS females from the rest of the experiment. The analysis is therefore continued on the basis of the simple analysis of variance of Table 60.

Since the two mean squares for overall deviation and for heterogeneity respectively do not differ significantly they may be combined to provide the estimate of error variation for calculating the standard error of the D etc. The pooled sum of squares is 0.50191 for 8 df so giving a mean square of 0.06274. V_D, the variance of D, is found by multiplying this by c_{DD} from the matrix in Table 57, V_H as the error variance multiplied by c_{HH}, V_{E1} using c_{11} as the multiplier and V_{E2} using c_{22}. Then $V_D = 0.06274 \times 12.3944 = 0.7776$. From this we find $s_D = \sqrt{V_D} = 0.8818$ as the standard error of D, and similarly for s_H etc. The standard errors apply to the estimates of D etc. from the individual parts of the experiment. The four parts are combined in the overall estimates which will thus have variances one-quarter as large as those of the estimates from the individual parts of the experiment, and standard errors half as large. The standard errors of the individual and of the overall estimates are shown in Table 58. It will be observed that the negative value found for H from the BS females is only just over twice its standard error, which since the latter is based on no more than 8 df is not fully significant though it might again be thought suggestive of some unexplained inconsistency or anomaly in the results from this part of the experiment.

The standard errors found in this way are, however, open to some suspicion for in using a common estimate of error variance in calculating V_D, V_H, V_{E1} and V_{E2} we are assuming that D, H, E_1 and E_2 themselves have the same error variance (the differences in the error variances of their estimates arising solely from the differences in the coefficients with which they appear in V_{1F2} etc.). This obviously cannot be so for the overall estimate of D is some eight times as large as the corresponding E_2 and even E_1 is over six times as large as E_2. We have thus probably underestimated the standard errors of D and E_1 while overestimating that of E_2. In principle, estimates of the standard errors which do not suffer from this disability can be found empirically from comparisons among the four estimates of each quantity obtained from the different parts of the experiment (see Nelder, 1953). Thus we can use the ordinary formula for calculating a variance to find V_D empirically as:

$$\tfrac{1}{3}[(2\cdot8813^2 + 3\cdot8533^2 + 2\cdot5123^2 + 2\cdot9550^2) - \tfrac{1}{4}(2\cdot8813 + 3\cdot8533$$
$$+ 2\cdot5123 + 2\cdot9550)^2] = 0\cdot3240$$

there being 3 df among the four independent estimates of D. Then $s_D = \sqrt{V_D}$ $= \sqrt{0\cdot3240} = 0\cdot5694$. (The standard error of D as estimated from the overall data will again be half of this.) The standard errors of the estimates of the four quantities found empirically in this way are set out at the right of Table 58. As expected the value for s_{E2} is lower than its counterpart obtained using the common estimate of error variation and that for s_{E1} is higher. Unexpectedly, however, that for s_D is lower, but we should remember that the empirical standard errors are each based on only 3 df and so are themselves subject to high sampling variation and not to be taken as very precise estimates. Indeed, unless an experiment is highly subdivided, standard errors obtained empirically in this way will always be based on too few comparisons to be really trustworthy.

The value found empirically for s_H in the present experiment is very high, thus reflecting the disparity among the four estimates of H. By comparison with this s_H, the negative value of H from the BS females is obviously of no significance, and we should perhaps not dwell on it further. It is clear from the analysis that there is good evidence of D type variation, but no good evidence of dominance. It is also clear that, since $E_2 > \tfrac{1}{10}E_1$, there are greater sources of non-heritable variation at work between cultures than within cultures, i.e. $E_b > 0$, as indeed is no more than expected.

28. WEIGHTED ANALYSIS

The method just described for obtaining from second degree statistics estimates of the components of variation has the advantage of relative simplicity in use. In particular, when it has once been worked out a c-matrix can be used in the analysis of all experiments yielding the same set of statistics. Thus the c-matrix of Table 57 can be used wherever the available statistics comprise V_{1F2}, V_{1S3}, W_{1S23}, V_{2S3}, E_1 and E_2. In the same way if the experiment had been carried out with plants capable of self-pollination it might have yielded values for V_{1F2}, V_{1F3}, V_{2F3}, W_{1F23}, E_1 and E_2. The appropriate c-matrix could thus have been worked out and would have been available for use with the data from all subsequent experiments of this same type. A number of matrices are already available in the literature for use in this way (see for example, Mather, 1949a; Mather and Vines, 1952).

The method has, however, three disadvantages. First, it ignores the sampling variation appearing in a lower rank variance as a reflection of a higher rank variance. For example, we saw in Section 25 that V_{1F3} is basically $\tfrac{1}{2}D + \tfrac{1}{16}H$

$+E_b$ but it must have added to it

$$\frac{1}{m}V_{2F3} = \frac{1}{m}(\tfrac{1}{4}D + \tfrac{1}{8}H + E_w)$$

to allow for the sampling variation in the distribution of the F_3 family means, arising from the variation within these F_3 families. Only by ignoring the sampling variation component can a standard matrix be worked out once and for all, as m is likely to vary from experiment to experiment whether a result of deliberate action or unforeseen circumstance. If m is reasonably high the effect of ignoring the sampling variation will be small; but if m is not large, ignoring the sampling variation could lead to a not unimportant distortion of the results.

The second disadvantage is that the statistics are all given equal weight in the estimation of the components of variation. This implies that the statistics are themselves known with equal precision, which though achievable in an experiment designed with this specific end in view, will not generally be the case. And if the statistics are not known with equal precision the more precise statistics should obviously be given greater weight in the analysis than the less precise.

Thirdly, the values of the statistics used as material for estimating the components of variation may themselves be correlated, whereas the method of analysis set out in the previous section assumes that they are not. Thus in Cooke et al.'s experiment V_{1F2} was found from the parents of the BIP families, and the sternopleural numbers of these parents and the means of these BIP families were used in finding W_{1S23}. So the values of V_{1F2}, V_{1S3} and W_{1S23} must in some measure be correlated and it would obviously be desirable for the method of estimation to take this into account.

The way of making allowance for differing precisions of the individual statistics and for correlations among them has been developed by Hayman (1960a) and Nelder (1960). The process of weighted estimation involves the setting up of matrices which are individual to each experiment, and the introduction of terms to represent the sampling variation in the statistics of lower rank thus presents no difficulty. We have already seen examples of weighted estimation in earlier chapters and it will be illustrated further by reference to the experiment of Cooke et al. already discussed. The only statistics affected by the inclusion of an item to cover sampling variation are V_{1S3} and E_2. The former now takes the form

$$V_{1S3} = \tfrac{1}{4}D + \tfrac{1}{16}H + E_b + \tfrac{1}{10}V_{2S3} = \tfrac{1}{4}D + \tfrac{1}{16}H + E_b$$
$$+ \tfrac{1}{10}(\tfrac{1}{4}D + \tfrac{3}{16}H + E_w),$$

where the second part of the expression reflects the effect of the sampling variation. The coefficient of $\tfrac{1}{10}$ is used because the family means were based on 10 males or females as the case might be. It should be noted, however, that in analysing the 'overall' statistics this coefficient would become $\frac{1}{40}$ as each of these were the averages of four values, one from each part of the experiment and each based on 10 males or females.

We can thus set down the coefficients of D, H, E_b and E_w in the expressions for the statistics V_{1F2} etc. observed in the experiment in the form of Table 61. Now as Hayman (1960a) points out, in matrix notation D, H, E_w, E_b can be regarded as a column vector $\mathbf{g}$, V_{1F2} etc. as a row vector $\mathbf{v}$ and the coefficients as a matrix $\mathbf{M}$, determined of course by the genetical structure and physical layout of the experiment. If we then construct the transpose $\mathbf{M}'$ of $\mathbf{M}$ by interchanging the first row and first column, second row and second column etc., the operation of setting up the left sides of the equation giving unweighted least squares estimates, as described in the previous section, is the premultiplication of $\mathbf{M}$ by $\mathbf{M}'$ to give $\mathbf{M}'\mathbf{M}$. This is then inverted to give the c-matrix of Table 57, which is thus $(\mathbf{M}'\mathbf{M})^{-1}$. The right sides of the equation of estimation are obtained at the same time and in the same way as the left sides. They thus represent $\mathbf{M}'\mathbf{v}$, just as the left sides are $\mathbf{M}'\mathbf{M}$. Finally, since the estimates of D, H etc. are obtained by multiplying the values of $S(Dy)$ etc. by the c-matrix, we now see that they can be represented in general terms as $\mathbf{g} = (\mathbf{M}'\mathbf{M})^{-1}\mathbf{M}'\mathbf{v}$ and the expected values of the statistics are then found as $E\mathbf{v} = \mathbf{Mg}$.

TABLE 61
Matrix of coefficients relating the components of variation to the statistics observed in the *Drosophila* experiment of Cooke *et al.*

		D	H	E_w	E_b
	V_{1F2}	$\frac{1}{2}$	$\frac{1}{4}$	1	·
	V_{1S3}	$\frac{1}{4}+\frac{1}{40}$	$\frac{1}{16}+\frac{3}{160}$	$\frac{1}{10}$	1
$\mathbf{v}$	W_{1S23}	$\frac{1}{4}$	·	·	·
	V_{2S3}	$\frac{1}{4}$	$\frac{3}{16}$	·	1
	E_1	·	·	1	·
	E_2	·	·	$\frac{1}{10}$	1

A dot indicates a coefficient of zero

The calculation of weighted least squares estimates of the components of variation follows the same general pattern except that the weights are introduced into the derivation of both sides of the equation of estimations. If we use $\mathbf{V}$ to denote the error variance-covariance matrix of the $\mathbf{v}$ (not to be confused with the c-matrix which is the variance-covariance matrix of the estimates $\hat{\mathbf{g}}$) we find

$$\hat{\mathbf{g}} = (\mathbf{M}'\mathbf{V}^{-1}\mathbf{M})^{-1}\mathbf{M}'\mathbf{V}^{-1}\mathbf{v}.$$

In words, the transpose $\mathbf{M}'$ is first multiplied by the invert of $\mathbf{V}$ and the product so obtained is used to premultiply first $\mathbf{M}$, to give the left side of the equation of estimation, and then $\mathbf{v}$ to give the right side. The left side, $\mathbf{M}'\mathbf{V}^{-1}\mathbf{M}$ is then inverted to give $(\mathbf{M}'\mathbf{V}^{-1}\mathbf{M})^{-1}$ which is the weighted c-matrix, and this is used to premultiply the right side of the equation, $\mathbf{M}'\mathbf{V}^{-1}\mathbf{v}$, to give the weighted estimates of $\mathbf{g}$ just as was done in the unweighted estimation.

The matrix $\mathbf{V}$ takes care both of the differences in precision of the statistics, V_{1F2} etc. and of their correlations. Let us, however, look first at the simpler case where they are uncorrelated. The sampling variance of a variance V is $2V^2/N$ where N is the number of degrees of freedom from which V is found. That of the covariance W is $(W^2 + V_1 V_2)/N$ where V_1 and V_2 are the variances of the groups of observations whose covariance is W. Since we are assuming the absence of correlations between the statistics, the error variance matrix $\mathbf{V}$ will be diagonal, containing nothing but a series of elements of this kind down the main diagonal. Inverting such a diagonal matrix results simply in a further diagonal matrix whose elements are the reciprocals of the elements of $\mathbf{V}$. Being thus the reciprocals of the sampling variances of the statistics the weights so obtained are the amounts of information (see e.g. Mather, 1946a, 1967a) about these statistics and $\mathbf{V}^{-1}$ may be described as an information matrix. All that is then necessary to obtain weighted estimates of the components in such a case is to follow the procedure described in the previous section but, when obtaining the equation of estimation, to multiply through not by the coefficients of D, H etc. in turn but by the products of these coefficients and the amounts of information or weights, i. Thus, take for example $V_{1F2} = \frac{1}{2}D + \frac{1}{4}H + E_w$ whose weight will be $i = N/2V_{1F2}^2$. Its contribution to the left side of the first equation of estimates will be

$$\frac{i}{2}(\tfrac{1}{2}D + \tfrac{1}{4}H + E_w),$$

the multiplier being the product of the coefficient of D and the weight i. Similarly the contributions of $V_{2S3} = \frac{1}{4}D + \frac{3}{16}H + E_w$ to the left sides of the three relevant equations would be

$$\frac{i}{4}(\tfrac{1}{4}D + \tfrac{3}{16}H + E_w)$$

$$\frac{3i}{16}(\tfrac{1}{4}D + \tfrac{3}{16}H + E_w)$$

$$i(\tfrac{1}{4}D + \tfrac{3}{16}H + E_w)$$

where $i = N/2V_{2S3}^2$. The same multipliers would, of course, be applied to the numerical value observed for V_{2S3} to obtain its contribution to the right sides of the equation, which by derivation from the notation of the previous section we can denote as $S(iDy)$, $S(iHy)$ etc.

This procedure is, of course, the same as that used for estimating m, $[d]$ and

[h] from the generation means in Chapters 4 and 5. The arithmetic example used in the earlier section fully illustrates the calculation, so obviating the need for a further illustration here. Two points may, however, be noted. In the earlier case the components m, [d] and [h] were estimated as a means to a different end, namely that of testing the goodness of fit of the hypothesis that there was no non-allelic interaction with the observed results. For this test the values of the generation means expected from the estimates of m etc. were compared with those observed, the sum of squares of differences between observation and expected (each square being multiplied by the weight appropriate to that mean) being distributed as a χ^2. Although in the present case the estimation of D, H, E_w and E_b is a primary aim, the same χ^2 test of goodness of fit is available and serves to test whether there is any evidence for dependence of the genes in either transmission (i.e. linkage) or action (i.e. non-allelic interaction), or of course both.

The second point relates to the weights themselves. In the examples of the earlier chapters, these were found as the reciprocals of the variances of the means as observed empirically in the experiments. The same procedure could be adopted in the estimation of D, H, etc. from V_{1F2} etc. if the experiment included sufficient replicates to give reliable empirical values for the variances of V_{1F2} etc. This will, however, seldom be the case, for experiments of this kind are demanding of time and space and few have even the four replicates of Cooke *et al.*'s *Drosophila* experiment. We therefore must generally fall back on finding the variances of V_{1F2} etc. as $2V^2/N$ or, for covariances, $(W^2 + V_1 V_2)/N$, as noted above. These variances should, however, be found using not the values observed for V_{1F2} etc. but the values expected for them, based on the weighted estimates of D, H, E_w and E_b (Hayman, 1960a). Thus finding the best estimates of the components of variation depends on using weights which themselves depend on the estimates of the components. We must therefore proceed by iteration, computing the weights first from the values observed from the statistics, using them to find estimates of D etc., recomputing the weights using the values expected for the statistics on the basis of these first estimates of D etc., using these new weights to find improved estimates of the components and so on, the iterative process continuing until its repetition fails to produce improvement in the estimates and the test of goodness of fit.

This process can be illustrated by the example used by Hayman. His data came from part of an experiment with *Nicotiana rustica* and consisted of the two inbred parents, their F_1, F_2, F_3 and F_4 generations. The parents were crossed reciprocally to give the two reciprocal F_1s, from each of which was grown an F_2, F_3 and F_4. The plants were grown in two blocks, the plots within a block each comprising 5 plants. Each block included 5 plots of each parent, 5 of each F_1, 5 of each F_2, 1 plot of each of 10 F_3s from each of the reciprocals and 1 plot of each of 50 F_4s from each reciprocal, the 50 being obtained by selfing 5 plants from each of 10 F_3 families the year before. The data from parents and F_1s were pooled in finding E_1 and E_2. Thus E_1 is based on $N = 160$ (where N is the number of degrees of freedom), 5×4 from each parent in each block and similarly $5 \times 4 \times 2 \times 2$ from the F_1s, the factors of 2 representing the two blocks and the two reciprocals. Similarly E_2 is based on $N = 32, 4 \times 2 \times 2$ from the parents and $4 \times 2 \times 2$ from the F_1s. With five plots of each reciprocal F_2 in each block, V_{1F2} is based on $N = 5 \times 4 \times 2 \times 2 = 80$. Similarly V_{1F3} is based on $N = 9 \times 2 \times 2 = 36$ and V_{2F3} on $N = 10 \times 4 \times 2 \times 2 = 160$. Turning to the F_4, there are 10 groups each of 5 families from each reciprocal, so V_{1F4} stems from $N = 9 \times 2 \times 2 = 36$, V_{2F4} from $N = 10 \times 4 \times 2 \times 2 = 160$ and V_{3F4} from $N = 10 \times 5 \times 4 \times 2 \times 2 = 800$. The numbers of degrees of freedom recorded by Hayman for V_{2F4} and V_{3F4} do not, however, agree exactly with these expectations (see Table 62), doubtless because some F_4 plants and, it would appear likely, 7 F_4 families either failed in the experiment or were excluded for other reasons.

Turning next to the effect of sampling variation as the expectations for the various statistics, with 5 plants per plot we expect

$$V_{1F3} = \tfrac{1}{2}D + \tfrac{1}{16}H + E_b + \tfrac{1}{5}V_{2F3}$$
$$= \tfrac{1}{2}D + \tfrac{1}{16}H + E_b + \tfrac{1}{5}(\tfrac{1}{4}D + \tfrac{1}{8}H + E_w).$$
$$\text{Similarly } V_{2F4} = \tfrac{1}{4}D + \tfrac{1}{32}H + E_b + \tfrac{1}{5}(\tfrac{1}{8}D + \tfrac{1}{16}H + E_w).$$

Now V_{1F4} is the variance of the means of the F_4 groups, each of which comprises 5 families, since 5 plants were selfed from each of the F_3 families the year before. So

$$V_{1F4} = \tfrac{1}{2}D + \tfrac{1}{64}H + \tfrac{1}{5}V_{2F4}$$
$$= \tfrac{1}{2}D + \tfrac{1}{64}H + \tfrac{1}{5}(\tfrac{1}{4}D + \tfrac{1}{32}H + E_b) + \tfrac{1}{25}(\tfrac{1}{8}D + \tfrac{1}{16}H + E_w).$$

Finally, while, of course, $E_1 = E_{2w}$, sampling variations make $E_2 = E_b + \tfrac{1}{5}E_w$. Putting these together with the expectations for V_{1F2}, V_{2F3} and V_{3F4} which are unaffected by sampling variation gives the **M** matrix as set out in Table 63.

Three covariances W_{1F23}, W_{1F34} and W_{2F34}, could have been calculated but were, in fact, excluded by Hayman from this analysis. Furthermore, since V_{1F2} was found from a set of plants which did not include the F_2 individuals selfed in the previous year to give the F_3s recorded in the experiment, and since the F_3 plants used in the previous year as parents of the F_4s similarly constituted no part of the F_3 observations, there is no sampling correlation between V_{1F2} and V_{1F3}, V_{1F3} and V_{1F4}, or V_{2F3} and V_{2F4}. So the estimates of the different variances may be treated as uncorrelated. Thus the **V** matrix is diagonal and the weights may be obtained simply as the reciprocals of the sampling variances in the way described above.

Two characters were observed in the experiment; plant height and time of first flowering. The values observed for the statistics, V_{1F2} etc. in respect of plant height are set out in Table 62, together with the values of N on which they are based and the weights found directly from these. As for example in the case

TABLE 62
Weighted analysis of the data in plant height from Nicotiana rustica (after Hayman, 1960a)

| | Observed | N | First weight | Expected after iteration | | | Unweighted expectation |
				1	2	5	
V_{1F2}	447·03	80	0·000 200 1	416·06	418·51	419·81	411·74
V_{1F3}	278·19	36	0·000 232 6	398·71	441·10	439·74	394·32
V_{2F3}	236·52	160	0·001 430 1	250·84	251·87	252·39	254·13
V_{1F4}	437·68	36	0·000 094 0	311·35	372·84	368·45	370·26
V_{2F4}	266·39	153	0·001 078 1	244·13	262·97	263·16	216·26
V_{3F4}	170·77	770	0·013 201 3	168·19	168·52	168·64	175·35
E_1	83·55	160	0·011 460 8	85·61	85·16	84·90	96·58
E_2	90·71	32	0·001 944 5	89·55	84·90	86·58	38·19

| | Estimates after iteration | | | Unweighted analysis |
	1	2	5	
D	516·0	638·8	629·1	655·5
H	289·9	55·9	81.5	−50·5
E_w	85·6	85·2	84·9	96·6
E_b	72·5	67·9	69·6	18·9
s_D	129·3	112·3	124·6	103·2
s_H	326·7	290·8	312·1	386·0
s_{Ew}	9·2	8·7	8·7	41·0
s_{Eb}	23·5	19·8	19·7	38·3
$\chi^2_{(4)}$	5·87	3·67	3·68	—

of V_{1F2}, its variance is

$$\frac{2V^2}{N} = \frac{2 \times 447·03^2}{80}$$

so giving

$$i = \frac{80}{2 \times 447·03^2} = 0·000\,200\,1.$$

The use of these weights gives the values of D, H, E_w and E_b and their standard errors as shown in the first column of the lower part of the table from which are obtained the expectations for V_{1F2} etc., shown above them. These afford the means of calculating the weights for the second round of the iterative procedure. Now the weights applied to V_{1F2} have become

$$i = \frac{80}{2 \times 416·06^2}$$

and so on. The results of the second calculation are set out in the next column

TABLE 63
Matrix of coefficients relating the components of variation to the statistics observed in Hayman's experiment with *Nicotiana*

		g		
	D	H	E_w	E_b
V_{1F2}	$\frac{1}{2}$	$\frac{1}{4}$	1	
V_{1F3}	$\frac{11}{20}$	$\frac{7}{80}$	$\frac{1}{5}$	1
V_{2F3}	$\frac{1}{4}$	$\frac{1}{8}$	1	
V_{1F4}	$\frac{111}{200}$	$\frac{39}{1600}$	$\frac{1}{25}$	$\frac{1}{5}$
V_{2F4}	$\frac{11}{40}$	$\frac{7}{160}$	$\frac{1}{5}$	1
V_{3F4}	$\frac{1}{8}$	$\frac{1}{16}$	1	
E_1			1	
E_2			$\frac{1}{5}$	1

of the table. Hayman actually carried out five such rounds of calculations and his final results are shown next in the table, those for the third and fourth rounds being omitted as unimportant for our present purpose. χ^2 testing of goodness of fit is shown for each of the rounds of calculation, the χ^2 being based on 4 df since 4 parameters have been estimated from the 8 quantities, V_{1F2} etc. Finally, in the last column of the table are given values of D etc. and the expectations for V_{1F2} etc. based on them as found by the simple unweighted method described in the previous section.

It will be seen that there is little difference between the final estimates of the components of variation and those obtained by two rounds of calculation. Clearly the iteration could have stopped at this point. The second round of calculation was, however, desirable as the estimates it yields show marked changes in D and H from those resulting from the first round, though the standard errors of D and H are altered but little. It is also worthy of note that the estimates yielded by the simple unweighted analysis for D, H and E_w agree quite well with results of the weighted analysis, in the case of D and H being in fact nearer to the values obtained in the latter after two (or five) rounds of calculation than those obtained from the first round. E_b is, however, greatly underestimated by the unweighted analysis. Similarly the unweighted analysis yields quite reasonable values for s_D and s_H but greatly overestimates s_{Ew} and s_{Eb}. Unweighted analysis gave a negative value for H, but this was insignificant and so is not disturbing. Weighted analysis looks superior in that it gave a positive value for H, but again this is insignificant so that in fact neither analysis produces any evidence for dominance. It might be noted, too, that there is no reason why weighted analyses should not also on occasion yield an insignificant negative estimate for H, or indeed any other of the components.

In respect therefore of the genetical components of variation, (though not in respect of the environmental components) the simple unweighted analysis emerges well from this comparison. The weighted analysis must, nevertheless, always offer a great advantage over the unweighted for it yields a test of goodness of fit which the latter does not. The χ^2 shows that the fit is satisfactory and that there is thus no evidence of disturbance from linkage or non-allelic interaction. Such confidence in the adequacy of the treatment could only be obtained by more indirect means, if at all, from unweighted analysis.

29. THE GENERAL CASE

The structure of Hayman's data was such that there was no sampling correlation between the statistics from which the components of variation were estimated. This will not generally be true. Turning, as an example, to Cooke *et al.*'s experiment with *Drosophila*, it will be recalled from the description in Section 27 that the flies from which V_{1F2} was calculated were the parents of the BIP families from whose means V_{1S3} was found. There must thus be a sampling correlation between these two statistics. Furthermore, since W_{1S23} was found from the same F_2 flies and S_3 means, it must be correlated for the same statistical reason with both V_{1F2} and V_{1S3}. The matrix is therefore no longer diagonal, as with Hayman's experiment, but takes the form of Table 64 in respect of the genetical variances and covariances (Cooke *et al.*, 1962).

TABLE 64

The **V** matrix (sampling variances and covariances) for the *Drosophila* experiment (Cooke *et al.*)

$$
\mathbf{v} \begin{cases} V_{1F2} \\ V_{1S3} \\ W_{1S23} \\ V_{2S3} \end{cases} \begin{bmatrix} 2V_{1F2}^2/N & 2W_1^2S_{23}/N & 2V_{1F2}.W_{1S23}/N & \cdot \\ 2W_{1S23}^2/N & 2V_{1S3}^2/N & 2V_{1S3}.W_{1S23}/N & \cdot \\ 2V_{1F2}.W_{1S23}/N & 2V_{1S3}.W_{1S23}/N & (W_{1S23}^2 + V_{1F2}V_{1S3})/N & \cdot \\ \cdot & \cdot & \cdot & 2V_{2S3}^2/N' \end{bmatrix}
$$

N is the number of degrees of freedom on which are based V_{1F2}, V_{1S3}, W_{1S23}
N' is the number of degrees of freedom on which is based V_{2S3}
(for the values of N and N' see Table 56)

Since this matrix is not diagonal it cannot be inverted and used by simply finding the reciprocals of the diagonal elements and taking these as weights, as was done in Hayman's analysis described in the previous section. On the contrary, its inversion involves a fairly heavy calculation, to give the information matrix, which is, of course, also not diagonal, and which must be premultiplied by the transpose of the **M** matrix before their product is postmultiplied by **M** to give the left sides of the equation of estimation and by **v** to give the right sides. Thus in the general case, weighted analysis becomes very heavy computationally and indeed must be expected to require the use of a computer. Cooke *et al.* did, in fact, carry out the weighted analysis of the *Drosophila* experiment with the aid of a computer.

Before we consider their results, however, three points should be made. First, they introduced a further statistic into their calculations by finding the variance of the means of the F_2 cultures. Since this should be no more than the sampling variation to be expected from the variance within F_2 cultures, except that its non-heritable element will be that appropriate to the comparison between the means of sets of 10 flies (the number of individuals of the set counted from each culture – see Section 27) rather than that appropriate to comparisons between flies from the same culture, this newly introduced statistic has the expectation $\frac{1}{20}D + \frac{1}{40}H + E_2$. Its introduction serves, of course, to provide a further degree of freedom for the test of goodness of fit of the expectations based on the estimates of the components of variation with the values observed for the various statistics.

Secondly, Cooke et al. estimate E_1 and E_2 as the non-heritable variances within and between bottles, as indeed Mather (1949a) did in his method of unweighted analysis. In this latter case no complication was introduced thereby; but in weighted analysis where items are brought into e.g. V_{1S3}, to take care of the sampling variation arising from the variance within families, V_{2S3}, Hayman's practice of estimating E_w and E_b as the actual non-heritable components is much to be preferred.

Thirdly, the number of iterations to be carried out is relatively unimportant where a computer is available, but some means of deciding when to stop is obviously necessary. This Cooke et al. achieved by finding the ratios of the squares of the changes in the estimates of the components of variatioh to the respective variances of these estimates, and continuing the iteration until the greatest of these fell below a predecided value.

The results of Cooke et al.'s weighted analysis is shown in Table 65. These estimates of the components of variation should be compared with those from the unweighted analysis as set out in Table 58. The overall estimates from the unweighted analysis are repeated in Table 65 to facilitate this comparison with the weighted estimates. Three points emerge from the comparison, whether of overall estimates or estimates for the four parts of the experiment. First, the weighted estimates of the heritable components of variation are in the main a little smaller than their unweighted counterparts: the differences are, however, not large. Second, despite this tendency for the weighted estimate of D and H to be smaller than the unweighted, the two categories agree strikingly in the picture they reveal; including the anomalous behaviour of the BS female part of the experiment. Third, the standard errors of the weighted estimates are generally little more than half the size of those from the unweighted analysis, so drawing attention to the gain achieved in precision by the more exact procedure. This increased precision brings out even more strongly the anomalous features of the data from the BS females, for it now shows the estimate of H obtained from these flies to be significantly negative – an impossible result for a quadratic quantity. This could, of course, be held to throw doubt upon the genetical theory upon which the analysis is based; but in

TABLE 65

Weighted analysis of Cooke et al.'s *Drosophila* experiment

	B × S		S × B		Overall	Overall unweighted
	Males	Females	Males	Females		
D	$2{\cdot}488 \pm 0{\cdot}443$	$3{\cdot}978 \pm 0{\cdot}468$	$2{\cdot}308 \pm 0{\cdot}442$	$2{\cdot}750 \pm 0{\cdot}457$	$2{\cdot}897 \pm 0{\cdot}227$	$3{\cdot}048 \pm 0{\cdot}441$
H	$0{\cdot}094 \pm 1{\cdot}122$	$-3{\cdot}871 \pm 1{\cdot}060$	$2{\cdot}843 \pm 1{\cdot}049$	$2{\cdot}548 \pm 0{\cdot}998$	$0{\cdot}378 \pm 0{\cdot}530$	$0{\cdot}955 \pm 1{\cdot}090$
E_1	$2{\cdot}695 \pm 0{\cdot}144$	$2{\cdot}564 \pm 0{\cdot}134$	$2{\cdot}231 \pm 0{\cdot}124$	$1{\cdot}953 \pm 0{\cdot}110$	$2{\cdot}362 \pm 0{\cdot}065$	$2{\cdot}313 \pm 0{\cdot}122$
E_2	$0{\cdot}428 \pm 0{\cdot}062$	$0{\cdot}392 \pm 0{\cdot}056$	$0{\cdot}336 \pm 0{\cdot}052$	$0{\cdot}317 \pm 0{\cdot}049$	$0{\cdot}370 \pm 0{\cdot}028$	$0{\cdot}371 \pm 0{\cdot}090$
$\chi^2_{(3)}$	$9{\cdot}276$	$3{\cdot}638$	$1{\cdot}454$	$2{\cdot}676$		

the absence of any supporting evidence from the remainder of the experiment, it would seem wiser to regard it as reflecting some unexplained anomaly that has arisen in this group of observations.

The sum of squares of differences between observation and expectation based on the estimates in the unweighted analysis becomes the χ^2 of the weighted analysis and in becoming a χ^2 it permits an immediate test of goodness of fit. This advantage stems directly from weighted analysis and cannot be achieved without it. Though the sum of squares from the unweighted analysis was for 2 df in each part of the experiment, the χ^2 is for 3 df because of the introduction of the variance of F_2 culture means as an additional statistic in the weighted analysis. The χ^2 for the BS males is significant at the 5 % level, but none of the others is significant. Nor is the sum of the χ^2s from the four parts of the experiment ($\chi^2_{(12)} = 17{\cdot}04$, $P = 0{\cdot}20 - 0{\cdot}10$). There is thus no reason to doubt the adequacy, for interpreting these results, of the simple genetical hypothesis taken as the basis of this analysis – apart, of course, from the negative value yielded for H by the BS female which has already been the subject of comment.

In that the estimates it yields of the components of variation are subject to larger standard errors, the unweighted analysis is less efficient than the weighted. Its efficiency in any given case can be measured by the ratio of the sampling variance of the estimates from the weighted analysis to the sampling variance of the unweighted estimates. The efficiency to be expected from unweighted analysis is discussed by Cooke et al. for experiments like that with Drosophila and also for experiments like another carried out with Nicotiana rustica. The expected efficiency of estimation was, in all cases, good for the genetical components of variation, being better for D than for H: in fact in no case was the expected efficiency of the unweighted estimate below 85 % for D, generally being well over 90 %, and the expected efficiency of the corresponding estimate of H was in all cases greater than 65 %, lying, indeed, usually between 70 % and 80 %. The expected efficiency of estimation of the non-heritable components of variation was lower, but in only one example was it below 50 % and the non-heritable components are, in any case, of lesser interest to the geneticist than the heritable. Thus not only is there no indication that unweighted analysis gives misleading or biased estimates of the components: its expected efficiency is not unacceptably low in either of the two types of experiment that have been analysed. Clearly in the absence of computer facilities it is an acceptable alternative to weighted analysis. As Cooke et al. say 'despite its shortcomings, the simple unweighted treatment may be expected to yield reasonable estimates of the components of variation, though the values found for their standard error cannot be regarded as fully reliable'. The greatest advantage of the weighted treatment lies in the χ^2 test of goodness of fit to which it leads and which cannot be obtained from unweighted analysis.

One last comparison remains to be made before we leave the Drosophila experiment. In a later paper, Cooke and Mather (1962) record the mean numbers of sternopleural chaetae in the 27 types of female and 18 types of male that can be produced by combining the X, II and III chromosomes, taken as units, of the B and the S lines in all possible ways. From such results it is possible to obtain direct estimates of the three d's, one from each chromosome, and the three corresponding hs in the females and two in the males where, of course, h for the X chromosome has no meaning, since heterozygotes cannot be produced. It is also possible to obtain estimates of parameters representing non-allelic interaction of the i, j, l and more complex types, but we shall not allow this to detain us as these interactions proved to be of little significance for the comparison we are seeking to make.

The estimates of the ds and the hs obtained by Cooke and Mather from males and females are given in Table 66. One point should be made. Since these are linear quantities, obtained by linear comparisons among means of sternopleural chaeta numbers, they take sign, and the convention has been adopted of designating the effect as positive where the higher chaeta number is associated with the chromosome from the B stock and the lower chaeta number with its homologue from S. It will be seen that the d_1 (the main effect of chromosome X) was small, with d_2 (relating to chromosome II) larger and d_3 (relating to chromosome III) much the largest of all. There was little evidence of dominance.

TABLE 66
The ds and hs associated with the three major chromosomes in the *Drosophila* experiment (Cooke and Mather, 1962)

Parameter	d_1	d_2	d_3	h_1	h_2	h_3
Females	0·064	0·669	1·136	−0·009	0·027	0·007
Males	−0·123	0·538	1·141	—	0·041	0·034

Now $D = S(d^2)$ and $H = S(h^2)$. We can synthesize from Table 66 values of D and H for comparison with those obtained by the analyses described above. We find

$$\hat{D} = 0·064^2 + 0·669^2 + 1·136^2 = 1·74 \text{ for females}$$

and $\qquad \hat{D} = (-0·123)^2 + 0·538^2 + 1·141^2 = 1·61 \text{ for males.}$

H is very small indeed for both sexes.

Now comparison with Table 65 shows that the value obtained for D, by analysis of V_{1F2} etc., is larger than that now obtained by synthesis. Indeed, the overall D from weighted analysis is 2·90 by comparison with an overall synthetic value of just under 1·70. Why the discrepancy, which must be at any rate on the border of significance?

In the first place Cooke and Mather show that the values of d_1 found respectively from females and males are almost surely over-low estimates; but this of itself will not explain the discrepancy. Nor, as Cooke and Mather further show, will it be explained by neglect of the parameters measuring non-allelic interaction, significant though some of these are. The explanation is most likely to be found elsewhere. The chromosomes were manipulated and treated as units in Cooke and Mather's experiment, each d being in fact the total effect of the polygenic combination carried by the corresponding chromosome taken as a unit. Then if the combination is a balanced one, recombination between the homologues could release variation (see Chapters 1 and 7). Now such variation could take place in Cooke *et al.*'s experiment from which the analytical values were obtained, but it could not in Cooke and Mather's experiment from which the synthetic values were obtained. Thus the values found for D in the two ways could differ, synthetic D being the lower in so far as recombination had released variation initially hidden in balanced polygenic combinations. Tests described by Cooke and Mather gave good evidence that the small value of d_1 did not indicate lack of activity in the X chromosome, which had in fact initially carried a balanced combination and that recombination released variation so leading to a larger value of d_1 and, by derivation, to a larger value of D in the later generations from which the analytical value of D had been obtained. There was no such release of variation by recombination from chromosome III; but then there was less reason to expect it since the value found initially for d_3 was large and did not provide the same suggestion of a balanced polygenic combination as did the smaller value of d_1.

The comparison of the synthetic and analytical Hs is less informative. Though the values of H for the SB males and females are significant in the weighted analysis, there is no consistent evidence of dominance from the analysis. There is thus no good reason to regard the synthetic and analytical Hs as being in conflict, though the value of the latter is generally higher than that of the former. Should this difference, however, be real, it could be explained in just the same way as the difference in the Ds. There is evidence that dominance is ambidirectional for chaeta characters (Breese and Mather, 1957, 1960, and see p. 274). Then as recombination of a balanced combination released variation of the D type it could also upset an initial balance of dominance and release variation of the H type. This possibility was not, however, tested further by Cooke and Mather.

7 *Interaction and linkage*

30. THE EFFECT OF INTERACTION

The contributions to the genetic variation of gene pairs at different loci are additive, in the way assumed in the previous chapter, only if the genes at the various loci are independent in both action and inheritance. Independence in action means, of course, that the increment added to the character in question by the substitution of one allele for another at a given locus is uninfluenced by the remaining genes under consideration. Independence in inheritance means that the two alleles at one locus are equally likely to be transmitted from parent to offspring together with a given allele at the second locus. Apart from special cases, therefore, dependence of genes in inheritance means linkage. We will consider the consequences for the components of variation of these two kinds of dependence in turn.

The effects of non-allelic interaction (or epistasis as it is sometimes called) on the mean expression of the character in the different generations has been considered in Chapter 5. Moving on to the second degree statistics, we note first that the variances of true breeding parental lines and their F_1s are unaffected by genic interactions for they reflect only non-heritable differences. Turning to the F_2 between two such true breeding lines and considering two unlinked pairs of genes, A–a and B–b (Table 19), we note that the mean is $\frac{1}{2}h_a + \frac{1}{2}h_b + \frac{1}{4}l_{ab}$ and find the variance as

$$V_{1F2} = \frac{1}{2}(d_a + \frac{1}{2}j_{ab})^2 + \frac{1}{2}(d_b + \frac{1}{2}j_{ba})^2 + \frac{1}{4}(h_a + \frac{1}{2}l_{ab})^2 + \frac{1}{4}(h_b + \frac{1}{2}l_{ab})^2 + \frac{1}{4}i_{ab}^2$$
$$+ \frac{1}{8}(j_{ab}^2 + j_{ba}^2) + \frac{1}{16}l_{ab}^2.$$

The effect of the i interaction appears as a separate item, but the main effects of the js appear in relation to the ds and of l in relation to the hs. The appearance of $(d + \frac{1}{2}j)$ and $(h + \frac{1}{2}l)$ is not surprising when we observe from the margin of Table 19 that the average departure from the mid-parent of individual homozygotes for an allele at one locus is $(d + \frac{1}{2}j)$ and that of the heterozygote is $(h + \frac{1}{2}l)$.

177

Proceeding to F_3 we find

$$V_{1F3} = \tfrac{1}{2}(d_a + \tfrac{1}{4}j_{ab})^2 + \tfrac{1}{2}(d_b + \tfrac{1}{4}j_{ba})^2 + \tfrac{1}{16}(h_a + \tfrac{1}{4}l_{ab})^2$$
$$+ \tfrac{1}{16}(h_b + \tfrac{1}{4}l_{ab})^2 + \tfrac{1}{4}i_{ab}^2 + \tfrac{1}{32}(j_{ab}^2 + j_{ba}^2) + \tfrac{1}{256}l_{ab}^2;$$

$$V_{2F3} = \tfrac{1}{4}(d_a + \tfrac{1}{4}j_{ab})^2 + \tfrac{1}{4}(d_b + \tfrac{1}{4}j_{ba})^2 + \tfrac{1}{8}(h_a + \tfrac{1}{4}l_{ab})^2$$
$$+ \tfrac{1}{8}(h_b + \tfrac{1}{4}l_{ab})^2 + \tfrac{5}{16}i_{ab}^2 + \tfrac{7}{64}(j_{ab}^2 + j_{ba}^2) + \tfrac{1}{32}l_{ab}^2;$$

$$W_{1F23} = \tfrac{1}{2}(d_a + \tfrac{1}{2}j_{ab})(d_a + \tfrac{1}{4}j_{ab}) + \tfrac{1}{2}(d_b + \tfrac{1}{2}j_{ba})(d_b + \tfrac{1}{4}j_{ba})$$
$$+ \tfrac{1}{8}(h_a + \tfrac{1}{2}l_{ab})(h_a + \tfrac{1}{4}l_{ab}) + \tfrac{1}{8}(h_b + \tfrac{1}{2}l_{ab})(h_b + \tfrac{1}{4}l_{ab})$$
$$+ \tfrac{1}{4}i_{ab}^2 + \tfrac{1}{16}(j_{ab}^2 + j_{ba}^2) + \tfrac{1}{64}l_{ab}^2.$$

In the F_3 variances the $(d + \tfrac{1}{2}j)$ and $(h + \tfrac{1}{2}l)$ of F_2 have been replaced by $(d + \tfrac{1}{4}j)$ and $(h + \tfrac{1}{4}l)$, again corresponding to the average departures from the mid-parent of homozygotes and heterozygotes in the F_3 generation. And, as might be expected, the corresponding expressions in the covariance of F_2 and F_3 are the geometric mean of their counterparts in the F_2 and F_3 variances, i.e. $(d + \tfrac{1}{2}j)(d + \tfrac{1}{4}j)$ and $(h + \tfrac{1}{2}l)(h + \tfrac{1}{4}l)$. It is no surprise that in V_{1F4}, V_{2F4} and V_{3F4} we find terms in $(d + \tfrac{1}{8}j)$ and $(h + \tfrac{1}{8}l)$, and we may be confident that in W_{1F34} and W_{2F34} they will be in $(d + \tfrac{1}{4}j)(d + \tfrac{1}{8}j)$ and $(h + \tfrac{1}{4}l)(h + \tfrac{1}{8}l)$.

These expressions relate only to two gene pairs. The expressions in d and h require generalization in two ways. In the first place, in so far as further genes are involved in similar interactions with A–a and B–b, we substitute $(d_a + \tfrac{1}{2}Sj_a.)$ for $(d_a + \tfrac{1}{2}j_{ab})$ where $Sj_a.$ is the sum of j_{ab}, j_{ac} etc. and $(h_a + \tfrac{1}{2}Sl_a.)$ for $(h_a + \tfrac{1}{2}l_{ab})$ etc. in F_2 statistics in order to cover the interactions of these further genes with A–a and B–b. We then generalize the expressions over all genes showing digenic interaction by writing

$$V_{1F2} = \tfrac{1}{2}D + \tfrac{1}{4}H + \tfrac{1}{4}I + \tfrac{1}{8}J + \tfrac{1}{16}L$$

where $D = S(d_a + \tfrac{1}{2}Sj_a.)^2$, $H = S(h_a + \tfrac{1}{2}Sl_a.)^2$, $I = S(i^2)$, $J = S(j^2)$ and $L = S(l^2)$ with, similarly

$$V_{1F3} = \tfrac{1}{2}D + \tfrac{1}{16}H + \tfrac{1}{4}I + \tfrac{1}{32}J + \tfrac{1}{256}L$$
$$V_{2F3} = \tfrac{1}{4}D + \tfrac{1}{8}H + \tfrac{5}{16}I + \tfrac{7}{64}J + \tfrac{1}{32}L$$
$$W_{1F23} = \tfrac{1}{2}D + \tfrac{1}{8}H + \tfrac{1}{4}I + \tfrac{1}{16}J + \tfrac{1}{64}L$$

Each expression should of course have an appropriate E added to it.

Apart from the terms in I, J and L which we might expect generally to be smaller than the terms in D and H, these expressions are the same as those already found for V_{1F2} in the previous chapter, the effects of interaction being accommodated by changing the definition of D from $S(d_a^2)$ to $S(d_a + wSj_a.)^2$ and that of H from $S(h_a^2)$ to $S(h_a + wSl_a.)^2$, the coefficient w changing with the generation, being $\tfrac{1}{2}$ in F_2, $\tfrac{1}{4}$ in F_3, $\tfrac{1}{8}$ in F_4 and so on. The test of non-allelic interactions is thus provided by a test of the homogeneity of D and H over generations. The terms in I, J and L are a distraction in such a test, but probably not a serious one. If they are neglected in the formulation of the variances and covariances these being written exactly as in the last chapter, but with the redefined D and H, the terms in I, J and L will tend to merge with E and to make E inhomogeneous too. Then in so far as the assumption of homogeneity of the compound E exaggerates the apparent inhomogeneity of D and H, the evidence for interaction will tend to be exaggerated. It should be observed, however, that unlike D and H, the effects of interaction not accommodated by the definitions of D and H but appearing in terms in I, J and L, are not homogeneous within generations. To this extent they may be a disturbing influence in the recognition of interaction by a test based on change in the definitions of the parameters between, while assuming their constancy within, generations.

Turning to biparental progenies of the third generation we find for the two gene case

$$V_{1S3} = \tfrac{1}{4}(d_a + \tfrac{1}{2}j_{ab})^2 + \tfrac{1}{4}(d_b + \tfrac{1}{2}j_{ba})^2 + \tfrac{1}{16}(h_a + \tfrac{1}{2}l_{ab})^2$$
$$+ \tfrac{1}{16}(h_b + \tfrac{1}{2}l_{ab})^2 + \tfrac{1}{16}i_{ab}^2 + \tfrac{1}{64}(j_{ab}^2 + j_{ba}^2) + \tfrac{1}{256}l_{ab}^2;$$
$$V_{2S3} = \tfrac{1}{4}(d_a + \tfrac{1}{2}j_{ab})^2 + \tfrac{1}{4}(d_b + \tfrac{1}{2}j_{ba})^2 + \tfrac{3}{16}(h_a + \tfrac{1}{2}l_{ab})^2$$
$$+ \tfrac{3}{16}(h_b + \tfrac{1}{2}l_{ab})^2 + \tfrac{3}{16}i_{ab}^2 + \tfrac{7}{64}(j_{ab}^2 + j_{ba}^2) + \tfrac{15}{256}l_{ab}^2;$$
$$W_{1S23} = \tfrac{1}{4}(d_a + \tfrac{1}{2}j_{ab})^2 + \tfrac{1}{4}(d_b + \tfrac{1}{2}j_{ba})^2 + \tfrac{1}{16}i_{ab}^2.$$

Here the main terms are in $(d + \tfrac{1}{2}j)$ and $(h + \tfrac{1}{2}l)$ just as they are in F_2. This is not surprising because, taken overall, the nine genotypes appear with the same frequencies in this generation as they do in F_2. It means, however, that comparison of the variances and covariances of these biparental families with the variance of F_2 offers no test of interaction through its effect on the terms in D and H, for the definition of these remains the same. But if sib-mating is continued for a further generation the interaction changes. The overall distribution of the genotypes in the S_4 generation is no longer that of F_2 and S_3, and this is reflected in the definition of D and H which depends on terms of the types $(d + \tfrac{3}{8}j)$ and $(h + \tfrac{3}{8}l)$ respectively. Heterogeneity of D and H once again provides a test for non-allelic interactions.

Since we are assuming A–a and B–b to be unlinked, it is immaterial to the variances and covariances of F_2, F_3, S_3 and other such derived generations, whether the initial cross was AABB × aabb or AAbb × aaBB. With interactions present this is not, however, true of the statistics from the families obtained by backcrossing the F_1 to the parents. Indeed we have already seen in Chapter 5 how the expressions for the means of the backcrosses differ according to whether the genes are associated or dispersed in the parents. These effects appear as differences in the signs of the interaction parameters in the expectation of the means. They appear similarly in the summed variances of the two backcrosses which are

$$V_{B1} + V_{B2} = \tfrac{1}{2}(d_a + \tfrac{1}{2}j_{ab} \mp \tfrac{1}{2}j_{ba})^2 + \tfrac{1}{2}(d_b \mp \tfrac{1}{2}j_{ab} + \tfrac{1}{2}j_{ba})^2$$
$$+ \tfrac{1}{2}(h_a \mp \tfrac{1}{2}i_{ab} + \tfrac{1}{2}l_{ab})^2 + \tfrac{1}{2}(h_b \mp \tfrac{1}{2}i_{ab} + \tfrac{1}{2}l_{ab})^2$$
$$+ \tfrac{1}{8}(j_{ab} \pm j_{ba})^2 + \tfrac{1}{8}(i_{ab} \pm l_{ab})^2,$$

where in the case of the double signs the upper one applies with associated genes (AABB × aabb) and the lower one with dispersed genes (AAbb × aaBB).

In the backcrosses the effect of interaction in inflating or reducing the summed variances, by comparison with the variances expected in its absence, depends on the distribution of the genes between the parents as well as on the signs of the interaction parameters. In F_2 and its derived generations inflation or reduction of the statistics depends only on these signs. If we consider two gene pairs and we look at

$$V_{1F2} = \tfrac{1}{2}(d_a + \tfrac{1}{2}j_{ab})^2 + \tfrac{1}{2}(d_b + \tfrac{1}{2}j_{ba})^2 + \tfrac{1}{4}(h_a + \tfrac{1}{2}l_{ab})^2 +$$
$$\tfrac{1}{4}(h_b + \tfrac{1}{2}l_{ab})^2 + \tfrac{1}{4}i_{ab}^2 + \tfrac{1}{8}(j_{ab}^2 + j_{ba}^2) + \tfrac{1}{16}l_{ab}^2$$

we see that while i will always tend to increase the variance, j will tend to increase it when positive and decrease it when negative, at least until it attains a value 1·6 times that of d (Mather, 1967b). Similarly l will increase the variance when of the same sign as h but decrease it when of the opposite sign to h, again at least until it reaches a value of 1·6h. As we have observed in Chapter 5, all the classical types of non-allelic interactions are, as indeed they must be, definable in terms of the relations between d_a, d_b, h_a, h_b, i_{ab}, j_{ab}, j_{ba}, and l_{ab}. In the classical complementary interaction, which gives a 9 : 7 ratio in a mendelian F_2, all these parameters have the same numerical value, with the ds and js both positive, and the hs, i and l having the same sign as one another, either all positive or all negative. Thus complementary interaction must inflate the variance of F_2 and also the variances of F_3, S_3 and later generations, though to varying degrees depending especially on the coefficients of j and l in the terms contributing to D and H, respectively (Mather, 1967b).

Similarly, duplicate interaction, leading to the classical F_2 ratio of 15 : 1, requires the eight parameters to be all of equal numerical value, but now with the js negative and i and l of like sign to one another but opposite to that of h. Thus duplicate interaction will decrease the variances of F_2 and its derived generations at least until j itself attains a certain ratio in relation to d and l in relation to h. As we have seen, this critical ratio is not reached in V_{1F2} until j and l materially exceed d and h, respectively, and it becomes even higher in later generations.

These effects of complementary and duplicate interactions are illustrated in Fig. 11 which is taken from Mather (1967b). It is assumed that, sign apart, $d_a = d_b = h_a = h_b$, and $i_{ab} = j_{ab} = j_{ba} = l_{ab} = \theta d_a$ where θ is a measure of the intensity of the interaction. At $\theta = 1$ interaction is full, and according to the signs of the parameters the two genes show the classical complementary or duplicate relationships. $0 < \theta < 1$ implies partial interaction and $1 < \theta$ a super-interaction. With two genes in this relation, complementary interaction increasingly inflates V_{1F2} as θ rises from 0 while duplicate interaction decreases it to a minimum obtained when $\theta = 0.8$. Then V_{1F2} rises with θ until at the critical ratio $\theta = j/d = l/h = 1.6$ it re-attains the value it has in the absence of interaction, which it then exceeds as θ rises above 1.6.

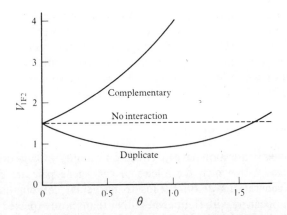

Figure 11 The effect on V_{1F2} of complementary and duplicate interactions between two gene pairs. θ measures the intensity of the interaction, which is absent when $\theta = 0$. At $\theta = 1$ interaction is complete and the two genes accord with the classical complementary or duplicate pattern, as the case may be. $0 < \theta < 1$ implies partial interaction and $1 < \theta$ super-interaction. (Reproduced by permission from Mather 1967b.)

With two genes it thus requires a high super-interaction of the duplicate type to restore to V_{1F2} its non-interactive level. In other words, with two genes one would expect a duplicate relation virtually always to reduce the variance of F_2. Since, however, with more than two genes in the system, the terms contributing to D and H take the forms $(d_a + \frac{1}{2}Sj_a.)^2$ and $(h_a + \frac{1}{2}Sl_a.)^2$ this critical value of θ falls reciprocally as the number of genes rises. Assuming all the ds equal both to one another and to all the hs, neglecting sign, and all the interaction parameters equal to one another and to θd, again neglecting sign, the value of θ at which, with duplicate interaction among all pairs of genes, V_{1F2} again reaches its non-interactive level is related to the number of genes in the system in Table 67. In general, with k genes in such a system the non-interactive value of V_{1F2} is regained when $\theta = 8/(2k+1)$ and V_{1F2} is at its minimal value when $\theta = 4/(2k+1)$.

TABLE 67
The effect on the variance of F_2 of duplicate interactions pairwise among k genes. The values of θ at which V_{1F2} is (a) minimal and (b) equal to its value in the absence of interaction are shown against the number of genes (and see in the text)

Number of genes	k	2	3	4	5	10	20
Value of θ (a)	$\dfrac{4}{2k+1}$	0·80	0·57	0·44	0·36	0·19	0·10
(b)	$\dfrac{8}{2k+1}$	1·60	1·14	0·89	0·73	0·38	0·20

In the classical complementary and duplicate relations $d_a = d_b$. As we have seen in Section 16, when $d_a \neq d_b$ the complementary interaction becomes the classical recessive epistasis, and the duplicate relation becomes the classical dominant epistasis. Thus recessive epistasis, like the complementary relation, is an interaction which inflates the variation, while dominant epistasis, like the duplicate relation, is one which tends to decrease it.

31. THE EFFECT OF LINKAGE

Where the genes in question have no effects on viability, linkage does not affect the frequencies with which the alleles of each gene pair are recovered in segregating generations. It only leads to particular combinations of these alternatives appearing with frequencies other than those expected on the basis of independence. On a scale where the increments added to the phenotype by the various genes are themselves additive, the total effect of a gene on the phenotypes of a family of given size will be the same, apart from sampling variation, no matter what its linkage relations may be: the relative frequencies of the particular combinations in which the alleles occur with other genes will have no effect, because every one which is over-common will be balanced by another being correspondingly rare. Linkage, therefore, can of itself have no effect on the mean measurements of segregating families, provided that the scale chosen for representing the phenotypes satisfies the scaling tests developed in Chapter 4. Equally then the operation of linkage does not vitiate these scaling tests.

Linkage, though not affecting the means, shows its effect in the second degree statistics, the variances and covariances used in partitioning the variation. Consider the simplest case of two genes A–a and B–b. When in the coupling phase with recombination value $p\,(= 1 - q)$ the ten genotypes of F_2 are expected with the frequencies shown in Table 68. This table also gives the phenotypic deviations of each class from the mid-parent, and the mean phenotypes of the corresponding F_3 families. It is not difficult to see from these data that the mean phenotype of F_2 is unaffected by the linkage, being $\frac{1}{2}(h_a + h_b)$ as before. But the sum of squares of deviations from the mid-parent is now

$$\tfrac{1}{4}[q^2(d_a + d_b)^2 + 2pq(d_a + h_b)^2 \ldots + q^2(-d_a - d_b)^2].$$

On subtracting $[\frac{1}{2}(h_a + h_b)]^2$ to correct for the departure of the F_2 mean from the mid-parent, the sum of squares of deviations from the mean, and with it the heritable variance (since the frequencies sum to unity), becomes

$$V_{1F2} = \tfrac{1}{2}[d_a^2 + d_b^2 + 2(1 - 2p)d_a d_b] + \tfrac{1}{4}[h_a^2 + h_b^2 + 2(1 - 2p)^2 h_a h_b].$$

Two new terms are to be observed in this expression, both involving the recombination value, in one case combined with the product $d_a d_b$ and in the

TABLE 68
Frequencies, F_2 phenotypes and mean F_3 phenotypes of the ten genotypic classes in an F_2 for two coupled genes

		AA	Aa	aa
The arrangement within each cell is: Frequency F_2 phenotype Mean F_3 phenotype	BB	q^2 $d_a + d_b$ $d_a + d_b$	$2pq$ $h_a + d_b$ $\frac{1}{2}h_a + d_b$	p^2 $-d_a + d_b$ $-d_a + d_b$
	Bb	$2pq$ $d_a + h_b$ $d_a + \frac{1}{2}h_b$	C $2q^2$ $h_a + h_b$ $\frac{1}{2}(h + h_b)$ $2p^2$ $h_a + h_b$ $\frac{1}{2}(h_a + h_b)$ R	$2pq$ $-d_a + h_b$ $-d_a + \frac{1}{2}h_b$
	bb	p^2 $d_a - d_b$ $d_a - d_b$	$2pq$ $h_a - d_b$ $\frac{1}{2}h_a - d_b$	q^2 $-d_a - d_b$ $-d_a - d_b$

All frequencies should be divided by 4

other with $h_a h_b$. With free recombination $p = \frac{1}{2}$, so that $1 - 2p = 0$ and these new terms vanish to leave the expressions obtained in the last chapter. When linkage is complete $p = 0$, giving $1 - 2p = 1$ and, aside from non-heritable variation, $V_{1F2} = \frac{1}{2}(d_a + d_b)^2 + \frac{1}{4}(h_a + h_b)^2$. The two genes then act as one. Even, however, where recombination occurs, recombinant types will be rare if p is small, and the genes will effectively act as one except in so far as selection may isolate one of the rare recombinants.

If we now write $D = d_a^2 + d_b^2 + 2(1 - 2p)d_a d_b$

and $\qquad H = h_a^2 + h_b^2 + 2(1 - 2p)^2 h_a h_b$

in place of the earlier definitions of these two components of heritable variation, we can again put $V_{1F2} = \frac{1}{2}D + \frac{1}{4}H + E$. It can then be shown that, as before, $V_{1F3} = \frac{1}{2}D + \frac{1}{16}H + E$ and $W_{1F23} = \frac{1}{2}D + \frac{1}{8}H$.

If the two genes are in repulsion, the expression for D is changed to $d_a^2 + d_b^2 - 2(1 - 2p)d_a d_b$, but, as might be expected, that for H is not changed. It should be noted, however, that the term $2(1 - 2p)^2 h_a h_b$ will be positive only if h_a and h_b are reinforcing one another by acting in the same direction. If they are opposing one other, this term must be negative. Thus opposition of the heterozygous increments resembles repulsion of the genes, and reinforcement resembles coupling in its effect on the variances and covariances (Fig. 12). It must nevertheless be remembered that opposition versus reinforcement is a

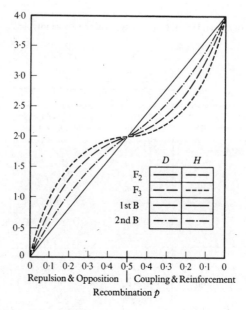

Figure 12 Change in the contribution made by two segregating genes to D and H in F_2, F_3 and backcrosses according to the genes' linkage relations. Calculations are based in $d_a = d_b = h_a = h_b = 1$.

physiological distinction, while repulsion versus coupling is a mechanical one.

When we turn to the mean variance of F_3 we find that it too can still be written in its old forms of $V_{2F3} = \frac{1}{4}D + \frac{1}{8}H + E$, but not only are D and H in this expression different from the corresponding items in the case of independent inheritance, they also differ from those given by V_{1F2}, etc., with linked inheritance. Taking the coupling case we find, in respect of these two genes,

$$V_{2F3} = \tfrac{1}{4}[d_a^2 + d_b^2 + 2(1-2p)^2 d_a d_b] + \tfrac{1}{8}[h_a^2 + h_b^2 + 2(1-2p)^2 (1-2p+2p^2)h_a h_b].$$

The repulsion case gives the same expression, but with the sign of the term in $d_a d_b$ reversed.

Then
$$D = d_a^2 + d_b^2 \pm 2(1-2p)^2 d_a d_b$$
and
$$H = h_a^2 + h_b^2 + 2(1-2p)^2 (1-2p+2p^2)h_a h_b,$$

the $\pm$ indicating the change in sign of the $d_a d_b$ term according to the phase of linkage. These expressions for D and H differ from those given by V_{1F2}, V_{1F3} and W_{1F23} in that $(1-2p)^2$ replaces $(1-2p)$ and $(1-2p)^2 (1-2p+2p^2)$ replaces $(1-2p)^2$ in the $d_a d_b$ and $h_a h_b$ terms, respectively. The definitions of D and H have changed, not with the generation, as when interaction was at issue, but with the rank of the variances.

In the same way V_{1S3} and W_{1S23} can still be written as $\frac{1}{4}D + \frac{1}{16}H + E$ and $\frac{1}{4}D$ respectively, with D and H having values as in V_{1F2}. When we come to the mean variance of biparental progenies we find that $V_{2S3} = \frac{1}{4}D + \frac{3}{16}H + E$, but D and H have changed from their V_{1F2} values just as they did in the case of V_{2F3}.

Turning to progenies raised by backcrossing, the expectations for which are given in Table 69, we find that $D = d_a^2 + d_b^2 + 2(1 - 2p)d_a d_b$ for coupling, and $D = d_a^2 + d_b^2 - 2(1 - 2p)d_a d_b$ for repulsion, H being given by $h_a^2 + h_b^2 + 2(1 - 2p)h_a h_b$ irrespective of linkage phase. In H the sign of the term in p will, of course, vary according to the opposition or reinforcement of the h_a and h_b increments. When D and H are redefined in this way the summed variances of the two first backcrosses still equal $\frac{1}{2}D + \frac{1}{2}H + 2E$, and the various combinations of variances of means of second backcrosses as well as covariances of means of second backcrosses with their first backcross parents also have the compositions shown in Table 49. Once again, however, the expressions for D and H change when we turn to the mean variance of the second backcross progenies. Though the formulae of Table 49 apply, D is now $d_a^2 + d_b^2 \pm 2(1 - 2p)(1 - p)d_a d_b$ and H is $h_a^2 + h_b^2 + 2(1 - 2p)(1 - p)h_a h_b$, the phase of linkage determining the sign of the recombination term in the expression for D.

TABLE 69
Linkage in backcrosses: frequencies and phenotypes of the classes

Coupling

	× AABB			× aabb	
	AA	Aa		Aa	aa
BB	q $d_a + d_b$	p $h_a + d_b$	Bb	q $h_a + h_b$	p $-d_a + h_b$
Bb	p $d_a + h_b$	q $h_a + h_b$	bb	p $h_a - d_b$	q $-d_a - d_b$

Repulsion

	× AAbb			× aaBB	
	AA	Aa		Aa	aa
Bb	p $d_a + h_b$	q $h_a + h_b$	BB	p $h_a + d_b$	q $-d_a + d_b$
bb	q $d_a - d_b$	p $h_a - d_b$	Bb	q $h_a + h_b$	p $-d_a + h_b$

All frequencies should be divided by 2

Fig. 12 shows the changes in value of D and H from F_2, F_3 and the first (1B) and second (2B) backcrosses with the recombination fraction, when $d_a = d_b = h_a = h_b = 1$. The effects of repulsion and coupling on D are paralleled by those of opposition and reinforcement on H. All the various Ds and Hs are equal when $p = 0$ or $p = 0.5$, i.e. for complete linkage or free assortment. At any other value of p, the values of D and H in F_3, and in 2B, approach more nearly to those given by free assortment than do the values of D and H in F_2 and 1B. The difference is greatest between $p = 0.20$ and $p = 0.30$.

In the absence of linkage, i.e. where $p = 0.5$, the heritable portion of V_{2F3} is half that of V_{1F2}, and the heritable portion of $V_{2B11} + V_{2B12}$ or $V_{2B21} + V_{2B22}$ is half that of $V_{B1} + V_{B2}$ (Table 49), $2V_{2F3}$ and V_{1F2} are compared in Fig. 13 over all recombination values in the various combinations of repulsion and coupling with opposition and reinforcement when $d_a = d_b = h_a = h_b = 1$. Fig. 14 shows the same comparison for $V_{B1} + V_{B2}$ and $2(V_{2B11} + V_{2B12})$ or $2(V_{2B21} + V_{2B22})$. Again it will be seen that the effect of linkage on the variances is greatest at the intermediate values of p, whatever the combination of linkage phase and dominance co-operation.

The presence of linkage may be detected by comparing the magnitudes of the heritable portions of the variances of different rank in F_2 and F_3 or in 1B and 2B. With linkage, $2V_{2F3}$ no longer equals V_{1F2}, and $2(V_{2B11} + V_{2B12})$ or $2(V_{2B21} + V_{2B22})$ no longer equals $V_{B1} + V_{B2}$. Before, however, turning to the application of this test, we must consider the general case of linkage of more than two genes.

Where three or more linked genes are involved, the formulae of Table 49 still

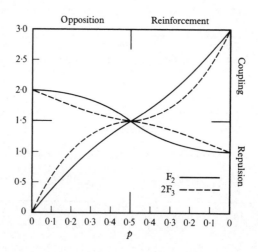

Figure 13 Change in the contribution ($\frac{1}{2}D + \frac{1}{4}H$) made to V_{1F2} and $2V_{2F3}$ by two segregating genes according to the genes linkage relations. Calculations are based on $d_a = d_b = h_a = h_b = 1$

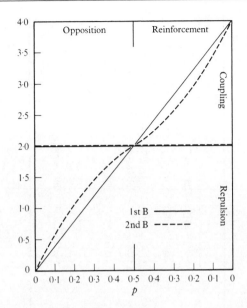

Figure 14 Change in the contribution ($\frac{1}{2} D + \frac{1}{2} H$) made by two segregating genes to $V_{B1} + V_{B2}$ and $2(V_{2B11} + V_{2B12})$ according to the genes' linkage relations. Calculations are based on $d_a = d_b = h_a = h_b = 1$.

hold, but with D and H redefined as shown in Table 70. S_C and S_R stand for the sums of all the items with the two genes in coupling and repulsion, respectively. It should be noted that p is the recombination value shown by the two genes in question, irrespective of any gene between them. In the same way, the phase of a linkage is independent of intermediate genes. Thus where there are three linked loci arranged as AbC/aBc, the recombination values are A–B; p_{ab}, B–C; p_{bc} and A–C; p_{ac} ($= p_{ab} + p_{bc} - 2cp_{ab}p_{bc}$ where c is the coincidence value). Also A and B and B and C are repulsed, but A and C are coupled. Then in F_2

$$D = d_a^2 + d_b^2 + d_c^2 - 2(1 - 2p_{ab})d_a d_b - 2(1 - 2p_{bc})d_b d_c + 2(1 - 2p_{ac})d_a d_c.$$

H is unaffected by linkage phase and so

$$H = h_a^2 + h_b^2 + h_c^2 + 2(1 - 2p_{ab})^2 h_a h_b + 2(1 - 2p_{bc})^2 h_b h_c$$
$$+ 2(1 - 2p_{ac})^2 h_a h_c,$$

but the remarks made above about the effect of linkage phase on the value of D will here apply to reinforcement and opposition relations.

The linkage of a number of genes will exert its maximum effect on the variances and covariances when all are coupled and all their heterozygous effects are reinforcing. All the terms containing p will then be positive. The effects of repulsion and opposition can never be so great, except where only

TABLE 70
D and H with linkage

Rank	Statistics	D	H
1	V_{1F2}, V_{1F3}, W_{1F23}, V_{1F4}, W_{1F34}, V_{1S3}, W_{1S23}	$S(d^2) + 2S_C[(1-2p)d_a d_b]$ $- 2S_R[(1-2p)d_a d_b]$	$S(h^2) + 2S[(1-2p)^2 h_a h_b]$
2	$\begin{cases} V_{2F3},\ V_{2F4},\ W_{2F34} \\[4pt] V_{2S3} \end{cases}$	$S(d^2) + 2S_C[(1-2p)^2 d_a d_b]$ $- 2S_R[(1-2p)^2 d_a d_b]$ ditto	$S(h^2) + 2S[(1-2p)^2(1-2p+2p^2)h_a h_b]$ $S(h^2) + 2S[(1-2p)^3(1-\tfrac{4}{3}p)h_a h_b]$
3	V_{3F4}	$S(d^2) + 2S_C[(1-2p)^3 d_a d_b]$ $- 2S_R[(1-2p)^3 d_a d_b]$	$S(h^2) + 2S[(1-2p)^2(1-2p+2p^2)^2 h_a h_b]$
1	$V_{B1} + V_{B2}$, $V_{1B11} + V_{1B12}$, $V_{1B21} + V_{1B22}$, $W_{1B1\cdot11} + W_{1B2\cdot22}$, $W_{1B1\cdot12} + W_{1B2\cdot21}$	$S(d^2) + 2S_C[(1-2p)d_a d_b]$ $- 2S_R[(1-2p)d_a d_b]$	$S(h^2) + 2S[(1-2p)h_a h_b]$
2	$V_{2B11} + V_{2B12}$, $V_{2B21} + V_{2B22}$	$S(d^2) + 2S_C[(1-2p)(1-p)d_a d_b]$ $- 2S_R[(1-2p)(1-p)d_a d_b]$	$S(h^2) + 2S[(1-2p)(1-p)h_a h_b]$

S_C = sum over all pairs of genes in coupling
S_R = sum over all pairs of genes in repulsion

two gene pairs are concerned, since more than two genes can be neither all repulsed nor all opposed to one another. The maximum effects of repulsion and opposition might be expected when adjacent genes are in repulsion and opposition. Even in such a case, however, the 1st, 3rd, 5th, etc. must be coupled and reinforcing, as must the 2nd, 4th, 6th etc. Inequality of the d and h increments of the various genes will also reduce the effect of linkage on the variances and covariances.

The effect of linkage on the value of a statistic could be zero even though linkage were in fact present; for the coupling and repulsion items, as well as the reinforcement and opposition items, could balance. The circumstances in which such a balance will be achieved must depend on the relative magnitudes of effect of the genes and on their recombination frequencies, as well as on their phasic relations. Even, however, where linkage items balance in D and H of the first rank they will not do so exactly in D and H of other ranks. The effect of linkage will still appear, though under such circumstances it may well be small.

These various considerations will, perhaps, be more easily seen from Fig. 15. This compares the heritable portions of V_{1F2} and $2V_{2F3}$ for the various arrangements of four genes. It is assumed that all the d and h increments are of unit magnitude (though the h increments differ in sign as indicated), and that there is no interference in recombination between different segments, i.e. that all coincidence values are 1. The three primary recombination values p_{ab}, p_{bc} and p_{cd} are further assumed all to be 0·125. Wherever V_{1F2} markedly exceeds or falls short of the value it would have in the absence of linkage, $2V_{2F3}$ occupies an intermediate position. In such cases D and H are changed in the same direction. But where D and H depart from their unlinked expectations in opposite ways, the pooled departure being therefore a relatively small one, $2V_{2F3}$ may depart more widely than V_{1F2} or it may even depart in the opposite direction. The difference between V_{1F2} and $2V_{2F3}$ must then be following that of the D item, which is itself tending towards the unlinked equilibrium value, since D preponderates over H in determining the magnitudes and changes of these variances.

Curiously enough with this tight linkage, the maximum change relative to the value of V_{1F2} in the repulsion series, and hence the easiest detection of the linkage, does not come with adjacent genes repulsed and opposed. It is in fact obtained with the arrangement RCR and oro where R indicates repulsion, C coupling, o opposition and r reinforcement of adjacent genes, the four genes being treated as three successive pairs. But at $p = 0·250$ the maximum change is found with RCR and ooo, and at $p = 0·375$ with the extreme arrangement RRR and ooo. On the coupling and reinforcement side, of course, the extreme situation CCC and rrr, always gives the greatest variance drop.

It is clear, therefore, that although a significant difference between the heritable portions of V_{1F2} and $2V_{2F3}$ will give evidence of linkage, the phasic balance cannot be inferred with full certainty from the direction of the change between F_2 and F_3 when dominance is present and the change is small. If D and

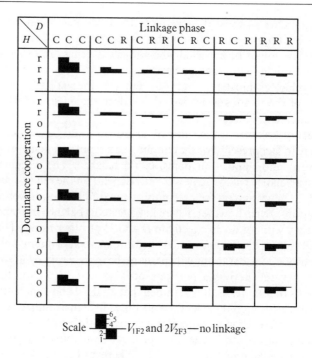

Scale ▇ V_{1F2} and $2V_{2F3}$—no linkage

Figure 15 The effect of linkage phase and dominance co-operation on the contribution made by four segregating genes (all *d*s and *h*s assumed to be 1), to V_{1F2}, left, and $2V_{2F3}$, right. The zero line is the value 3, contributed by four genes to both V_{1F2} and $2V_{2F3}$ when they are unlinked. Values above 3 are represented by solids above, and those below 3 by solids below the line, as indicated on the scale. The value for $2V_{2F3}$ is generally, but not always, between that for V_{1F2} and the unlinked value of 3. Adjacent genes are assumed to show $p = 0.125$ with no interference. C = coupling, R = repulsion, r = reinforcement, o = opposition. These relations are shown for adjacent genes so that, for example, CRC indicates ABcd/abCD. Similarly oro indicates that h_a and h_d have signs similar to one another but opposite to those of h_b and h_c, where a, b, c, d is the order in the chromosome.

H were assessed separately in F_3, as they are in F_2, the situation would become much clearer, for any counteracting effects that their changes might have relative to one another, would then become apparent. Such a separation can be made where F_4 data are available to do for F_3 what F_3 data do for F_2.

32. THE TEST OF LINKAGE

The effects of both interaction and linkage on variation in the descendants of a cross between true breeding lines is basically to change the constitution of *D* and *H* while leaving unchanged the coefficients with which they contribute to the various second degree statistics (though it should be observed that interaction has some side effects in components of variation which have no relation to D and H). With interaction the changes in D and H are from generation to generation. Thus for example in V_{1F2}, D depends on terms of the kind $(d_a + \frac{1}{2} Sj_a.)^2$ whereas in both V_{1F3} and V_{2F3} the corresponding item is $(d_a + \frac{1}{4} Sj_a.)^2$. In the covariance relating the two generations the terms are the geometric means of those appearing in the variances of the two generations.

The changes brought about in D and H by linkage show a different pattern. They follow not the generation but the rank of the statistics, the rank being, of course, the number of occasions on which recombination has had opportunity to affect the differences that the statistic in question is measuring. Thus variation in F_2 reflects the consequences of one round of recombination, at gametogenesis in F_1. V_{1F2} is therefore of rank one. Variation among the means of F_3 families also reflects this same single round of recombination in the F_1, since the mean expression of a segregating family is unaffected by the linkage relations of the genes that are segregating. So V_{1F3} is of rank one, as is W_{1F23}; but the mean variance of F_3 families reflects the effect of recombination at gametogenesis in F_2 as well as that in F_1. It is thus a second rank statistic and is correspondingly denoted as V_{2F3}. Similarly the mean variance of F_4 families will be of rank three and is denoted by V_{3F4}.

The test of linkage is thus basically a test of homogeneity of D and H over the statistics of different rank. The corresponding test of interaction on the other hand is a test of homogeneity of D and H over the statistics from different generations. In principle, therefore, it should be possible to detect the effects of interaction and linkage separately by appropriate analyses of the second degree statistics. In practice three difficulties arise. First, although rank and generation represent different classifications of the statistics, these classifications are not wholly independent of one another. Thus for example the F_3 generation yields statistics of both rank one and rank two, while the F_2 can yield only a rank one statistic. A test could therefore reveal heterogeneity over the generations, even though linkage was the only disturbing factor. The safeguard is that in such a case, there would also be residual heterogeneity within the F_3 generation, and this the analysis must be designed to reveal if it is to lead to confident interpretation. Similarly heterogeneity over rank can arise from interaction because of the correlations between rank and generation; but in such a case there should also be heterogeneity within rank, and this would be absent if linkage were the only disturbing factor. Again the analysis must provide a measure of the residual disturbance if any heterogeneity between ranks is to be interpreted with confidence as due to linkage. This matter has been discussed in some detail by Opsahl (1956), who has also shown that in at any rate certain specific cases the correlation between the effects of linkage and interaction may be nearly one-half, so emphasizing the need to be on guard against facile interpretations of disturbances in D and H as due to linkage on the one hand or interaction on the other. We will return to some of his results later.

The second possible difficulty is that changes in D and H over generations reflect the effects of only the j and l type interactions, for only j contributes to D and only l to H. The effects of i interactions will not therefore be expected to be detected by this test of interaction and they could consequently be falsely attributed to linkage, though they could mimic only repulsion linkage and so will be restricted in their nuisance value (Hayman and Mather, 1955).

Finally, while the test of linkage is relatively undemanding, there being never more ranks of statistics to take into account than segregating generations observed (and sometimes fewer), that for interaction can become very much more troublesome in its requirements, partly because of the problem of the i interaction already noted, partly because of the effects of those components of variation which depend only on the interaction parameters themselves and are separate from the changes in D and H, and partly because the covariances have to be separated from the variances in the test of heterogeneity of D and H since they conform with the variances of neither parent nor offspring generation in the definitions of D and H but lie between the two.

Fortunately, interaction can be detected and analysed by comparison of the generation means through the scaling tests as we have seen in Chapters 4 and 5. Indeed the scaling tests afford a more expeditious and precise means of detecting and interpreting interaction than does comparison of the second degree statistics for segregating generations. The value of the scaling tests cannot be too strongly emphasized as a preliminary to any analysis of second degree statistics, the more so as they further provide a means of searching for a transformation of scale by which the interaction can be removed or at any rate reduced, if such a transformed scale exists. Given that any interactions have been accommodated in this way, the test of linkage is placed on a correspondingly sounder basis.

In the absence of interaction linkage cannot be detected from a comparison of generation means; its action is revealed only when we turn to the second degree statistics. The method of procedure can be illustrated by reference to the observations of Cooke et al. (1962) on the inheritance of the number of sternopleural chaetae in Drosophila melanogaster already partly analysed in Chapter 6. This example is particularly apposite to the illustration of the test for linkage, since the segregating generations observed included only F_2 and S_3, the biparental progenies of the third generation. As we have already seen in Section 30, the effects of interaction on the definitions of D and H are the same in the variances of F_2 and S_3 and in their covariance. Any heterogeneity that might be observed can, therefore, be attributed to linkage with some confidence.

Of the four statistics involving D and H, V_{1F2}, V_{1S3} and W_{1S23} are of rank one and V_{2S3} is of rank two. If linkage is operative D and H in V_{2S3} will differ from D and H in the first rank statistics. A perfect fit must therefore be obtained for V_{2S3}, and indeed this would be secured by adjustment in either D or H alone. So the D and H of rank two cannot be assessed separately and there is in fact no need to assess them separately for the test of linkage: the important thing is that as a result of such adjustment the sum of squares of deviations of the observed statistics from their expectations, based on the best fitting Ds and Hs, will be reduced by this perfect fit in V_{2S3}. Since only one adjustment is needed to achieve this fit, only one degree of freedom is used up in separating the second rank D and H from their first rank counterparts. So in essence the test of linkage devolves into estimating the rank one D and H from V_{1F2}, V_{1S3} and W_{1S23} alone, assuming a perfect fit in V_{2S3}. The residual sum of squares is then calculated and compared with the sum of squares of deviation from expectation when a single D and a single H are estimated from all the data together. The comparison shows us how far the assumption of linkage, and with it the acceptance of change in D and H with rank, reduces the sum of squares of deviations, and so enables us to judge whether the effect is a significant one.

Part of the work for the test of linkage has been done in Sections 27 and 29. We must now undertake a similar calculation with V_{2S3} excluded, and we shall undertake this first using Mather's (1949a) simple unweighted analysis.

Our first task is to estimate D, H, E_1 and E_2 from the observations but excluding V_{2S3}. We thus have as data from, for example, the BS males (Table 56)

$$V_{1F2} = \tfrac{1}{2}D + \tfrac{1}{4}H + E_1 = 4\cdot8860$$
$$V_{1S3} = \tfrac{1}{4}D + \tfrac{1}{16}H + E_2 = 1\cdot0673$$
$$W_{1S23} = \tfrac{1}{4}D = 0\cdot5797$$
$$E_1 = 2\cdot6854$$
$$E_2 = 0\cdot4881$$

Using the methods described in Chapter 5, four least squares expressions can be found as

$$0\cdot375\,000\,D + 0\cdot140\,625\,H + 0\cdot500\,000\,E_1 + 0\cdot250\,000\,E_2$$
$$0\cdot140\,625\,D + 0\cdot066\,406\,H + 0\cdot250\,000\,E_1 + 0\cdot062\,500\,E_2$$
$$0\cdot500\,000\,D + 0\cdot250\,000\,H + 2\cdot000\,000\,E_1$$
$$0\cdot250\,000\,D + 0\cdot062\,500\,H \qquad\qquad + 2\cdot000\,000\,E_2$$

Inverting this least squares matrix by equating to 1, 0, 0, 0 etc. and solving, gives the c-matrix shown in Table 71. For the BS males

$$S(Dy) = 2\cdot8448 \quad \text{we find} \quad \hat{D} = 2\cdot1034$$
$$S(Hy) = 1\cdot2832 \qquad\qquad\quad \hat{H} = 4\cdot3002$$
$$S(E_1 y) = 7\cdot5514 \qquad\qquad\quad \hat{E}_1 = 2\cdot7123$$
$$S(E_2 y) = 1\cdot5554 \qquad\qquad\quad \hat{E}_2 = 0\cdot3804$$

The estimates obtained from the four parts of the experiment separately and those from the overall data are shown in Table 72. From these, the

TABLE 71
The c-matix for the unweighted analysis of the *Drosophila* experiment excluding V_{2S3}

D		H		E_1		E_2	
c_{DD}	14·315 790	c_{HD}	− 30·315 790	c_{1D}	0·210 526	c_{2D}	0·842 105
c_{DH}	− 30·315 790	c_{HH}	94·315 790	c_{1H}	− 4·210 526	c_{2H}	0·842 105
c_{D1}	0·210 526	c_{H1}	− 4·210 526	c_{11}	0·973 684	c_{21}	0·105 263
c_{D2}	0·842 105	c_{H2}	0·842 105	c_{12}	0·105 263	c_{22}	0·578 948

TABLE 72
The components of variation in the *Drosophila* experiment excluding V_{2S3}

	BS males	BS females	SB males	SB females	Overall
D	2·1034	4·2798	2·1999	2·8517 ± 0·3522	2·8496 ± 0·1761
H	4·3002	− 5·8698	4·3190	3·3335 ± 0·9043	1·5720 ± 0·4522
E_1	2·7123	2·4828	2·2568	1·8967 ± 0·0919	2·3372 ± 0·0460
E_2	0·3804	0·4579	0·3555	0·3125 ± 0·0708	0·3766 ± 0·0354

expectations and the deviations of the values observed from these expectations can be found as shown in Table 73. The sum of squares of the deviations of observed values from expected are shown in the bottom line of Table 73. It should be observed that, as explained in Section 27, the sum of squares in the 'overall' column must be multiplied by 4 to make it comparable to the others, so giving $0.001\,20 \times 4 = 0.004\,80$ for this item.

TABLE 73
Values expected for the statistics excluding V_{2S3} and the deviation of those observed (Table 56) from these expectations

	BS males	BS females	SB males	SB females	Overall
Expectations					
V_{1F2}	4·839 1	3·155 3	4·436 5	4·155 9	4·155 0
V_{1S3}	1·175 0	1·161 0	1·175 4	1·233 8	1·187 3
W_{1S23}	0·525 9	1·070 0	0·550 0	0·712 9	0·712 4
E_1	2·712 3	2·482 8	2·256 8	1·896 7	2·337 2
E_2	0·380 4	0·457 9	0·355 5	0·312 5	0·376 6
Deviations					
V_{1F2}	0·026 9	− 0·007 8	− 0·006 2	0·009 4	0·005 6
V_{1S3}	− 0·107 7	0·031 1	0·024 8	− 0·037 8	− 0·022 5
W_{1S23}	0·053 8	− 0·001 5 6	− 0·012 4	0·018 9	0·011 2
E_1	− 0·026 9	0·007 8	0·006 2	− 0·009 4	− 0·005 6
E_2	0·107 7	− 0·031 1	− 0·024 7	0·037 7	0·022 4
Sum of squares	0·027 54	0·002 30	0·001 46	0·003 39	0·001 20

With the exclusion of V_{2S3} from the calculation, four parameters have been estimated from five statistics – five parameters from six statistics if we like formally to bring in V_{2S3} and the joint estimate of the second rank D and H to which it can be regarded as leading. In either case, however, the sum of squares at the foot of each column of Table 73 is clearly based on one degree of freedom in contrast to the sums of squares in the analysis of Section 27 (see Table 59) each of which was based on two degrees of freedom. One degree of freedom has been used up by letting V_{2S3} become, so to speak, its own expectation, i.e. by allowing for the effects of linkage. Thus we can now divide each sum of squares of Table 59 into two parts, each for one degree of freedom, the first representing residual variation after allowing for linkage and the second representing the effect of linkage. The former is the sum of squares of Table 73 and the latter the difference between this and the sum of squares of Table 59. To take the BS males as an example, we observe that the sum of squares (for two degrees of freedom) from Table 59 is 0·3430, while that from Table 73 (for one degree of freedom) is 0·0275. Then linkage accounts for $0·3430 - 0·0275 = 0·3155$ with 0·0275 arising from residual causes. Linkage thus accounts for a much more substantial part of the variation than do the residual causes; but this evidence for linkage is not of itself compelling as each part of the variation is estimated from only one degree of freedom and a disparity much greater than that observed with these BS males would be needed before significance was attained.

All four parts of the experiment show a greater mean square (which, for one degree of freedom, is of course the same as the sum of squares) for linkage than for residual causes. Summing over all four parts of the experiment we find as sums of squares:

Linkage (by subtraction)
$$0·3155 + 0·0950 + 0.0510 + 0·0057 = 0·4672 \text{ for 4 df}$$
Residual (from Table 73)
$$0·0275 + 0·0023 + 0·0015 + 0·0034 = 0·0347 \text{ for 4 df}$$
Total (from Table 59)
$$0·3430 + 0·0973 + 0·0525 + 0·0091 = 0·5019 \text{ for 8 df}$$

So taking all four parts of the experiment together we find the mean square for linkage to be 0·1168 and that for residual variation to be 0·0087, giving a variance ratio of 13. With both mean squares based on four degrees of freedom such a variance ratio has a probability of between 0·05 and 0·01. The evidence for linkage has mounted up.

The analysis of variance may, however, be taken a stage still further by bringing in the sum of squares from the overall columns of Tables 59 and 73. In Section 27 the total sum of squares was broken into two parts, for the overall deviations (derived from the overall results and based on 2 df) and for heterogeneity (derived by differencing and based on 6 df). This was set out in

Table 60. A similar analysis can be carried out on the sums of squares from Table 73. The sum of squares from the overall deviations will now be based on 1 df and is 0·004 80. The difference between this and the sum of squares for overall deviations (based on 2 df) as found in Section 27 is 0·086 40 − 0·004 80 = 0·081 60 and is ascribable to linkage. Similarly the heterogeneity sum of squares, found from Table 73, is 0·027 54 + 0·002 30 + 0·001 46 + 0·003 39 − 0·004 80 = 0·029 89, is based on 3 df and reflects heterogeneity of residual variation. If this is subtracted from the heterogeneity item of the earlier analysis of variance (Table 60) we obtain 0·415 51 − 0·029 89 = 0·385 62 for 3 df reflecting heterogeneity of linkage effects.

The full analysis of variance obtained in this way is set out in Table 74. The items for the residual variation overall and its heterogeneity do not differ significantly and may therefore be amalgamated to give a pooled estimate of error variation of 0·008 67 for 4 df. Against this both the overall linkage and the heterogeneity of linkage effects are significant at the 5% level. There is thus evidence that linkage is at work. It is, however, to be noted that the mean square for linkage heterogeneity is somewhat larger than that for overall linkage itself. Thus if we accept that linkage is established we must accept further that its effects vary from one part of the experiment to another.

TABLE 74
Full analysis of variance of the *Drosophila* experiment

Item	SS	df	MS	VR	P
Linkage	0·081 60	1	0·081 60	9·4	0·05–0·01
Residual	0·004 80	1	0·004 80		
[(Sum of L and R)	0·086 40	2]			
Heterogeneity					
of linkage	0·385 62	3	0·128 54	14·8	0·05–0·01
of residual	0·029 89	3	0·009 96		
[(Sum of HL and HR)	0·415 51	6]			
Total	0·501 91	8			
Pooled residual variation	0·034 69	4	0·008 67		

The weighted analysis of the data from this experiment, allowing for linkage, is described by Cooke et al. (1962). It follows the general pattern of the weighted analysis neglecting linkage as described in Section 29, except, of course, that the D and H of V_{2S3} are not assumed to be the same as the other D and H in the experiment, thus allowing for any effects that linkage may produce. In the unweighted analysis just described, this change in D and H between the first and second rank statistics was accommodated by omitting V_{2S3} from the analysis (which is here described as the exclusive analysis to distinguish it from the inclusive analysis of Chapter 6), thus in effect allowing for the change with rank in D and H by taking the value observed for V_{2S3} as its own expectation. The process is not quite so simple in the weighted analysis because, as will be recalled from Section 29, the contribution to the value of V_{1S3} from the sampling variation arising from V_{2S3} was allowed for in the formulation of the expectation or V_{1S3}. Thus the second rank values of D and H appear not only in V_{2S3} but also, albeit in a minor way, in V_{1S3}. The weighted exclusive analysis cannot thus allow for the change in D and H with rank merely by excluding V_{2S3} but must specifically allow for an estimate of the second rank components of variation. Since, however, the second rank D and H appear in the same combination, i.e. $\frac{1}{2}D + \frac{3}{16}H$, in both V_{2S3} and in the item in V_{1S3} reflecting sampling variation stemming from V_{2S3}, these two second rank components cannot be estimated separately and only one degree of freedom is therefore used up in allowing for linkage, just as in the unweighted analysis. It will be recalled, too, that the variance of the means of the F_2 cultures was also brought into this weighted analysis by Cooke *et al.*, so giving an extra statistic and hence an extra degree of freedom.

The χ^2s testing goodness of fit after the inclusive weighted analysis are given in Table 65. These are repeated in Table 75, together with the corresponding χ^2 testing goodness of fit after the exclusive analysis. Each inclusive χ^2 stems from 3 df, and each exclusive χ^2 from 2 df, since one more component of variation (the combination $\frac{1}{2}D + \frac{3}{16}H$ of the two second rank components) has been estimated. The difference between the corresponding inclusive and exclusive χ^2s is thus itself a χ^2 for 1 df providing a test of linkage. These linkage χ^2s are also set out in the upper part of Table 75 for the various parts of the experiment. The linkage χ^2 for the BS males is easily significant at the 5 % level and almost so at the 2 % level. That for the BS females has a probability between 0·10 and 0·05 and is thus somewhat suggestive, if not formally significant. Neither of the χ^2s from the SB part of the experiment is even suggestively large. The exclusive χ^2s, testing for residual variation after the

TABLE 75
The analysis of χ^2 from the weighted analysis of the *Drosophila* experiment

Part of experiment		BS males	BS females	SB males	SB females
Inclusive	$\chi^2_{(3)}$	9·276	3·638	1·454	2·676
Exclusive	$\chi^2_{(2)}$	4·105	0·582	0·102	2·660
Linkage	$\chi^2_{(1)}$	5·171	3·056	1·352	0·016

Analysis of linkage χ^2

	df	χ^2	P
Overall linkage	1	0·719	0·5–0·3
Heterogeneity of linkage	3	8·876	0·05–0·02
Total	4	9·595	

elimination of linkage, are significant neither individually nor jointly. There is no evidence of residual causes of variation, and we should indeed remember that comparisons of D and H based on F_2 and S_3 cannot yield evidence of interaction.

The linkage χ^2s can be analysed further. Cooke et al. give the linkage χ^2 obtained from the overall data of the experiment as 0·719. We can thus obtain the analysis shown in the lower part of Table 75 by pooling the four linkage χ^2s, one from each part of the experiment, to give a total χ^2 and subtracting from this the overall χ^2 to leave a χ^2 testing heterogeneity of linkage. Again, as with the unweighted analysis (Table 74), this heterogeneity of linkage is significant, though now, in contrast to the unweighted analysis, the overall item is not. With both analyses therefore we have evidence of disturbances traceable to the effects of linkage but evidence which must be treated with caution since in one case (the unweighted analysis) the overall effect is blurred by significant heterogeneity between the parts of the experiment, and in the other (the weighted analysis) the evidence springs only from such heterogeneity.

The evidence for linkage from this experiment with *Drosophila* is not unambiguous. This does not of course mean that none of the genes involved are linked to one another. Indeed, as we noted in Section 29, Cooke and Mather (1962) obtained clear evidence of such linkage in the X chromosome using another technique. It should be borne in mind that linkage displays its effect in biometrical genetics not by producing new combinations of genes recognizable from individual phenotypes in proportions not otherwise expected, but by changing the relative magnitudes of variances and covariances which reflect the occurrence and frequencies of many different phenotypes and genotypes. As we have seen in Section 31 either tight linkage or very loose linkage may well escape detection: linkage, where the recombination values are of intermediate size, is most likely to be recognized because this produces the greatest change in the D and H components of variation and therefore in the variances and covariances. Furthermore, since the effect as observed is the summation of the consequences of all the pairs of linked genes, coupling and repulsion linkages can balance one another in their effects and so escape recognition.

At the same time, however, linkage can reveal itself clearly in biometrical analysis. Mather (1949a), in collaboration with Dr U. Philip, describes an experiment on the inheritance of ear-conformation in barley. The parents were the varieties Spratt and Goldthorpe, and ear-conformation was represented by a metric compounded of ear width, ear length and the length of the central six internodes of the ear. This metric was arrived at by the calculations of a discriminant function, described in detail by Mather (1949a) and referred to in Section 8. The experiment was set out in five randomized blocks with 117 plots per block. Of these, 100 were devoted to 100 different F_3 families derived from F_2 plants grown the year before, 10 were given to the F_2, 3 to F_1, and 2 to each parental variety. Each plot was intended to contain 10 plants but this number was not reached in every case and indeed in a few plots so many plants failed that it was deemed desirable to exclude the whole plot from the analysis. Block differences were significant. V_{1F3} and W_{1F23} were therefore calculated using appropriate block means, the sum of squares and of cross-products so obtained being pooled over blocks to give the overall figures. E_2 was estimated from parents and F_1 using deviations of plot means from the appropriate block means. V_{1F2}, V_{2F3} and E_1 were, of course, found from the appropriate families using variation round the plot means.

The mean measurements of the two parents were 327·9 and 300·6. The F_1 showed heterosis with a mean of 400·2. The F_2 and F_3 means were 362·3 and 343·7, respectively, both of which are rather higher than would be expected in the absence of non-allelic interaction. The departures were, however, too small to justify rescaling and, indeed as we shall see, there is no indication from the analysis into D, H and E that it has been materially upset by any inadequacy of scale that may have existed.

The various second degree statistics are set out separately for each block in Table 76, together with the overall or joint figures obtained by pooling over blocks. The overall values approximate to the means of the individual block values but do not exactly equal them because the failure of certain plots, as already noted, led to the blocks making slightly different contributions to the pool.

TABLE 76
Barley data – ear-conformation

Statistic	Block					Joint	Expectations	
	A	B	C	D	E		Inclusive	Exclusive
V_{1F2}	9492·0	11540·9	9179·5	8673·7	9926·4	9713·1	9326·2	9710·6
V_{1F3}	6289·7	5838·6	5935·1	7148·8	6013·4	6246·7	6241·6	6241·6
W_{1F23}	6934·6	6342·5	6742·5	7439·6	6696·8	6833·0	6840·5	6840·5
V_{2F3}	4513·1	4532·5	4342·4	4323·1	3867·9	4314·1	5082·9	4314·1
E_1	1443·8	736·9	1433·1	1720·9	747·6	1221·4	839·6	1224·0
E_2	278·6	57·9	306·8	403·2	54·9	219·1	224·2	224·2

The results include three statistics of rank one (V_{1F2}, V_{1F3}, W_{1F23}) and one of rank two (V_{2F3}). All these, together with the direct estimates of E_1 and E_2, are used in forming the inclusive equations of estimation of D, H and E_1 and E_2, and hence the inclusive c-matrix which is set out in the upper part of Table 77. The effects of linkage are accommodated by letting V_{2F3} become its own expectation, so to speak, and the exclusive equation of estimation will be based on V_{1F2}, V_{1F3}, W_{1F23} and the direct estimates of E_1 and E_2. The exclusive c-matrix is set out in the lower part of Table 77. The value of $S(Dy)$ etc. and the

TABLE 77
c-matrices for analysis of the barley experiment

	D		H		E_1		E_2
Inclusive							
c_{DD}	10·526 316	c_{HD}	−30·315 789	c_{1D}	1·157 895	c_{2D}	−1·684 211
c_{DH}	−30·315 789	c_{HH}	107·789 473	c_{1H}	−5·894 737	c_{2H}	4·210 526
c_{D1}	1·157 895	c_{H1}	−5·894 737	c_{11}	0·780 702	c_{21}	−0·105 263
c_{D2}	−1·684 211	c_{H2}	4·210 526	c_{12}	−0·105 263	c_{22}	0·789 474
Exclusive							
c_{DD}	10·526 316	c_{HD}	−30·315 789	c_{1D}	1·157 895	c_{2D}	−1·684 210
c_{DH}	−30·315 789	c_{HH}	107·789 473	c_{1H}	−5·894 737	c_{2H}	4·210 526
c_{D1}	1·157 895	c_{H1}	−5·894 737	c_{11}	0·947 368	c_{21}	−0·105 263
c_{D2}	−1·684 210	c_{H2}	4·210 526	c_{12}	−0·105 263	c_{22}	0·789 474

estimates derived from them are shown in Table 78 for both inclusive and exclusive analyses. The overall expectations, both inclusive and exclusive, are also set out in Table 76 and the analysis of variance, derived from the sums of squares of deviations of the separate block estimate, is given in Table 79.

TABLE 78
Results of analysis of barley data

	Inclusive	Exclusive		Inclusive	Exclusive
$S(Dy)$	12 474·9	11 396·4	D	10 388·8	10 388·8 ± 808·9
$S(Hy)$	4 212·1	3 672·8	H	13 169·0	13 169·0 ± 2588·4
$S(E_1y)$	15 248·7	10 934·6	E_1	839·6	1 224·0 ± 242·7
$S(E_2y)$	6 465·8	6 465·8	E_2	224·2	224·2 ± 221·5

TABLE 79
Analysis of variance of barley data

Item	SS	df	MS	$t_{[25]}$	P
Linkage	4 590 309	1	4 590 309	3·76	0·001
Residual interaction	14 329	1	14 329	0·21	0·90–0·80
Replication	7 755 458	24	323 144		
Total	12 360 096	26			
Pooled error	7 769 787	25	310 791		

Six statistics were observed for each of five blocks, making 30 deviations or 30 df in all. Of these, 4 are taken up in fitting the overall values of D, H, E_1 and E_2, leaving 26. Since there are five estimates of each statistic each row of Table 76 contributes 4 df to a pool of 24, for differences among the replicate estimates of the statistics. The remaining 2 of the 26 df in the analysis are assignable, 1 to the effect of linkage and 1 to the residual variation in D and H of the kind ascribable to genic interaction. This latter is labelled Residual interaction in the analysis. It would appear from the analysis of variance that there is good evidence of linkage but no evidence of any other variation in the value of D and H. Any shortcomings of the scale have thus failed to inflate seriously the item for Residual interaction. These shortcomings cannot therefore be having any material effect on the estimated values of D and H though they could affect the use of these values in, for example, predicting advances under selection.

The value of V_{2F3} is less than $\frac{1}{2}V_{1F2}$ even before E_1 has been deducted from each. There has thus been a fall in D and H from the first to the second rank statistics. This may well be due in part to the linked genes having reinforcing dominance, but since H contributes less to V_{1F2} and V_{2F3} than does D, it seems inescapable that the change must be due chiefly to the linkage being preponderantly in the coupling phase; that is to say, within the linked groups the $+$ alleles must be preponderantly associated on one chromosome and the $-$ alleles on the other. The means of the two parents do not, however, differ widely. This suggests further that at least two linked groups of genes are involved, coupling existing within each group but the two groups balancing one another in the parents, with each parent having the $+$ alleles of one linkage group and the $-$ alleles of the other.

When the effect of linkage has been removed the remaining 25 df (which may be pooled once the residual interaction has shown itself not to be significant) give an error mean square of $310\,791\cdot5$. From this the standard errors of D and H, as they appear in statistics of rank one, may be found using the exclusive c-matrix. There will of course be a factor of $\frac{1}{5}$ in the calculation because of the 5 replicates in the experiment. These standard errors are given in Table 78. It is clear that H as well as D is significant. The genes must therefore show dominance, and since the F_1 shows heterosis the dominance must be at any rate preponderantly towards high values. The estimate of H is, in fact, actually larger than that of D but obviously not significantly so. There is thus no ground for postulating overdominance, i.e. for postulating that $h > d$ for any gene.

Before leaving this example it is worth noting that the estimate of E_2 has a standard error nearly as large as itself. This is, of course, a consequence of the simple method of analysis, which, as noted in Chapter 6, gives equal weight to all the statistics used in the estimation of the components of variation, even though clearly E_2, for example, varies less in absolute terms than do the other statistics. But again, while weighted analysis by the methods already described would remove this anomaly, the results of the simple analysis cannot be regarded as grossly misleading since the evidence of linkage comes from changes in D and H and the estimate of E_2 is of but very secondary interest. To the geneticist the values of D and H are of much more importance than that of E_2.

In neither the *Drosophila* nor the barley experiments is the evidence for linkage obscured by interaction. Such evidence as there is for linkage in the *Drosophila* results cannot be attributed to side effects of interaction because, as we have seen, an experiment involving only the F_2 and S_3 as segregating generations is very insensitive to the consequences of interaction. In the barley the case is different, for here the sum of squares of deviations taken out by linkage is so large and the residuum attributable to interaction is so small that it would be utterly unreasonable to seek an interpretation that did not involve linkage.

Not all examples, however, permit such confident interpretation. Opsahl (1956) describes an experiment with *Nicotiana rustica* in which he observed both plant height and flowering-time in families raised by crossing F_2 plants back to both parental lines and to the F_1. Although part of this experiment followed the North Carolina type 3 pattern, it was not analysed in that way. Instead all the data were included in a component analysis of the more general kind that we have discussed in Chapter 6 and in various sections of the present chapter. The data on plant height provided evidence of dominance but not of interaction or linkage and are thus of no interest to us in the present connection. The analysis of flowering-time revealed a more complex situation. The scaling tests gave evidence of interaction, but no attempt was made to eliminate this by rescaling.

Now in his discussion of the correlation between the effects of linkage and interaction, Opsahl points out that a useful way to proceed is first to allow explicitly for linkage in the analysis, in which case any residual disturbance must be due to interaction; and then to allow explicitly for interaction in the experiment, when any residual disturbances must be due to linkage. This he does with his flowering-time results, and he obtains the two analyses of Table 80. The first analysis, making explicit provision for linkage, leaves a mean square for residual disturbance that is not only significant but as great as that for linkage. There can thus be no doubt of the presence of interaction. The

TABLE 80
Opsahl's analysis of flowering-time in *Nicotiana rustica*

A. Explicit allowance made for linkage	df	MS	P
Linkage	4	153·34	0·02–0·01
Residual disturbances	4	145·34	0·02–0·01
Error variation	12	32·35	
B. Explicit allowance made for interaction			
Interaction	3	193·85	0·01
Residual disturbances	5	122·63	0·05–0·02
Error variation	12	32·35	

The estimate of error variation is based on differences between duplicate statistics obtained from the two blocks of the experiment second analysis, explicitly providing for interaction, also leaves a mean square for residual disturbances (ascribable on the face of things only to linkage) which is at least suggestively large, though this time not so great as that unambiguously assignable to interaction in the first analysis.

The evidence points, therefore, to both interaction and linkage as expressing their effects in the determination of flowering-time in this experiment. The evidence for linkage is not, however, perhaps as clear as it appears, for Opsahl did not have enough statistics to fit the components of variation necessary to allow for interaction in a fully general way. Instead he made certain simplifying assumptions which would accommodate fully any duplicate or complementary interaction, yet left it possible that the effects of other forms of interaction would not be fully incorporated into the explicit item for interaction in the second part of the analysis. Any effects not so incorporated would, of course, appear as residual disturbances and thus become confused with the effects of linkage. The evidence for linkage is thus not wholly above suspicion, though one feels that it is rather better than Opsahl himself suggests. Opsahl recognizes and emphasizes that his approach may leave the position, in respect of linkage, uncertain. It will not, however, by any means always do so and it is of considerable prospective value in seeking to arrive at an understanding of complex situations where the test for linkage is complicated by the presence of interaction.

This problem of the unambiguous detection of linkage despite the presence of non-allelic interaction, and *vice versa*, has been discussed in detail by Perkins and Jinks (1970) who also brought genotype × environment interaction into their consideration. They used as illustrative material an experiment which included 21 different types of family from a cross between varieties 1 and 5 of *Nicotiana rustica* and which yielded 36 independent variances. The results of this unique experiment permitted the unambiguous detection of non-allelic interaction even though both linkage and genotype × environment interaction were present. It also pointed the way towards using the F_2 triple test cross (p. 151 *et seq.*) for detecting linkage and non-allelic interaction independently of each other.

In the F_2 triple test cross the comparison $(\bar{L}_{1i} + \bar{L}_{2i} - 2\bar{L}_{3i})$ provides an unambiguous test for non-allelic interaction. The contribution of two unlinked pairs of interacting genes, A–a and B–b to this comparison is

$$\tfrac{1}{24} i_{ab}^2 + \tfrac{1}{96}(2j_{ab}^2 + 2j_{ba}^2 + l_{ab}^2).$$

The sum of squares for this comparison is the sum of the n deviations squared for n degrees of freedom. The item for 1 degree of freedom corresponding with the correction term of a normal sum of squares has the expectation $\tfrac{1}{24} i_{ab}^2$ leaving a variance for $n-1$ degrees of freedom whose expectation is

$$\tfrac{1}{96}(2j_{ab}^2 + 2j_{ba}^2 + l_{ab}^2).$$

For many pairs of genes these two expectations become $\frac{1}{24}[i]^2$ and $\frac{1}{48}J$ $+\frac{1}{96}L$, respectively. This separation of the square of the additive × additive interaction component of the mean, $[i]$ from the contribution of additive × dominance and dominance × dominance interactions of the variances is independent of the distribution of the genes in the parents of the F_2 and of linkage.

Plant height in the F_2 triple test cross based on the cross of varieties 1 and 5 of *N. rustica* showed no non-allelic interactions (Section 27). Flowering time, which was also recorded, however, showed highly significant interactions. We have therefore, in Table 81 partitioned its sum of squares for epistasis for 40 degrees of freedom into the item for 1 degree of freedom (overall epistasis) and the remainder for 39 degrees of freedom (families × epistasis) and similarly for the corresponding replicate interaction sum of squares.

TABLE 81
Partitioning the non-allelic interaction in an F_2 triple test cross experiment with *Nicotiana rustica* for the character flowering time (basis single plant)

Item	SS	df	MS	χ^2	VR	P
1. Overall epistasis	140·16	1	140·16	9·87		0·01–0·001
					7·23	0·02–0·01
2. Epistasis × families	755·99	39	19·38	53·26		0·10–0·05
Replicates × 1	2·80	1	2·80	0·20		NS
Replicates × 2	475·37	39	12·19	33·49		NS
Within families	13 498·97	951	14·19			

The two replicate interactions are not significant when tested against the within family variance and the epistasis × families item just fails to reach significance. Nevertheless, we have tested the overall epistasis against its interaction with families as well as against the within family variance with the same highly significant result. We have therefore strong evidence of an $[i]$ type component, though no more than a suggestion of a J or L component.

Turning to linkage we note that in its absence the arrays of gametes produced by an F_1 and an F_2 are identical. Where, therefore, these gametes are mated at random as they are, for example, in an F_2 ($F_1 \times F_1'$) and the L_3 of a triple test cross ($F_2 \times F_1$) they produce identical populations of genotypes. In the absence of linkage the genetical variation within an F_2 and an L_3 is, therefore, expected to be the same irrespective of the nature of the gene action and interaction. By randomizing the F_2 and L_3 together in the same experiment their environmental components of variation will be the same, as will also the interaction between the environmental and genetic components of variation. The F_2 and L_3 will, therefore, have the same means and variances (Perkins and Jinks, 1970).

The modification of D and H in the presence of linkage for the L_3 is

$$D = Sd^2 + S_c[\tfrac{3}{2}(1 - 2p)d_ad_b + \tfrac{1}{2}(1 - 2p)^2 d_ad_b]$$
$$- S_R[\tfrac{3}{2}(1 - 2p)d_ad_b + \tfrac{1}{2}(1 - 2p)^2 d_ad_b]$$
$$H = Sh^2 + S[\tfrac{3}{2}(1 - 2p)h_ah_b + \tfrac{1}{2}(1 - 2p)^2 d_ad_b]$$

The linkage biases are, of course, less than in the corresponding D and H for the variance of the F_2 (Table 70) because of the additional round of recombination in the one F_2 parent of the L_3. Unless the linkages are between genes which are displaying non-allelic interactions the means of the F_2 and L_3 still have identical expectations in the presence of linkage (Section 25). To detect linkage we can, therefore, compare the L_3 and F_2 variances. In practice this can be best achieved by a variance ratio providing that the experiment is appropriately designed.

The F_2 is a single family from which we can obtain the within-family variance. The L_3 consists of n families each of size m (see Section 27). The variance of L_3 that we require is the sum of the within and between family components of variation, that is, $\sigma_w^2 + \sigma_b^2$. If we estimate the total variance within the L_3 it has the expectation

$$\sigma_w^2 + \frac{nm - m}{nm - 1}\sigma_b^2$$

for $nm - 1$ degrees of freedom. Only by making m equal to 1 can we, therefore, estimate the appropriate L_3 variance for direct comparison with the F_2 variance.

The relationship between F_2 and L_3 applies also to B_1 ($F_1 \times P_1$) and B_2 ($F_1 \times P_2$) families and the L_1 ($F_2 \times P_1$) and L_2 ($F_2 \times P_2$) generations of an F_2 triple test cross, respectively. By raising F_2, B_1 and B_2 families and an F_2 triple test cross we have, therefore, three independent variance ratio tests for linkage. Perkins and Jinks (1970) have reported such an analysis for plant height in the cross of varieties 1 and 5 of *N. rustica* with the results given in Table 82.

TABLE 82
Variance ratio tests for linkage based on comparisons of the total σ^2s for the families and generations specified.

Sources	Total σ^2	df	P
F_2	34·22	298	
L_3	37·59	398	NS
B_1	28·19	158	
L_1	29·16	398	NS
B_2	23·92	158	
L_2	29·46	298	0·02–0·002

Since the larger of the pair of variances is always the numerator of the variance ratio the corresponding probability must be doubled. In all three comparisons the generation from the triple test cross has the larger variance but it is significantly so for L_2 and B_2 only. Since the larger variances come from the generations with the smaller linkage bias any linkages must be predominantly in the repulsion phase.

33. GENOTYPE × ENVIRONMENT INTERACTION

In Sections 19 and 20 the specification and estimation of genotype × environment interactions, their minimization by scalar changes and the estimation of the contribution of macroenvironmental effects, such as treatment, locational and seasonal differences, to the means of generations have been described. In this section the approach will be extended to cover the contribution of genotype × environment interaction arising in response to microenvironmental differences within and between families raised in the same macroenvironment, to the variation within generations.

Where the various genotypes are distributed at random over all environments within a single macroenvironment the overall mean phenotype of each genotype will be independent of the genotype × environment interactions as represented by the g items (Section 19); though, of course, its sampling variation will reflect the magnitudes and relations of the gs. The effects of the interactions on the phenotypic variance can be illustrated from the case of a single gene difference, A–a in two environments, $+e_1$ and $-e_1$ (Table 34) using the approach of Mather and Jones (1958) and Jones and Mather (1958). The variance of a pure breeding line with genotype AA will be $(e_1 + g_{1da})^2$ while that of an aa line will be $(e_1 - g_{1da})^2$. Similarly, the variance of the F_1 heterozygote Aa produced by crossing two such lines will be $(e_1 + g_{1ha})^2$. In the F_2 of the same cross the three genotypes occur in the proportions $\frac{1}{4}:\frac{1}{2}:\frac{1}{4}$ and since $(e_1 + g_{1da})^2 + (e_1 - g_{1da})^2 = 2e_1^2 + 2g_{1da}^2$ the variance of the F_2 becomes

$$V_{1F2} = \tfrac{1}{2}d_a^2 + \tfrac{1}{2}e_1^2 + \tfrac{1}{2}g_{1da}^2 + \tfrac{1}{4}h_a^2 + \tfrac{1}{2}(e_1 + g_{1ha})^2$$
$$= \tfrac{1}{2}d_a^2 + \tfrac{1}{2}g_{1da}^2 + \tfrac{1}{2}h_a^2 + \tfrac{1}{2}g_{1ha}^2 + e_1^2 + e_1 g_{1ha}.$$

If we now extend this model to the two genes at each of two loci A–a and B–b in each of two environments $+e_1$ and $-e_1$, we have to consider four possible homozygous combinations: the association phases AABB and aabb and the dispersion phases AAbb and aaBB. Taking first the AABB homozygote, its variation over the two environments is given by

$$\tfrac{1}{2}(e_1 + g_{da} + g_{db})^2 + \tfrac{1}{2}(-e_1 - g_{da} - g_{db})^2$$

which reduces to

$$e_1^2 + g_{da}^2 + g_{db}^2 + 2g_{da}g_{db} + 2e_1 g_{da} + 2e_1 g_{db}.$$

The corresponding expectations for the remaining three homozygotes are for AAbb:

$$e_1^2 + g_{da}^2 + g_{db}^2 - 2g_{da}g_{db} + 2e_1g_{da} - 2e_1g_{db},$$

aaBB:

$$e_1^2 + g_{da}^2 + g_{db}^2 - 2g_{da}g_{db} - 2e_1g_{da} + 2e_1g_{db},$$

aabb:

$$e_1^2 + g_{da}^2 + g_{db}^2 + 2g_{da}g_{db} - 2e_1g_{da} - 2e_1g_{db}.$$

Using the approach of Perkins and Jinks (1970) we may now consider any number of such pairs of genes whether associated or dispersed. The general expectation for a homozygous line with the higher mean score (P_1) will then be

$$V_{P1} = e_1^2 + G_D + 2W_{gdigdk} + 2W_{e1gdi};$$

and for the homozygous line with the lower mean score (P_2)

$$V_{P2} = e_1^2 + G_D + 2W_{gdigdk} - 2W_{e1gdi},$$

which in any number (j) of environments becomes

$$V_{P1} = E_w + G_D + 2W_{gdigdk} + 2W_{ejgdi}$$

and
$$V_{P2} = E_w + G_D + 2W_{gdigdk} - 2W_{ejgdi}$$

respectively, where $E_w = S(e_j^2)$; $G_D = S(g_{di}^2)$; $W_{gdigdk} = \text{cov } g_{di}g_{dk}$ and $W_{ejgdi} = \text{cov } e_jg_{di}$.

By a similar progression from a single gene difference in two environments to any number of pairs of genes and environments the general expectation for the variance within F_1 families becomes

$$V_{F1} = E_w + G_H + 2W_{ghighk} + 2W_{ejghi}$$

where
$$G_H = S(g_{hi}^2) \text{ etc.}$$

Without introducing any further parameters we can now specify the expected variance of any generation whose genetic contribution can be accounted for by the additive and dominance effects of the genes. With complete randomization the environmental and genotype × environment components of variation appear only in the expectations of the within family variances although, of course, they appear as part of the sampling bias of the variances of higher rank.

In the expectations of all the within family variances five of the parameters, E_w, G_D, G_H, W_{ejgdi} and W_{ejghi} appear in the form

$$x(E_w + G_D + 2W_{ejgdi}) + y(E_w + G_D - 2W_{ejgdi}) \\ + z(E_w + G_H + 2W_{ejghi}),$$

where $x + y + z = 1$.

Hence we cannot estimate these five parameters no matter how many generations we raise following a cross between two inbred lines. We may, however, estimate various combinations of these parameters. For example,

$$(E_w + G_D + 2W_{ejgdi}) = GE_1$$
$$(E_w + G_D - 2W_{ejgdi}) = GE_2$$
$$(E_w + G_H + 2W_{ejghi}) = GE_3$$

or, alternatively,

$$(E_w + G_D) = \tfrac{1}{2}(GE_1 + GE_2)$$
$$W_{ejgdi} = \tfrac{1}{4}(GE_1 - GE_2)$$
$$(E_w + G_H + 2W_{ejghi}) = GE_3.$$

The expected within family variances in terms of GE_1, GE_2 and GE_3 are given in Table 83. Any nine of the statistics listed in the table are sufficient to solve for the genetic, environmental and interaction components of variance but, as we have seen, we cannot separate the environmental component from the interactions.

TABLE 83
Model specifying genetic, environmental and genotype × environment interaction contributions to the within family variance

Generation	D	H	$F^\dagger$	GE_1	GE_2	GE_3	W_{gdigdk}	W_{ghighk}	W_{gdighk}
P_1				1				2	
P_2					1			2	
F_1						1		2	
F_2	$\frac{1}{2}$	$\frac{1}{4}$		$\frac{1}{4}$	$\frac{1}{4}$	$\frac{1}{2}$		$\frac{1}{2}$	
B_1	$\frac{1}{4}$	$\frac{1}{4}$	$-\frac{1}{2}$	$\frac{1}{2}$		$\frac{1}{2}$	$\frac{1}{2}$	$\frac{1}{2}$	$\frac{1}{2}$
B_2	$\frac{1}{4}$	$\frac{1}{4}$	$\frac{1}{2}$		$\frac{1}{2}$	$\frac{1}{2}$	$\frac{1}{2}$	$\frac{1}{2}$	$-\frac{1}{2}$
F_3	$\frac{1}{4}$	$\frac{1}{4}$		$\frac{3}{8}$	$\frac{3}{8}$	$\frac{1}{4}$		$\frac{1}{8}$	
$F_2 \times P_1(L_1)$	$\frac{1}{8}$	$\frac{1}{8}$	$-\frac{1}{4}$	$\frac{1}{2}$		$\frac{1}{2}$	$\frac{1}{2}$	$\frac{1}{2}$	$\frac{1}{2}$
$F_2 \times P_2(L_2)$	$\frac{1}{8}$	$\frac{1}{8}$	$\frac{1}{4}$		$\frac{1}{2}$	$\frac{1}{2}$	$\frac{1}{2}$	$\frac{1}{2}$	$-\frac{1}{2}$
$F_2 \times F_1(L_3)$	$\frac{3}{8}$	$\frac{1}{4}$		$\frac{1}{4}$	$\frac{1}{4}$	$\frac{1}{2}$	$\frac{1}{2}$	$\frac{1}{2}$	
$*F_{2bip}$	$\frac{1}{4}$	$\frac{3}{16}$		$\frac{1}{4}$	$\frac{1}{4}$	$\frac{1}{2}$	$\frac{1}{2}$	$\frac{1}{2}$	
B_{11}	$\frac{1}{8}$	$\frac{1}{8}$	$-\frac{1}{4}$	$\frac{3}{4}$		$\frac{1}{4}$	$\frac{9}{8}$	$\frac{1}{8}$	$\frac{3}{8}$
B_{12}	$\frac{1}{8}$	$\frac{1}{8}$	$\frac{1}{4}$		$\frac{1}{4}$	$\frac{3}{4}$	$\frac{1}{8}$	$\frac{9}{8}$	$-\frac{3}{8}$
B_{22}	$\frac{1}{8}$	$\frac{1}{8}$	$\frac{1}{4}$		$\frac{3}{4}$	$\frac{1}{4}$	$\frac{9}{8}$	$\frac{1}{8}$	$-\frac{3}{8}$
B_{21}	$\frac{1}{8}$	$\frac{1}{8}$	$-\frac{1}{4}$	$\frac{1}{4}$		$\frac{3}{4}$	$\frac{1}{8}$	$\frac{9}{8}$	$\frac{3}{8}$
$B_1 \times F_1$	$\frac{3}{8}$	$\frac{1}{4}$	$-\frac{1}{4}$	$\frac{3}{8}$	$\frac{1}{8}$	$\frac{1}{2}$	$\frac{1}{8}$	$\frac{1}{2}$	$\frac{1}{4}$
$B_2 \times F_1$	$\frac{3}{8}$	$\frac{1}{4}$	$\frac{1}{4}$	$\frac{1}{8}$	$\frac{3}{8}$	$\frac{1}{2}$	$\frac{1}{8}$	$\frac{1}{2}$	$-\frac{1}{4}$
B_{1bip}	$\frac{1}{4}$	$\frac{3}{16}$	$-\frac{1}{4}$	$\frac{9}{16}$	$\frac{1}{16}$	$\frac{3}{8}$	$\frac{1}{2}$	$\frac{9}{32}$	$\frac{3}{8}$
B_{2bip}	$\frac{1}{4}$	$\frac{3}{16}$	$\frac{1}{4}$	$\frac{1}{16}$	$\frac{9}{16}$	$\frac{3}{8}$	$\frac{1}{2}$	$\frac{9}{32}$	$-\frac{3}{8}$
B_{1S}	$\frac{1}{4}$	$\frac{1}{8}$		$\frac{5}{8}$	$\frac{1}{8}$	$\frac{1}{4}$	$\frac{1}{2}$	$\frac{1}{8}$	$\frac{1}{4}$
B_{2S}	$\frac{1}{4}$	$\frac{1}{8}$		$\frac{1}{8}$	$\frac{5}{8}$	$\frac{1}{4}$	$\frac{1}{2}$	$\frac{1}{8}$	$-\frac{1}{4}$

* Usually denoted as F_3bip but this is inconsistent with the terminology that has had to be adopted for the backcross generations † $F = S(dh)$

Only three of the statistics V_{P1}, V_{P2} and V_{F1} differ uniquely in the presence of genotype × environment interactions, hence the heterogeneity of these three variances unambiguously detects their presence. In fact, these three variances will differ as a result of interactions much more than any other statistics, and so they are the most sensitive to their presence. In the presence of interactions there is no way in which the three statistics V_{P1}, V_{P2} and V_{F1} can be used to estimate the environmental dependent variances of the segregating generations as we can, for example, in the single gene case or in the absence of interactions.

So far we have been considering the situation where the individuals of all families and generations are distributed at random over a common range of environments within a single macroenvironment. This, however, is not a common situation and experimental convenience frequently rules it out. Indeed, in many higher animals members of the same litter are not readily separable before weaning, hence the litter rather than the individual is the unit of randomization. We must, therefore, consider an extension to the commoner situation where each family is raised in one or more compact groups, which are themselves distributed at random over the macroenvironment. The environmental differences affecting the variation of individuals within the family are then different from those affecting the variation in family means. These two sources of environmental variation, of course, correspond with the E_w and E_b components of variation of the earlier models (Chapter 6). We must, therefore, allow for the presence of both of these additive environmental components and for their individual interactions with the additive and dominance effects of the genes. Those involving E_w, the within family item (or within plot if families are divided among replicate plots each containing a number of individuals), are the same as in the completely randomized experiment. However, we must now introduce a new symbol to distinguish them both from the latter and from those involving E_b. This can be achieved by substituting G_{wD}, G_{wH}, $W_{ew_jg_{di}}$ and $W_{ew_ig_{hi}}$ for the G_D, G_H, $W_{e_jg_{di}}$ and $W_{e_jg_{hi}}$, respectively, of the completely randomized experiment (Table 83). All the properties of expectations for the completely randomized experiment apply to the within family expectations of the incompletely randomized experiment.

The new parameters required to specify the variances of family (or plot) means, exactly parallel those for the within family variances. Thus the parameters required are E_b, G_{bD}, G_{bH}, $W_{gb_{di}gb_{dk}}$, $W_{gb_{hi}gb_{hk}}$, $W_{eb_jgb_{di}}$ and $W_{eb_jgb_{hi}}$ which, apart from the subscript b to denote that they are between family effects, are identical with those described earlier. Examples of the expected within and between family variances in terms of these parameters are given in Table 84. For the P_1, P_2, F_1 and F_2 generations the expectations apply equally to replicated families or to replicate plots of the same family, since in these generations every family has the same expected mean and variance. For the F_3 generation, however, the expectations relate only to the family as the unit of

TABLE 84

The expected variance within and between plots (or families) in the presence of genotype × environment interactions

Generation	Within-plot* (or family)	Between-plot means (or family means)
P_1	$E_w + G_{wD} + 2W_{gwdigwdk} + 2W_{ewjgwdi}$	$E_b + G_{bD} + 2W_{gbdigbdk} + 2W_{ebjgbdi}$
P_2	$E_w + G_{wD} + 2W_{gwdigwdk} - 2W_{ewjgwdi}$	$E_b + G_{bD} + 2W_{gbdigbdk} - 2W_{ebjgbdi}$
F_1	$E_w + G_{wH} + 2W_{gwhigwhk} + 2W_{ewjgwhi}$	$E_b + G_{bH} + 2W_{gbhigbhk} + 2W_{ebjgbhi}$
F_2	$\frac{1}{2}D + \frac{1}{4}H + E_w + \frac{1}{2}G_{wD} + \frac{1}{2}G_{wH} + \frac{1}{2}G_{bD}$ $+ \frac{1}{4}G_{bH} + \frac{1}{2}W_{gwhigwhk} + W_{ewjgwhi}$	$E_b + \frac{1}{4}G_{bH} + \frac{1}{2}W_{gbhigbhk} + W_{ebjgbhi}$
F_3	$\frac{1}{4}D + \frac{1}{8}H + E_w + \frac{3}{4}G_{wD} + \frac{1}{4}G_{wH} + \frac{1}{4}G_{bD}$ $+ \frac{1}{8}G_{bH} + \frac{1}{8}W_{gwhigwhk} + \frac{1}{2}W_{ewjgwhi}$	$\frac{1}{2}D + \frac{1}{16}H + E_b + \frac{1}{2}G_{bD} + \frac{1}{8}G_{bH}$ $+ \frac{1}{8}W_{gbhigbhk} + \frac{1}{2}W_{ebjgbhi}$

* Although the contributions of E_w, G_{wD}, G_{wH}, $W_{ewjgwdi}$ and $W_{ewjgwhi}$ are given separately they cannot be estimated; only compounds of these such as $\overline{GE}_1$, $\overline{GE}_2$ and $\overline{GE}_3$ (see Table 83) can be separated.

randomization, since families do not have the same expected mean and variance.

Two features of these expectations are worthy of note. First, while the within family parameters appear only in the within plot or the within family variance, the between family parameters appear in the expectations of both the within and between family (or plot) variances. Second, because G_{bD} and G_{bH} appear in the within family variances without E_b it is possible to separate the additive environment component of the family differences from the genotype × environment interactions. It is still impossible, however, to achieve the same separation for the corresponding within family components. Hence, in an incompletely randomized experiment in which the individuals in each generation are divided among different plots (which may or may not coincide with family units) a total of 17 parameters, three of which are compound, are required to specify the expected variances assuming an additive-dominance model. Nine generations of an appropriate kind must, therefore, be raised in order to estimate all the parameters of the model.

It should be noted that the heterogeneity of the variances of the P_1, P_2 and F_1 generations provide an equally sensitive and unambiguous test for the presence of genotype × environment interaction for complete and incomplete randomization and at the within and between plot or family level (Tables 83 and 84).

The problem of specifying the contribution of genotype × environment interactions to the within and between family variances in the presence of linkage has been examined by Mather and Jones (1958). The expected variances, in a form appropriate for fitting to observed variances, have been described by Perkins and Jinks (1970) for the case of complete randomization. They found that 29 parameters were required to specify completely the contributions of the genetic, environmental, interactive and linkage components of an additive-dominance model to the variances of the generations that can be derived by selfing, backcrossing and sib-mating following a cross between two inbred lines and raised in a completely randomized experiment. However, only 15 parameters could be estimated, 11 of which were compounds of two or more of the 29 parameters whose contributions are completely correlated in all the expectations and are hence inseparable. A detailed account of the definitions of these parameters and the procedure for detecting and separating the contributions of genotype × environment interactions and linkage when both are present are discussed and illustrated by Perkins and Jinks (1970).

The specification of genotype × environment interactions that has been developed and illustrated for the crosses between pairs of inbred lines can be extended to other mating systems without involving any new problems or principles. Some of the expectations for randomly mating populations obtained in this way are given by Mather and Jones (1958).

In Section 19 we showed that the genotype × environment interaction components of generation means are often linear functions of the additive environmental components. Hence we could substitute

$$g_d = b_d e_j$$

and

$$g_h = b_h e_j$$

in many of the expectations.

If the same empirical relationships apply to the microenvironmental differences within and between families (or plots) when grown in the same environment, then

$$G_{wD} = b_{wd}^2 E_w \qquad (G_D = b_d^2 E_w)$$
$$G_{wH} = b_{wh}^2 E_w \qquad (G_H = b_h^2 E_w)$$
$$G_{bD} = b_{bd}^2 E_b \qquad G_{bH} = b_{bh}^2 E_b$$

where

$$b_{wd} = \frac{W_{gwdiewj}}{E_w} \qquad b_{bd} = \frac{W_{gbdiebj}}{E_b}$$

$$b_{wh} = \frac{W_{gwhiewj}}{E_w} \qquad b_{bh} = \frac{W_{gbhiebj}}{E_b}.$$

We could therefore rewrite the expectations in Tables 83 and 84 in terms of the bs for use where a linear relationship is known to be present or can be assumed. However, since we cannot estimate E_w independently of G_{wD} (or G_D in the case of complete randomization) we can never test this assumption nor estimate b_{wd} or b_{wh} (b_d or b_h), though we can of course do so for the between family or plot components.

Nevertheless, though we cannot demonstrate a linear relation the means and variances of the families which constitute the L_1 and L_2 generations of an F_2

triple test cross provide a unique opportunity for detecting an interaction between the microenvironment and the additive and dominance action of the genes (Perkins and Jinks, 1971). We can combine the means and variances of L_1 and L_2 family variances which have the same F_2 parents to give half the sum and half the difference for the means and variances, that is, for the ith F_2 parent $\frac{1}{2}(\bar{L}_{1i} + \bar{L}_{2i})$, $\frac{1}{2}(\bar{L}_{1i} - \bar{L}_{2i})$, $\frac{1}{2}(\sigma_{L1i}^2 + \sigma_{L2i}^2)$ and $\frac{1}{2}(\sigma_{L1i}^2 + \sigma_{L2i}^2)$. In the absence of genotype × environment there is no covariance and hence no correlations between $\frac{1}{2}(\bar{L}_{1i} + \bar{L}_{2i})$ and $\frac{1}{2}(\sigma_{L1i}^2 + \sigma_{L2i}^2)$ and $\frac{1}{2}(\bar{L}_{1i} - \bar{L}_{2i})$ and $\frac{1}{2}(\sigma_{L1i}^2 - \sigma_{L2i}^2)$. In the presence of genotype × environment interactions the covariances between these statistics have the expectations: $\frac{1}{4}dg_de$ and $-hg_d^2 + hg_h^2 + \frac{1}{4}hg_he$, respectively for a single gene difference. The first has a value only if the interactions involve the additive gene action while the second depends primarily on interactions involving dominance. Extending these expectations to many gene loci introduces new terms but does not alter these conclusions.

Perkins and Jinks (1970) have carried out this analysis for plant height in the triple test cross based on the F_2 of the cross between varieties 1 and 5 of *N. rustica* (see Sections 27 and 32). The correlations for the 40 sums was -0.259 and that for the 40 differences -0.024 each for 38 degrees of freedom. Neither is significant. There is, therefore, no evidence of interactions with the microenvironment.

8 Randomly breeding populations

34. THE COMPONENTS OF VARIATION

Where the initial cross is between true breeding lines, and where self-mating, intercrossing and backcrossing can be practised at will, we have seen that a great multiplicity of statistics are available for estimating the genetic and environmental components of the generation means and variances. Most organisms will, however, by their own special properties set limits to the range of statistics which can be obtained. In such plants as wheat, oats and barley, self-pollination is easily secured, whereas crossing of any kind is tedious. Data will therefore generally come from measurement of F_2 and F_3 families. With self-incompatible and dioecious plants, and with most animals, self-mating is precluded. The statistics available will then be those from successive backcross generations and from F_2 and biparental progenies. Either can lead to a satisfactory analysis.

The value of true breeding parental lines to experiments of this kind is, of course, very great. Using such lines, simple Mendelian theory can be applied to the transmission of the genes, even though they cannot be followed individually in transmission. In particular, providing that there is no selective elimination, we know the relative frequencies of different kinds of gametes and of different kinds of zygotes in any generation derivable from crosses between such lines. A further advantage following from the use of true breeding lines is their availability for the direct assessment of environmental and genotype × environment interaction sources of variation and for providing tests of the adequacy of scale and for the presence of non-allelic interactions.

Partitioning the variation into its components is, therefore, much easier where true breeding lines are used in the initial cross. It is, however, possible to make a partition even where such lines are not available. If the genetic differences between individuals within two stocks are small compared with those between the stocks, the stocks may be treated as true breeding without serious error. The genetic variation within them is confounded with the non-heritable variation, and this may indeed be of very small consequence in crosses such as those between species.

213

Where a species is capable both of easy self-fertilization and of easy crossing, every member of any population may be itself treated as an F_1 between hypothetical parents. Self-pollination will yield an F_2, which may be used in its turn to give both F_3s and biparental progenies. The components of variation may be estimated from such families even in the absence of the hypothetical parents or of a sufficiently large F_1, from which a direct estimate of the environmental component of variation could be obtained. With a clonally propagatable plant, this absence of parents would not be felt, since a clone of any individual of any generation could be used to give the necessary estimates of non-heritable variation. There would remain the question of the adequacy of an additive-dominance model. This would be an adequate model on a scale on which the fall from the F_2 mean to the mean of all F_3s is twice that from the mean of all F_3s to the mean of all F_4s (see Chapter 4). By these means the genetical properties of the various individuals of any population could be determined by treating each individual as a unique F_1. By combining and comparing a sample of such individuals we could obtain a picture of the genetical properties and composition of the population as a whole.

In respect of a single gene difference, A–a with frequencies u_a and v_a ($= 1 - u_a$) the individuals in the population will be of three kinds AA, Aa and aa, and if the population is randomly mating they will be present in frequencies u_a^2; $2u_a v_a$; v_a^2. After one generation of selfing we shall have families whose means are d_a, $\frac{1}{2}h_a$ and $-d_a$ and heritable variances 0, $(\frac{1}{2}d_a^2 + \frac{1}{4}h_a^2)$ and 0 occurring in these frequencies, respectively. The means and variances correspond with the expectations of the P_1, F_2 and P_2 generations, respectively, of two inbred lines and the F_2 produced from a cross between them. For a number of independent genes, therefore, the mean variance within the families will be

$$S(uvd^2) + \tfrac{1}{2}S(uvh^2) = \tfrac{1}{4}D + \tfrac{1}{8}H_1$$

where $D = S(4uvd^2)$ and $H_1 = S(4uvh^2)$ which are the diallel forms of D and H_1 described in Chapter 9.

The expectations of the mean variances within families for further generations of selfing can, therefore, be obtained directly from those of the selfing series. We also have already seen (Section 25) that the general form of these expectations for generation n and rank j is

$$(\tfrac{1}{2})^j D + (\tfrac{1}{2})^{j-1} (\tfrac{1}{4})^{n-j} H \quad \text{where} \quad D = S(d^2), \quad H = S(h^2).$$

To these expectations must be added the standard environmental components of variation of the selfing series appropriate to the experimental design used. We can allow for the fact that only a proportion $2uv$ of the randomly mating population initiate a selfing series by using the diallel forms of D and H_1. The general expectation, then, holds for successive generations of selfing a randomly mating population by equating the initial population to an F_2 ($n = 2$) and the first generation produced by selfing to an F_3 ($n = 3$), at the same time considering only rank 2, or higher statistics ($j \geqslant 2$) since we are considering the mean variances within families, in the equivalent to the F_3, as the highest rank. That is, the expectation applies to $n \geqslant 3, j \geqslant 2$ and $n > j + 1 \geqslant 3$. Within similar restrictions on n we can derive the general expectation of the rank 2 parent offspring covariances, which equal

$$\tfrac{1}{4}D + (\tfrac{1}{2})^{2n-2} H_1,$$

where n refers to the parental generation. By using the covariances, two successive generations of selfing provide sufficient statistics to estimate D, H_1, E_w and E_b.

The success of this method of analysing a population of unknown composition and of unknown gene frequencies depends on enforcing a situation, by the initial selfing of the individual, where the gene frequencies become known and gene behaviour predictable from simple Mendelian principles. The same type of analysis can be put to good use in at least two ways in the later stages of an experiment which itself began with a cross between true breeding lines. First, if each F_2 plant is regarded as an F_1, its F_3 as the corresponding F_2, and so on, the F_2 population can be analysed plant by plant and its composition compared with that expected on the basis of the first analysis which will have already been obtained in the experiment and in which the F_2 was used as a whole for the understanding of the properties of the F_1 and the parents which gave rise to it. Secondly, lines raised from F_2 by selective breeding can be analysed in this way in later generations in order to observe the effects of the selection on the decay of variability. In both cases, the parental lines and the F_1 can be employed to give direct estimates of E_w and the question of the appropriate model on the chosen scale will already have been settled.

Still other means must be employed where it is desired to undertake the analysis of a population of unknown constitution, whose constituent individuals cannot be self-fertilized. Scaling problems and choice of an adequate model can usually be settled only by ancillary experiments or considerations, unless it is proposed to undertake the whole analysis by first producing a number of inbred lines, intercrossing these and then proceeding as before – a task which will usually be prohibitive if only by reason of its requirements in time. Where two or more true breeding lines are available, the population may be analysed by a triple test cross in a way to which we will return later (Section 40). Otherwise the population must be treated as it stands.

The early biometricians developed the method of correlation between relatives for this purpose of analysing populations over whose breeding no control could be exercised. This method has been thoroughly reviewed from the standpoint of Mendelian inheritance by Fisher (1918) but it is not confined in its application to the analysis of heritable variation. It has been extensively used for the partitioning of non-heritable variation into its various components (Lush, 1943; Wright, 1934a). Its use depends on the fact that where a correlation r exists between two variables, a proportion r^2 of the variation of one variate may be accounted for by reference to variation in the other, leaving

$1 - r^2$ as residual variation for which other causes must be sought. In this way an analysis of variance may be arrived at.

When used for the separation and analysis of heritable variation the method of correlation between relatives is, as would be expected, related to the methods already developed for the special case of crosses between true breeding stocks. It leads, in fact, to generalized formulae for the genetic components and at the same time shows us the special advantages of the type of data discussed in Chapters 4, 5 and 6. We will begin by assuming that an additive-dominance model is adequate on the scale chosen for taking measurements.

As we have previously noted, in a population mating at random, in which two alleles A and a of a gene occur with frequencies u_a and $v_a (= 1 - u_a)$, respectively, in the gametes taken as a whole, the three genotypes AA, Aa and aa will occur with the average frequencies $u_a^2 : 2u_a v_a : v_a^2$ in each generation. Then, if as before the contributions made by the three genotypes to the characters are d_a, h_a and $-d_a$, respectively, the mean will be

$$(u_a - v_a) d_a + 2u_a v_a h_a \qquad \text{(see Table 85)}.$$

The contribution which this gene makes to the variance of the population will be

$$(u_a^2 + v_a^2) d_a^2 + 2u_a v_a h_a^2 - [(u_a - v_a) d_a + 2u_a v_a h_a]^2.$$

This simplifies to

$$2u_a v_a [d_a^2 + 2 (v_a - u_a) d_a h_a + (1 - 2u_a v_a) h_a^2],$$

which can be recast in the form:

$$2u_a v_a [d_a^2 + 2 (v_a - u_a) d_a h_a + (v_a - u_a)^2 h_a^2 + 2u_a v_a h_a^2]$$

or
$$2u_a v_a [(d_a + (v_a - u_a) h_a)^2 + 2u_a v_a h_a^2].$$

Where the genes are independent in action and unlinked in transmission, the total heritable variation will be the sum of the series of such items, one from each gene, namely

$$S2u_a v_a [d_a + (v_a - u_a) h_a]^2 + S4u_a^2 v_a^2 h_a^2.$$

If we now put

$$D_R = S4u_a v_a [d_a + (v_a - u_a) h_a]^2 \quad \text{and} \quad H_R = S16u_a^2 v_a^2 h_a^2$$

the heritable component of the total variation, V_R, becomes

$$\tfrac{1}{2} D_R + \tfrac{1}{4} H_R.$$

Apart from sampling variation, this heritable variation will be constant from generation to generation.

When $u = v = \tfrac{1}{2}$ for all genes, as in the F_2 of a cross between two true breeding lines, these general expressions for D_R and H_R reduce to Sd^2 and Sh^2, and the heritable variance itself reduces, as it should, to that already found for the variance of an F_2. Thus if, and only if, $u = v = \frac{1}{2}$ the contribution made by d and h to the heritable variation may be separated in the analysis. Otherwise D_R contains some effect of h, and H_R is correspondingly less than the summed effects of all the squared h deviations. D_R may be greater or less than Sd^2 according to the relative frequencies of the more dominant and more recessive alleles (Figs. 16 and 17). In general, therefore, D_R is not the additive variation in

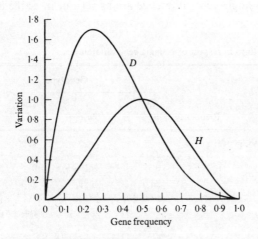

Figure 16 Change in the contributions made by a gene to D_R and H_R according to the frequency of its dominant allele in a randomly breeding population. In calculating the variation it was assumed that $d = h = 1$.

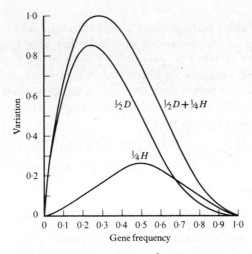

Figure 17 Change in the contribution ($\frac{1}{2}D_R + \frac{1}{4}H_R$) made by a gene to the variation of a randomly breeding population according to the frequency of its dominant allele. The separate items $\frac{1}{2}D_R$ and $\frac{1}{4}H_R$ are also shown. Calculations are based on $d = h = 1$.

the sense used in the earlier chapters, although it is frequently referred to as such. It is, in fact, the additive variation in the statistical sense, rather than in the genetical sense we have adopted.

Within a randomly mating population we can find various degrees of relationships between its members that can be used as a basis for partitioning the total variation. The most obvious is the grouping of the individuals into families. The nine possible types of matings in a randomly mating population in respect of a single gene difference are as set out in Table 85.

TABLE 85
The possible matings in respect of a single gene difference

Male parents		Frequency	Genotype AA u_a^2	Aa $2u_a v_a$	aa v_a^2	
	Genotype		u_a^4	$2u_a^3 v_a$	$u_a^2 v_a^2$	Frequency
			d_a	$\frac12 d_a + \frac12 h_a$	h_a	Mean
Female	AA	u_a^2	0	$\frac14(d_a - h_a)^2$	0	Variance
parents			$2u_a^3 v_a$	$4u_a^2 v_a^2$	$2u_a v_a^3$	
	Aa	$2u_a v_a$	$\frac12 d_a + \frac12 h_a$	$\frac12 h_a$	$-\frac12 d_a + \frac12 h_a$	
			$\frac14(d_a - h_a)^2$	$\frac12 d_a^2 + \frac14 h_a^2$	$\frac14(d_a + h_a)^2$	
			$u_a^2 v_a^2$	$2u_a v_a^3$	v_a^4	
	aa	v_a^2	h_a	$-\frac12 d_a + \frac12 h_a$	$-d_a$	
			0	$\frac14(d_a + h_a)^2$	0	

Overall mean $(u_a - v_a)d_a + 2u_a v_a h_a$

These nine possible matings are of six kinds in respect of the genotypes of the parents and of the progenies to which they give rise and they are recognizable as equivalent to parental, F_1, F_2 and backcross generations whose means and variances are known (see Chapters 4 and 6).

The mean variance within these six kinds of families follows directly from this table as the summed products of the frequencies of each kind of family and their within family variance, thus

$$u_a^4 \cdot 0 + 4u_a^3 v_a \cdot \tfrac14 (d_a - h_a)^2 + 2u_a^2 v_a^2 \cdot 0 + 4u_a^2 v_a^2 \cdot (\tfrac12 d_a^2 + \tfrac14 h_a^2)$$
$$+ 4u_a v_a^3 \cdot \tfrac14(d_a + h_a)^2 + v_a^4 \cdot 0,$$

which reduces to

$$u_a v_a d_a^2 + u_a v_a (v_a - u_a) dh + (u_a v_a - u_a^2 v_a^2) h_a^2.$$

On summation over all independent genes it yields

$$\tfrac14 D_R + \tfrac{3}{16} H_R$$

as the heritable component of the within family variance, V_{2SR}.

The difference between this and the total variation, $\frac{1}{2}D_R + \frac{1}{4}H_R$ is, of course, the heritable variation between family means, V_{1SR}. We can derive this variance directly from the frequencies and family means in the table, as

$$u_a^4 d_a^2 + 4u_a^3 v_a \left(\tfrac{1}{2}d_a + \tfrac{1}{2}h_a\right)^2 + 2u_a^2 v_a^2 h_a^2 + 4u_a v_a^3 \left(-\tfrac{1}{2}d_a + \tfrac{1}{2}h_a\right)^2 + 4u_a^2 v_a^2 \left(\tfrac{1}{2}h_a\right)^2$$
$$+ v_a^4 d_a^2 - [(u_a - v_a)d_a + 2u_a v_a h_a]^2$$

the last term being the correction for the mean. On summation over all independent genes this reduces to

$$\tfrac{1}{4}D_R + \tfrac{1}{16}H_R.$$

This is also the heritable covariance of full-sibs, W_{FSR} and its expectation may be obtained directly as a covariance. There is only one complication to bear in mind. In a family from, for example, the AA × Aa, equal numbers of AA and Aa offspring are expected. Then, of pairs of sibs taken at random from such a family $\frac{1}{4}$ will be both AA, $\frac{1}{4}$ will be both Aa and $\frac{1}{2}$ will include a sib of each kind. Such families occur with a frequency of $2u_a^3 v_a$ (Table 85) and so will contribute $\frac{1}{4} \cdot 2u_a^3 v_a d_a^2 + \frac{1}{4} \cdot 2u_a^3 v_a h_a^2 + \frac{1}{2} \cdot 2u_a^3 v_a d_a h_a$ to the sum of cross-products. The correction term will be the square of the overall mean for the population, i.e. $[(u_a - v_a)d_a + 2u_a v_a h_a]^2$.

Another type of relationship that may be found in a randomly mating population, and can provide a basis for partitioning the total variation, is that of half-sibs. These are progenies with one parent in common, the other parent being different. The covariance of half-sibs is the same as the variance of the means of half-sib groups produced by allowing each type of parent, for example AA, to mate with a random sample of the three types of parent in proportion to their frequencies in the population. There are, of course, three such half-sib groups produced by common parents of types AA, Aa and aa, respectively. Each type of half-sib group will itself occur with frequencies determined by the frequencies of the common parents AA, Aa and aa in the population, i.e. u_a^2, $2u_a v_a$ and v_a^2. The means and frequencies of the half-sib groups are therefore

Common parent	AA	Aa	aa	Overall mean
Frequency	u_a^2	$2u_a v_a$	v_a^2	
Mean of half-sib group	$u_a d_a + v_a h_a$	$\frac{1}{2}(u_a - v_a)d_a + \frac{1}{2}h_a$	$-v_a d_a + u_a h_a$	$(u_a - v_a)d_a + 2u_a v_a h_a$

Hence the heritable covariance of half-sibs, W_{HSR} (variance of half-sib group means) is

$$u_a^2(u_a d_a + v_a h_a)^2 + 2u_a v_a\left[\tfrac{1}{2}(u_a - v_a)d_a + \tfrac{1}{2}h_a\right]^2 + v_a^2(-v_a d_a + u_a h_a)^2$$
$$- [(u_a - v_a)d_a + 2u_a v_a h_a]^2 = \tfrac{1}{2}u_a v_a[d_a + (v_a - u_a)h_a]^2,$$

which summing over independent genes gives $\frac{1}{8}D_R$.

An alternative way of partitioning the total variation is to determine how much of this variation can be related to the variation among the parents in the previous generation. This can be estimated by the parent–offspring covariance. Its expectation may be derived in a number of ways. In one way the family structure of the offspring can be ignored and we need only consider the frequency with which the two types of gametes, A and a, combine with the three types of parents AA, Aa and aa, occurring with relative frequencies $u_a^2 : 2u_a v_a : v_a^2$. The population of gametes, with which those of each parent will combine in giving the next generation, will carry A and a with the relative frequencies u_a and v_a. The offspring of AA parents will, therefore, be AA in u_a of cases, and Aa in v_a of cases; the former having an expression of d_a and the latter of h_a. Taking the three types of parents in turn we can in this way build up a table showing the frequencies of the various relations of expression in parent and offspring (Table 86).

TABLE 86
The frequencies of the various offspring–parent relationships

Parents			Offspring		
Frequency	Genotype		AA	Aa	aa
			d_a	h_a	$-d_a$
u_a^2	AA	d_a	u_a^3	$u_a^2 v_a$	—
$2u_a v_a$	Aa	h_a	$u_a^2 v_a$	$u_a v_a$	$u_a v_a^2$
v_a^2	aa	$-d_a$	—	$u_a v_a^2$	v_a^3

The sum of cross-products of parent and offspring will therefore be

$$(u_a^3 + v_a^3) d_a^2 + 2 (u_a^2 v_a - u_a v_a^2) d_a h_a + u_a v_a h_a^2$$

from which must be deducted a term correcting for the means of parents and offspring. This will be the same as that used in calculating the variances, since the mean measurement of all offspring is the same as that of all parents. The contribution of this gene to the covariance of parent and offspring thus becomes

$$[u_a^3 + v_a^3 - (u_a - v_a)^2] d_a^2 + 2 [u_a^2 v_a - u_a v_a^2 - 2u_a v_a (u_a - v_a)] d_a h_a + (u_a v_a - 4u_a^2 v_a^2) h_a^2,$$

which reduces to

$$u_a v_a [d_a + (v_a - u_a) h_a]^2$$

and summing over all independent genes this becomes $W_{POR} = \frac{1}{4} D_R$.

The treatment of the non-heritable variation in a randomly mating population depends on whether we are dealing with data collected by simple observation of natural populations or with experimental populations maintained under controlled conditions and in the latter case on the degree of control and the experimental design used.

35. GENIC INTERACTION

The effects of interaction between non-allelic genes on the composition of the variation in the F_2 and other generations derived from a cross between two true-breeding lines were examined in Section 30. These effects were twofold: in the first place terms in $I = Si^2$, $J = Sj^2$ and $L = Sl^2$ appeared in the expressions for the heritable variation; and second, interaction items were introduced into D and H, which became $D = S(d_a + wSj_a.)^2$ and $H = S(h_a + wSl_a.)^2$, where the value of w depends on the generation, (F_2, F_3, S_3 etc.) under discussion. Non-allelic interaction has a similar two-fold effect on the composition of the variation in randomly breeding populations (Mather, 1974). The heritable variance of the population, for example, no longer has the structure

$$V_R = \tfrac{1}{2}D_R + \tfrac{1}{4}H_R, \text{ but becomes } V_R = \tfrac{1}{2}D_R + \tfrac{1}{4}H_R + \tfrac{1}{4}I_R + \tfrac{1}{8}J_R + \tfrac{1}{16}L_R,$$

while the covariance of parent and offspring now appears as $W_{POR} = \tfrac{1}{4}D_R + \tfrac{1}{16}I_R$. These, together with the structures of other statistics commonly derived from randomly breeding populations are collected together in Table 87.

TABLE 87
The structure of variances and covariances in randomly breeding populations

Statistic	Structure
Variance of population	$V_R = \tfrac{1}{2}D_R + \tfrac{1}{4}H_R + \tfrac{1}{4}I_R + \tfrac{1}{8}J_R + \tfrac{1}{16}L_R + E_w + E_b$
Covariance of parent and offspring	$W_{POR} = \tfrac{1}{4}D_R \quad + \tfrac{1}{16}I_R$
Covariance of full sibs	$W_{FSR} = \tfrac{1}{4}D_R + \tfrac{1}{16}H_R + \tfrac{1}{16}I_R + \tfrac{1}{64}J_R + \tfrac{1}{256}L_R + E_b$
Covariance of half sibs	$W_{HSR} = \tfrac{1}{8}D_R \quad + \tfrac{1}{64}I_R$
Variance of sibship means	$V_{1SR} = \tfrac{1}{4}D_R + \tfrac{1}{16}H_R + \tfrac{1}{16}I_R + \tfrac{1}{64}J_R + \tfrac{1}{256}L_R + E_b$
Mean variance of sibships	$V_{2SR} = \tfrac{1}{4}D_R + \tfrac{3}{16}H_R + \tfrac{1}{16}I_R + \tfrac{7}{64}J_R + \tfrac{15}{256}L_R + E_w$

E_w is the non-heritable variance within a sib-ship family. E_b reflects the effects of any common factors in the environments of members of the same family (the common family environment).

As in F_2 and its derived generations, the terms in I, J and L appearing in first rank variances and covariances of this table take coefficients which are the products of the coefficients of the relevant main terms. In other words, the coefficient of I is the square of the coefficient of D, that of L is the square of the coefficient of H, and that of J is the product of the coefficients of D and H. This relation no longer holds, however, when we turn to variances and covariances of higher rank, i.e. in this table to the mean variance of sibships, V_{2SR}.

Not surprisingly, the composition of the individual components of variation, D_R, etc., are more complex than their counterparts in an F_2 and its derivatives. One such complexity we have already seen, the appearance of contributions from h in D_R. This springs from the inequality of gene frequencies in a randomly breeding population, and as we have already noted it vanishes when $u = v$, as in an F_2, and D_R then reduces to D. To this must now be added the further complexities arising from the genic interactions. The derivation of the components of variation in a randomly breeding population allowing for the presence of non-allelic interaction as well as inequality of gene frequencies is algebraically tedious rather than difficult, and it will not be set out in detail. One prospective complexity can, however, be eliminated. In discussing the composition of D and H where interaction is present, we used a coefficient w for the contribution of j to D and of l to H, because their values change with the generation. This is no longer necessary when considering a randomly breeding population, for w always takes its F_2 value of $\frac{1}{2}$ in the statistics derived from observations made directly on a natural population.

Taking first the case of two gene pairs, where the phenotypes produced by the nine possible genotypes are as set out in Table 19 on page 83, and for brevity writing $u_a v_a$ as Π_a and $u_a - v_a$ as Δ_a, we find

$$D_R = 4\Pi_a [(d_a + 2\Pi_b j_{ab} + \Delta_b i_{ab}) - \Delta_a (h_a + \Delta_b j_{ba} + 2\Pi_b l_{ab})]^2$$
$$+ 4\Pi_b [(d_b + 2\Pi_a j_{ba} + \Delta_a i_{ab}) - \Delta_b (h_b + \Delta_a j_{ab} + 2\Pi_a l_{ab})]^2$$
$$H_R = 16\Pi_a^2 (h_a + \Delta_b j_{ba} + 2\Pi_b l_{ab})^2 + 16\Pi_b^2 (h_b + \Delta_a j_{ab} + 2\Pi_a l_{ab})^2$$
$$I_R = 16\Pi_a \Pi_b (i_{ab} - \Delta_b j_{ab} - \Delta_a j_{ba} + \Delta_a \Delta_b l_{ab})^2$$
$$J_R = 64\Pi_a \Pi_b^2 (j_{ab} - \Delta_a l_{ab})^2 + 64\Pi_a^2 \Pi_b (j_{ba} - \Delta_b l_{ab})^2$$
$$L_R = 256\Pi_a^2 \Pi_b^2 l_{ab}^2$$

When $u = v = \frac{1}{2}$ giving $\Pi = \frac{1}{4}$ and $\Delta = 0$, these expressions of course reduce to those for D, H, I, J and L as set out for F_2 on page 177; and when $i = j = l = 0$, I_R, J_R and L_R vanish while D_R and H_R reduce to those used in the previous section. Generalizing these two-gene expressions we find the definitions set out in Table 88 for the case of any number of genes. For comparison the counterpart F_2 expressions are also set out in the table.

It will be observed from the table that any variance or covariance carrying a term in D_R also includes one in I_R, and terms in J_R and L_R similarly appear with H_R. Separation of I_R from D_R, and J_R and L_R from H_R is thus not easy. In principle we can find $W_{POR} - 2W_{HSR} = (\frac{1}{4}D_R + \frac{1}{16}H_R) - 2(\frac{1}{8}D_R + \frac{1}{64}I_R) = \frac{1}{32}I_R$, but the two covariances must be estimated with some precision for this relation to provide a useful test for the presence of i-type interaction. And there must be no comparable disturbance from other factors such as assortative mating, selection, maternal effects or effects of the family environment. Similarly,

$$V_{2SR} - 3V_{1SR} = 8W_{HSR} - 2W_{POR} = \frac{1}{16}J_R + \frac{3}{64}L_R,$$

TABLE 88

Definitions of the components of variation in a randomly breeding population. The special case of F_2, where all gene frequencies are equal, is included for comparison.

Component	Randomly breeding population	F_2
D_R	$S_a 4\Pi_a [\,(d_a + 2S_b\Pi_b j_{ab} + S_b\Delta_b i_{ab})$	
	$\quad - \Delta_a (h_a + S_b\Delta_b j_{ab} + 2S_b\Pi_b l_{ab})]^2$	$S_a (d_a + \tfrac{1}{2}S_b j_{ab})^2$
H_R	$S_a 16\Pi_a^2 (h_a + S_b\Delta_b j_{ba} + 2S_b\Pi l_{ab})^2$	$S_a (h_a + \tfrac{1}{2}S_b l_{ab})^2$
I_R	$S_{ab} 16\Pi_a\Pi_b (i_{ab} - \Delta_b j_{ab} - \Delta_a j_{ba} + \Delta_a\Delta_b l_{ab})^2$	$S_{ab} i_{ab}^2$
J_R	$S_{ab} 64[\Pi_a\Pi_b^2 (j_{ab} - \Delta_a l_{ab})^2 + \Pi_a^2\Pi_b (j_{ba} - \Delta_b l_{ab})^2]$	$S_{ab}(j_{ab}^2 + j_{ba}^2)$
L_R	$S_{ab} 256\Pi_a^2\Pi_b^2 l_{ab}^2$	$S_{ab} l_{ab}^2$

S_a indicates summation over all genes
S_b indicates summation over all genes interacting with A–a
S_{ab} indicates summation over all pairs of interacting genes.

but V_{2SR} and V_{1SR} will not carry the same amount of non-heritable variation and some external estimates of their respective components must be available if we are to obtain a useful estimate of $J_R + \tfrac{3}{4}L_R$, even apart from the need for precision in the estimation of the five statistics that are used and for a known lack of sources of comparable disturbance.

In short, the direct observation of natural populations offers little scope for the detection of non-allelic interaction and the separation of its effects from those of the d and h contributions to the variation made by the genes taken as individuals; though, of course, knowledge of the ways in which interactions contribute to the various statistics may be of assistance in interpreting the implications of the relations found among these statistics, even if it is only by introducing a necessary note of caution into the conclusions that are drawn.

36. NATURAL POPULATIONS

In utilizing the expectations for the heritable component of variation to analyse natural populations it is usually necessary to make assumptions about the non-heritable sources of variation. The simplest situation that can be envisaged is where the individuals are distributed at random over the environment irrespective of whether they belong to the same or to different full-sib families or half-sib groups and where parents and offspring are independently scattered over the environment. In such a situation the environmental contribution to the total variation and to the mean variation within families will be E_w. The covariance between full-sibs, half-sibs, and parents and offspring will contain no environmental component.

Where, as in most animals, the progeny of single matings initially share a common location, even though at a later stage of development they may be dispersed over the whole environment, and this common location is chosen or maintained by one or both parents, this simple situation cannot apply.

Members of the same family will be more alike than members of different families because they will have shared a common location and the variance arising from the common environmental effects will inflate the full-sib and parent–offspring covariances.

Under some circumstances, for example, where the same mother has progeny by different fathers, the common element in the environment provided by the mother can inflate the half-sib covariance. Where, however, half-sibs are related through a common father and have different mothers the inflation of the covariance due to the common environment is likely to be smaller, unless, of course, the male parent is the dominant influence in choosing and maintaining the family environment.

In a natural population it is usually assumed, rather than proven, that random mating is occurring, and the confidence that we can place on any analysis must depend on the reasonableness of this assumption, and the absence of any compelling evidence to the contrary.

The most widely investigated natural populations are human, where mating is not strictly at random and members of the same family normally share a common environment. Thus the expectations based on the assumption of random mating will hold only approximately and the parent–offspring and full-sib covariances will be affected by the common family environment. The simplest system that is likely to be adequate must, therefore, include at least two environmental components, E_w for the variation within families and E_b for the effect of the different environments of different families. Hence we require at least four observed statistics to solve for the four parameters of the model, and, of course, we would need more if we wished to test the adequacy of the model. However, approximate solutions have frequently been obtained from the more readily obtainable statistics by ignoring the effects of the common family environment. For example:

The total variance V_R will equal $\frac{1}{2}D_R + \frac{1}{4}H_R + E_w$.

The parent–offspring covariance W_{POR} will equal $\frac{1}{4}D_R$.

The full-sib covariance W_{FSR} will equal $\frac{1}{4}D_R + \frac{1}{16}H_R$.

From which it follows that

$$D_R = 4W_{POR}$$
$$H_R = 16(W_{FSR} - W_{POR})$$
$$E_w = V_R + 2W_{POR} - 4W_{FSR}.$$

The corresponding correlations, which are the form in which the statistics are usually presented, are

$$r_{PO} = \frac{W_{POR}}{\sqrt{(V_P . V_O)}} = \frac{W_{POR}}{\sqrt{V_R^2}} = \frac{\frac{1}{4}D_R}{\frac{1}{2}D_R + \frac{1}{4}H_R + E_w}$$

$$r_{FS} = \frac{W_{FSR}}{\sqrt{(V_O . V_O)}} = \frac{W_{FSR}}{\sqrt{V_R^2}} = \frac{\frac{1}{4}D_R + \frac{1}{16}H_R}{\frac{1}{2}D_R + \frac{1}{4}H_R + E_w}$$

so that

$$\tfrac{1}{4}D_R = r_{PO}(\tfrac{1}{2}D_R + \tfrac{1}{4}H_R + E_w)$$

and

$$\tfrac{1}{4}D_R + \tfrac{1}{16}H_R = r_{FS}(\tfrac{1}{2}D_R + \tfrac{1}{4}H_R + E_w).$$

The general solution is

$$\frac{H_R}{D_R} = \frac{4(r_{FS} - r_{PO})}{r_{PO}} \quad \text{and} \quad \frac{E_w}{D_R} = \frac{1 - 2r_{PO}}{4r_{PO}} - \frac{H_R}{D_R} r_{PO}.$$

Fisher (1918) records that Pearson and Lee's data yield

$$r_{PO} = 0.4180 \text{ and } r_{FS} = 0.4619 \text{ for human cubit measurement.}$$

Substituting we find

$$\hat{H}_R = 0.4201 \hat{D}_R \quad \text{and} \quad \hat{E}_w = -0.0069 \hat{D}_R.$$

If we express the relative contributions of the components to the total variation as proportions, that is we put

$$V_R = \tfrac{1}{2}D_R + \tfrac{1}{4}H_R + E_w = 1.00$$

then,

$$\tfrac{1}{2}\hat{D}_R = 0.8360 \quad \text{and} \quad \hat{D}_R = 1.6720$$

$$\tfrac{1}{4}\hat{H}_R = 0.1756, \qquad \hat{H}_R = 0.7024$$

$$\hat{E}_w = -0.0116, \qquad \hat{E}_w = -0.0116.$$

In other words, the non-heritable variation must be very small, our estimate becoming negative through sampling error, while H_R is nearly half as large as D_R. So far we have ignored the common environmental effects because these data provide no means of correcting for them. We can examine the bias created by assuming their absence in circumstances where they are present.

In their presence:

The estimate of D_R, as we have obtained it, is in fact $D_R + 4E_b'$.

The estimate of H_R is $H_R + 16(E_b - E_b')$.

The estimate of E_w is $E_w + 2E_b' - 3E_b$.

E_b is the common environmental component of full-sibs and E_b' the common environmental component of parents and offspring. Since we might expect E_b to be greater than E_b' it is easy to see how our estimates of D_R and H_R might be inflated relative to E_w and how the estimate of the latter might become zero or even negative.

The assumption of random mating is also not fully justified as there is a marital correlation of 0.1 between the cubit measurements of mates. If this correlation is due to a similarity in genotype between mates, it must result in higher values for both r_{PO} and r_{FS} than would be obtained with the same genetic structure under a system of random mating. Our estimate of E_w could, therefore, be too low for this reason also. However, with such a low level of correlation between mates, only part of which will be due to genetic correlation and hence effective, the estimates are not going to be made seriously inaccurate by the assumption of random mating.

If we assume that $E_b = E_b'$ and that both are equal to the common environmental component of the half-sib covariance, we can, by including other observable statistics, estimate D_R, H_R, E_w and E_b and hence remove the assumption of no common environmental effects. In practice, however, an alternative approach is more useful, particularly with human populations, which exploits the special properties of twins.

37. TWINS

If twins arise at random in the population the total variation, V_R, among a sample of twins assuming random mating and an additive, dominance model of the heritable variation will be $\frac{1}{2}D_R + \frac{1}{4}H_R + E_w + E_b$. Where we have monozygotic twins raised together (MZT) the variation within the pair (σ_w^2) will be E_w. All of the remaining variation will be between families (σ_b^2).

Where monozygotic twins have been raised apart (MZA) and these are a random sample of all twins, the difference between individuals belonging to the same twin pair will still be entirely non-heritable but it will now include both within and between family components, that is, E_w and E_b. Providing that the separated twins are distributed at random among families in the population the variation within twin pairs will be $E_w + E_b$ and the variation between pairs (σ_b^2) will be $\frac{1}{2}D_R + \frac{1}{4}H_R$.

Thus with monozygotic twins raised apart we can separate the environmental and heritable sources of variation independently of the model assumed for the latter. Such observations do not, however, enable us to separate additive and dominance variation with these data. If we combine data from MZT and MZA we may now separate E_w from E_b and both from $\frac{1}{2}D_R + \frac{1}{4}H_R$ but we again cannot separate D_R and H_R.

Shields (1962) reports a measure of Neuroticism in man for 29 pairs of monozygotic female twins raised together, 26 pairs raised apart and 14 pairs of male twins raised apart which have been analysed by Jinks and Fulker (1970). The mean variances within families $\bar{V}_F = \sigma_w^2$ and the variances of family means $V_{\bar{F}} = \frac{1}{2}\sigma_w^2 + \sigma_b^2$ are as follows.

		Females	Males	Expectations
MZT	$V_{\bar{F}}$	11·0819	—	$\frac{1}{2}D_R + \frac{1}{4}H_R + E_b + \frac{1}{2}E_w$
	$\bar{V}_F$	8·1207	—	E_w
MZA	$V_{\bar{F}}$	14·5608	14·7307	$\frac{1}{2}D_R + \frac{1}{4}H_R + \frac{1}{2}E_b + \frac{1}{2}E_w$
	$\bar{V}_F$	9·6635	5·0000	$E_b + E_w$

The estimates of the total variances V_R, i.e. the total sum of squares divided by the total degrees of freedom, of the MZT and MZA samples are not significantly different and this one would expect since unless the sample sizes are small both should, on our model, be estimates of $\sigma_w^2 + \sigma_b^2 = \frac{1}{2}D_R + \frac{1}{4}H_R + E_w + E_b$. Furthermore, the mean scores in the three samples do not differ, being 9·72, 11·86 and 10·71 respectively. We are, therefore, justified in regarding them as three samples drawn from the same population and we can regard the males and females as replicate estimates of the same statistics. For twins raised apart therefore,

$$V_F = \frac{1}{2}D_R + \frac{1}{4}H_R + \frac{1}{2}E_b + \frac{1}{2}E_w = 14\cdot6458$$

and

$$\bar{V}_F = E_b + E_w = 7\cdot3318$$

hence,

$$\frac{1}{2}D_R + \frac{1}{4}H_R = 10\cdot9799$$

so that the proportion of heritable variation, the broad heritability, is 0·60 or 60%.

If we now combine the twins raised together and apart we have four statistics and three parameters since $\frac{1}{2}D_R$ and $\frac{1}{4}H_R$ are inseparable. We can, therefore, obtain least squares estimates by the normal procedures (see Sections 14 and 27). These are

$$\frac{1}{2}\hat{D}_R + \frac{1}{4}\hat{H}_R = 10\cdot0291$$
$$\hat{E}_w = 8\cdot7546$$
$$\hat{E}_b = -2\cdot0568$$

We can now calculate the expected values of the four statistics and we have one degree of freedom for comparing the observed and expected values. From replicate statistics (females versus males) we have an error variance for two degrees of freedom against which to test the significance of the discrepancy between the observed and expected values.

Source		Observed	Expected	Deviation
MZT	V_F	11·0819	12·3496	−1·2677
	$\bar{V}_F$	8·1207	8·7546	−0·6339
MZA	V_F	14·6458	13·3781	+1·2677
	$\bar{V}_F$	7·3318	6·6978	+0·6339

The SS of deviations = 4·0152 for 1 df.

The SS for sex differences = 10·8871 for 2 df.

However, for the two of the four statistics we are working with the averages of two replicates hence the replicate error for the deviations is

$$\frac{1}{2}\left(\frac{10\cdot8871}{2} + \frac{10\cdot8871}{4}\right) = 4\cdot0827.$$

Clearly, the deviations are not significant and the model fits adequately. The deviation mean square and the error mean square may now be pooled to give an error variance of 4.0602 for 3 df. By multiplying this by the appropriate coefficients on the leading diagonal of the inverted matrix (see Section 27) we obtain the error variances of each of the estimates of the three components and hence their standard errors. These are

$$\tfrac{1}{2}\hat{D}_R + \tfrac{1}{4}\hat{H}_R = \quad 10{\cdot}0291 \pm 2{\cdot}0400 \quad t_{(3)} = 4{\cdot}92 \quad P = 0{\cdot}01{-}0{\cdot}02.$$

$$\hat{E}_w = \quad 8{\cdot}7546 \pm 1{\cdot}9116 \quad t_{(3)} = 4{\cdot}58 \quad P = 0{\cdot}02$$

$$\hat{E}_b = \quad -2{\cdot}0568 \pm 2{\cdot}5488 \quad t_{(3)} = 0{\cdot}81 \quad P = 0{\cdot}40{-}0{\cdot}50.$$

Thus although the amount of replication is lamentably small we can see that the estimates of $\tfrac{1}{2}D_R + \tfrac{1}{4}H_R$ and of E_w are significantly greater than zero while the negative E_b is not. This result is in good agreement with that obtained from twins raised apart.

Before considering dizygotic twins there are two general points of interest arising from the present analysis. First, while we have assumed random mating and additive and dominance variation to arrive at the expectations of the heritable variation, the process of fitting the model and the estimates obtained would be unchanged irrespective of the mating system and the nature of the gene action. Thus we could substitute any expectation in place of $\tfrac{1}{2}D_R + \tfrac{1}{4}H_R$ in the model without affecting the outcome and without biasing the separation of heritable and environmental sources of variation. Second, examination of the deviations of observed and expected values of the four statistics after fitting $\tfrac{1}{2}D_R + \tfrac{1}{4}H_R$, E_w and E_b shows that they are identical in magnitude but differ in sign between the statistics from twins raised together and twins raised apart. Thus the 1 df for testing the significance of these deviations is, in fact, testing the difference between the total variance components for the two types of twins which have identical expectations on the model. This test is, therefore, equivalent to comparing the two total variances as a variance ratio and it is not surprising that the two tests agree in showing no significant difference.

We can extend the analysis indefinitely by including samples from other types of families, for example, dizygotic twins and normal sibs etc. Assuming random mating the statistics obtainable from dizygotic twins and full-sibs have the same expectations which are the standard within- and between-family variances. By including either or both pairs of statistics we are able to separate additive and dominance effects providing, of course, the assumption of random mating is a reasonable one.

Shields (1962) gives the Neuroticism scores for 16 pairs of female dizygotic twins raised together (DZT) and Jinks and Fulker have presented a combined analysis of these with the data from the monozygotic twins. The six observed statistics after pooling males and females as before, and their expectations on the random mating model are

Source		Observed	Expectations assuming $H_R = 0, E_b = 0$	Model
MZT	$V_{\bar{F}}$	11·0819	13·3926	$\tfrac{1}{2}D_R + \tfrac{1}{4}H_R + E_b + \tfrac{1}{2}E_w$
	V_F	8·1207	8·1605	E_w

MZA	$V_{\overline{F}}$	14·6458	13·3926	$\frac{1}{2}D_R + \frac{1}{4}H_R + \frac{1}{2}E_b + \frac{1}{2}E_w$
	$\overline{V}_F$	7·3317	8·1605	$E_b + E_w$
DZT	$V_{\overline{F}}$	11·7828	11·0645	$\frac{3}{8}D_R + \frac{5}{32}H_R + E_b + \frac{1}{2}E_w$
	$\overline{V}_F$	13·8552	12·8167	$\frac{1}{4}D_R + \frac{3}{16}H_R + E_w$

Fitting the full model confirms that E_b is not significantly different from zero and it also reveals that H_R is not significant. A D_R, E_w model may, therefore, be fitted leaving 4 df for testing the adequacy of the model against the replicate error. The least squares estimates are

$$\hat{D}_R = 18·6248$$
$$\hat{E}_w = 8·1605$$

leading to the expected values of the six statistics shown above.

The mean square for the deviation of observed and expected values of the statistics is 4·5964 for 4 df and the replicate error mean square is now 4·5363 for 2 df. It is clear, therefore, that the D_R, E_w model accounts for the observed values of the statistics. The pooled mean squares may, therefore, be used to obtain standard errors of D_R and E_w in the normal way. These show that D_R and E_w are highly significant.

$$\hat{D}_R = 18·6248 \pm 3·1245 \quad t_{(6)} = 5·96 \quad P = 0·01-0·001$$
$$\hat{E}_w = 8·1605 \pm 1·3530 \quad t_{(6)} = 6·03 \quad P = 0·01-0·001.$$

We can, of course, obtain approximate maximum likelihood estimates of D_R and E_w by weighting the observed statistics by their amounts of information (Section 28). This method gives

$$\hat{D}_R = 18·3380 \pm 4·9884 \quad c = 3·68 \quad P < 0·001$$
$$\hat{E}_w = 7·7199 \pm 1·2755 \quad c = 6·05 \quad P < 0·001$$

which agree very well with the simpler method previously given. Indeed the estimate of the proportion of the total variation, $\frac{1}{2}D_R + E_w$ due to heritable causes, which equals 0·54 or 54%, is very close to the earlier estimate based on monozygotic twins alone. The test of the fit of the model in the weighted analysis leads to an approximate $\chi^2_{(4)} = 1·3717$ ($P = 0·8$), which again confirms the adequacy of the simple model.

The four degrees of freedom for testing the adequacy of the model may be partitioned into two parts. Two degrees of freedom may be regarded as testing the effect of omitting H_R and E_b from the full model and two are for testing the equality of the total variance components of the three types of twins, which are expected to be identical on the model. Since the remainder mean square after fitting the D_R, E_w model is equal to the replicate error this confirms that H_R and E_b are not significantly different from zero and that the total variance components do not differ significantly. The latter conclusion is confirmed by the homogeneity of the three total variances. Thus the three types of twins are random samples from the same population and, therefore, subject to the same heritable and environmental sources of variation. Hence, these results provide no evidence that monozygotic twins, for example, are less exposed to environmental differences than dizygotic twins as has frequently been suggested.

In this and in the previous analysis of monozygotic twins alone we could improve the estimates of the components and increase the power of the tests of significance by repeating the estimation using the iterative weighted least squares procedures described in Section 28. With such a good fit to the model as is shown by these data, however, the improvement can only be marginal. It is more important to consider the validity of the assumptions implicit in the model, namely, no genotype × environment interaction, no genotype–environment correlation, random mating and no non-allelic interactions.

As we have seen when the full D_R, H_R, E_w and E_b model is fitted, the test of its goodness of fit is a test of the equality of the total variances. It can, therefore, only test the validity of any assumptions whose failure would have led to inequality of these variances.

The total variances for MZ and DZ twins whether raised together or apart are expected to be the same in the presence of genotype × environment interaction, assortative mating and non-allelic interaction. None of these can, therefore, be detected by failure of the test of goodness of fit of the models which omit them and we must look elsewhere for ways of detecting them.

In the presence of genotype × environment interactions all four total variances have the expectation

$$\tfrac{1}{2}D_R + \tfrac{1}{4}H_R + E_w + E_b + \tfrac{1}{2}G_{wDR} + \tfrac{1}{4}G_{wHR} + \tfrac{1}{2}G_{bDR} + \tfrac{1}{4}G_{bHR}$$

where the interaction components are defined as in Section 33. Nevertheless, monozygotic twins provide a powerful test for such interactions. Since the difference between a pair of monozygotic twins is solely environmental in origin, in the absence of interactions the magnitude of this difference should be independent of the genotypes of the twin pairs. Our measure of the genotypic differences between twins is the difference between family means of MZ twins raised apart. We can, therefore, test the assumption of no genotype × environment interaction by testing for the independence of the means (or sums) and differences of twin pairs where the twins have been raised apart (Jinks and Fulker, 1970). Although there is occasionally a basis for cross classifying twins with families, for example, where one of each pair of twins remains in its 'own' home and the other is fostered, in general this will not be so. Hence, in general, the sign that is allocated to the twin difference will be arbitrary and we will take its absolute value and ignore sign.

The means and differences for twin pairs may be examined for evidence of non-independence by plotting one against the other for all twin pairs. Non-independence would then show itself by a departure from a random scatter giving a linear or more complex relationship. Equally, we can detect non-independence by calculating the correlation between sums and differences over all pairs or by fitting linear or curvilinear regressions.

The 40 pairs of MZA scored for Neuroticism give a correlation of $r = 0.06$ for 38 df and scatter diagrams confirm the absence of any relationship between mean and difference for these data. There is, therefore, no evidence of genotype × environment interaction.

In the absence of monozygotic twins raised apart MZTs provide a similar but less satisfactory test because the differences include only within-family environmental effects (E_w) and the means include the common environmental effects that arise from sharing the same family environment (E_b). Nevertheless, if the scatter diagram and any statistical analysis show no relationship we can still conclude that genotype × environment interactions are absent but only, of course, in respect of the within family environment. An ambiguity in interpretation, however, arises when the tests reveal a relationship because, no matter how unlikely this may be, it could arise not from genotype × environment interaction but because of the non-independence of the within- and between-family environmental components. For the Neuroticism data the MZTs confirm the absence of interactions ($r = 0.15$ for 27 df).

For assortative mating an additional heritable component $(A/1 - A)D_R$ appears in the expectations for the total variance but always with the same coefficient as D_R. The total variances of MZT, MZA, DZT and DZA, therefore, all become

$$\tfrac{1}{2}D_R + \tfrac{1}{2}\left(\frac{A}{1-A}\right)D_R + \tfrac{1}{4}H_R + E_w + E_b$$

hence its presence cannot be detected by the test of goodness of fit of the model. The parameter A is the proportion of the correlation between mates that is expected to be attributable to the D_R component of the total variation and it may be estimated directly as the product of this correlation and the narrow heritability (Fisher, 1918).

The total variance in a randomly mating population in the presence of non-allelic interactions has been given by Mather (1974) and its derivation is given on page 221. This is the expected total variance of MZT, MZA, DZT and DZA. The presence of non-allelic interactions cannot, therefore, be detected by the failure of a model which excludes them and unlike genotype × environment interactions and assortative mating there is no alternative direct approach to detecting their presence using twins.

Of the four assumptions we recognized at the outset, the failure of only one, no genotype–environment correlation, can lead to inequality of the total variances and hence should be detectable by the test of goodness of fit of the D_R, H_R, E_w and E_b model (Jinks and Fulker, 1970). Three ways in which such a correlation can arise have, however, been recognized with quite different consequences (Eaves et al., 1977). The first of these is where a genotype selects its own environment. If superior genotypes seek out or create for themselves advantageous environments this will generate a genotype–environment correlation that will be inseparable from the genotypic effects because the components which define their contributions will always appear together in the expectations with the same coefficients. This, however, need not concern us since the genotypes role in choosing or creating its environment makes the latter an extension of the phenotype.

A second source of correlation is sibling effects (Eaves, 1976a). This is where competition or co-operation between siblings makes every sibling an important component of the environment of the other siblings. One might expect sibling effects to be more acute for twins than for pairs of non-twin siblings born a few years apart. Any appropriate model leads to the expectation that the contribution from this source, which can be negative for competition and positive for co-operation, will differ between MZ and DZ twins and between twins raised together and apart. They can, therefore, be detected as differences in the total variances.

The final source of genotype–environment correlation is cultural transmission (Eaves, 1976b). This arises when the environment shared by members of the same family depends on the phenotype of their parents. It produces a correlation between the genetical and environmental difference between families for twins raised by their natural parents. Where, however, they are raised in 'foster' homes this correlation will be reduced. Indeed, if fostering is random, an assumption we will return to later, this correlation should be reduced to zero. It can therefore, be detected by comparing the total variances of twins raised together and apart (Jinks and Fulker, 1970).

Repeatedly in this discussion we have seen the importance for our analyses of twins raised apart. The extent to which they allow us to achieve our objectives, however, rests on the validity of the assumption that each twin of a pair is distributed at random among the family environments present in the population. We can test whether 'foster' homes are a random sample of family environments by comparing their mean and variance for any particular measure with a random sample of 'own' homes. The measure could be physical or an index such as socio-economic class. On the other hand, a biological measure could be used of the kind described in Section 19. This might, for example, be the mean phenotype of the parents, natural or foster, who provide the home environment, for the character of interest.

These analyses can tell us whether foster homes are a random sample but not whether separated twins are allocated to them at random, i.e. whether any attempts to match the adopted child and the foster home have been successful in producing a so-called 'placement' effect. In order to test for such an effect we would have to look for a correlation between the family environment discussed earlier. Only if the correlations were non-significant could we confidently conclude that the separated twins provide a valid estimate of the total environmental effects and that they could fulfill the role allocated to them in the analyses we have described.

Although twins raised apart or the equivalent of DZA, full-sibs reared apart, are essential for the analyses described so far they are rarely available and most data typically consist of monozygotic and dizygotic twins raised together.

Loesch's observations on total finger ridge counts on 60 male and 50 female MZT and 62 male and 49 female DZT are an example (Loesch, 1979). The variances for females and males separately are as follows:

		Females	Males	Model
MZT	V_F	2412	1734	$\frac{1}{2}D_R + \frac{1}{2}H_R + E_b + \frac{1}{2}E_w$
	$\overline{V}_F$	84	113	E_w
DZT	V_F	1532	1433	$\frac{3}{8}D_R + \frac{5}{32}H_R + E_b + \frac{1}{2}E_w$
	$\overline{V}_F$	1533	1139	$\frac{1}{4}D_R + \frac{3}{16}H_R + E_w$

The mean number of finger ridges is higher for males than for females and the total variance for MZ twins is higher for females than for males ($P = 0.03$). Sexes, therefore, differ and they must either be analysed separately or sex differences must be allowed for in our models. The total variances of MZ and DZ twins do not differ significantly for females ($P = 0.70 - 0.60$) or males ($P = 0.40 - 0.30$). There is, therefore, no evidence of sibling effects but other sources of genotype–environment correlations and genotype × environment interactions cannot be ruled out.

With MZTs the relationship between twin difference and twin mean only tests for interactions involving E_w. As there is no significant relationship in either sex these interactions are not important. To proceed further with the analysis we must assume that interactions with E_b and cultural transmission are absent.

No solution to the D_R, H_R, E_w and E_b model can, however, be obtained. We have the choice of three perfect fit solutions. We may assume either that E_b or H_R are making no significant contribution to the variation or, alternatively, estimate E_w and the two compound parameters ($\frac{1}{4}D_R + \frac{3}{16}H_R$) and ($\frac{1}{4}D_R + \frac{1}{16}H_R + E_b$). If we assume $E_b = 0$ then the perfect fit estimates of D_R and H_R that we obtain have the following biases:

$$\hat{D}_R = D_R + 6E_b$$
$$\hat{H}_R = H_R - 8E_b.$$

Similarly, if we assume $H_R = 0$

$$\hat{D}_R = D_R + \frac{3}{4}H_R$$
$$\hat{E}_b = E_b - \frac{1}{8}H_R.$$

In both cases therefore, D_R will be inflated and H_R or E_b underestimated if we assume that E_b or H_R are zero when they are not. On the other hand, by assuming $H_R = 0$, half the estimate of D_R, $\frac{1}{2}\hat{D}_R = \frac{1}{2}D_R + \frac{3}{8}H_R$ is only slightly larger than the total heritable component of variation, $\frac{1}{2}D_R + \frac{1}{4}H_R$. Hence, by making this assumption a reasonable separation of the heritable and non-heritable sources of variation can be achieved even when H_R is not equal to zero.

Where one of the two parameter models fit, for example, E_w and E_b, D_R and

E_w, D_R and E_b, etc., and all of the others fail we can be confident of the results. If, however, all of the two parameter models fit equally well no unambiguous conclusion is possible. If, on the other hand, all two parameter models fail we cannot improve on the analysis which assumes $H_R = 0$.

Martin *et al.* (1981) have fitted these models to Loesch's MZT and DZT data on finger ridge counts using the iterative, weighted least squares procedure described in Section 28. The three parameter model with $H_R = 0$ gave negative values for E_b. The latter cannot, therefore, be large hence a D_R, E_w model was tried and was found to be satisfactory for both sexes.

The estimates of D_R and E_w and the χ^2 for two degrees of freedom testing the goodness of fit of the model were as follows:

	Females	Males
D_R	4812·35	3692·26
E_w	84·78	114·84
$\chi^2_{(2)}$	1·59	0·54

The estimates are highly significant ($P < 0.001$) and the χ^2s are non-significant. From the relative magnitudes of D_R and E_w it is clear that finger ridge count is highly heritable, indeed the narrow heritability, which equals the broad heritability in these data, is 0·97 for females and 0·94 for males.

A most interesting development in the use of twins has been proposed by Nance and Corey (1976). The progeny of MZ twins are genetically half-sibling families raised apart and the covariance of half-siblings provides a direct estimate of $\frac{1}{8}D_R$ (Section 34). More importantly, however, because the monozygotic twins may be the maternal or the paternal parent and their progenies may be divided into males and females the families of MZ twins provide some of the most sensitive tests for maternal effects and sex linkage (Sections 50, 52 and Haley *et al.*, 1981). These data, however, like twin data are at their most powerful when supplementing the commoner types of relationships found in natural populations such as parent–offspring, full siblings, cousins etc. and because they are more powerful the more complex sources of variation that we have briefly referred to such as assortative mating and genotype–environment correlations, become amenable to analysis (Eaves *et al.*, 1977, 1978).

38. EXPERIMENTAL POPULATIONS

With an experimental population steps can be taken to ensure that the assumption of random mating and assumptions about the distributions of non-heritable sources of variation are satisfied. Within limits imposed by the nature of the material we can then choose between a number of alternative breeding programmes that will guarantee random mating and between a number of experimental designs that will ensure that individuals, parts of families or whole families are randomized as single units over the environment with predictable consequences for the non-heritable components of the variances and covariances.

There are four basic breeding programmes that will ensure random mating, these are:

(i) Randomly mated biparental progenies (BIPs)
(ii) North Carolina design 1
(iii) North Carolina design 2
(iv) Diallels.

(i) The first design is the simplest. With hermaphroditic plants this involves using plants as male or female parents and crossing them in pairs chosen at random to yield $\frac{1}{2}n$ progeny families. Where, as in most animals and a few plants there are two sexes, equal numbers of males and females must be included in the sample and each male must be mated with a single randomly chosen female.

In species where a single mating regularly leads to more than one offspring per family the total variation among the offspring from all matings may be partitioned into two parts, the variation within families and the variation between families, respectively. If we approach the analysis through the analysis of variance we have

Item	df	Expected MS	
Between families	$\frac{1}{2}n - 1$	$\sigma_w^2 + m\sigma_b^2$	$= mV_F$
Within families	$\frac{1}{2}n(m-1)$	σ_w^2	$= V_F$

where $m = $ family size; $\sigma_w^2 = \frac{1}{4}D_R + \frac{3}{16}H_R + E_w (V_{2SR})$; $\sigma_b^2 = \frac{1}{4}D_R + \frac{1}{16}H_R + E_b$ = covariance of full-sibs ($V_{1SR} = W_{FSR}$).

Where births are regularly single this partitioning is possible only if the same matings are repeated at least once.

If this mating system is repeated in successive generations we merely obtain successive estimates of σ_w^2 and σ_b^2 with the same expectations in terms of the model. With this design, therefore, we cannot estimate all the parameters in the simple model without further assumptions, controls or additional statistics.

The assumptions would have to be: no dominance ($H_R = 0$) and no common family environment ($E_b = 0$), whereupon

$$D_R = 4\sigma_b^2$$

and

$$E_w = \sigma_w^2 - \sigma_b^2.$$

If H_R and E_b did not equal zero these would estimate

$$D_R + \frac{1}{4}H_R + 4E_b$$

and

$$E_w + \frac{1}{8}H_R - E_b$$

respectively, i.e. we would overestimate the genetic relative to the environmental component of variation.

Two kinds of controls could assist the estimation of the parameters. First, where the material is amenable single individuals can be randomized over the whole environment irrespective of family relationships. E_b then becomes zero and we need only assume $H_R = 0$ to obtain a solution for D_R and E_w. Second, individual families may be split at random and raised independently in different replications. This allows us to introduce into the analysis of variance an item for replicates or replicate × between families. Either will provide an estimate of E_b. We can then estimate D_R, E_w and E_b by assuming $H_R = 0$.

Additional statistics can be obtained in two ways: first, from the parent–offspring covariance W_{POR}; and second from the selfed progenies of the parents, the so-called augmented BIP design (Kearsey, 1970). The former could be obtained for any material that is randomly mated but its usefulness depends on a number of factors. In plants, independent randomization of the individuals in successive generations will guarantee that the parent–offspring covariance will contain no component due to a common family environment. However, since successive generations are raised in different environments this covariance will be biased if these differences lead to genotype × environment interactions between generations. In animals, the effect of any common environment between parents and offspring may be minimized by separating the two as soon as it is practicable, but it cannot be wholly eliminated, particularly in mammals. Thus we must either assume that any residual effect of the common environment is unimportant or introduce a new environmental parameter into the model to allow for the common element.

An alternative procedure which is available for self-fertile plants is to grow the selfed progeny of each parent simultaneously with, and randomized among, the progenies produced by random mating of the parents. These selfed progenies must, of course, be eliminated from any mating system devised to maintain the randomly breeding population. Their inclusion in the design provides, however, three further statistics, the variation within and between families produced by selfing and the covariance of parents selfed on offspring produced by random mating. Unfortunately, the expectations of these three statistics cannot be given in terms of the general form of D_R and H_R but require a special form which introduces new parameters into the model (see Section 41). Thus this apparently simple method does not overcome the difficulty. Nevertheless, appropriately analysed, the comparison of outcrossed and inbred family means provides a sensitive test for non-additive, heritable variation (Kearsey, 1970).

We may illustrate the analysis by reference to the data of Kearsey (1965). This consists of the flowering-times of biparental progenies from a natural population of *Papaver dubium* which had been maintained in the experimental field for one year by random mating. 43 families were grown as single randomized plants in each of two replicate blocks at two different sowing times in the same season. Each family was represented by five plants in each block in each sowing. The observed mean squares for each item in the analysis of variance are given for the two sowing times in Table 89.

TABLE 89
The analysis of variance of the biparental progenies from a population of *Papaver dubium*

Item	Sowing 1			Sowing 2		
	df	MS	P	df	MS	P
Between families	42	157·13	<0·001	42	66·92	<0·001
Blocks	1	15·05	NS	1	14·52	NS
Blocks × families	42	50·60	NS	42	18·44	NS
Within families	329*	49·61		334	22·69	

* In this and subsequent analyses the df are less than expected due to the loss of a few plants.

Neither blocks nor blocks × families are significant and we can pool them with the within-family item to estimate σ_w^2 and σ_b^2. Since the plants are randomized individually $E_b = 0$, hence:

$$\sigma_b^2 = \tfrac{1}{4}D_R + \tfrac{1}{16}H_R$$
$$\sigma_w^2 = \tfrac{1}{4}D_R + \tfrac{3}{16}H_R + E_w$$

and our estimates for the two sowings are

	Sowing 1	Sowing 2
$\hat{\sigma}_b^2 =$	10·75	4·42
$\hat{\sigma}_w^2 =$	49·63	22·69

In both cases the statistic with the larger expectation (σ_w^2) is the larger. We can, of course, solve for only two parameters in each sowing, therefore, we must assume that $H_R = 0$ whereupon

	Sowing 1	Sowing 2
$\hat{D}_R =$	43·00	17·68
$\hat{E}_w =$	38·88	18·27

It has been shown above (page 235) that if H_R is not zero then D_R is inflated by $\tfrac{1}{4}H_R$ and E_w by $\tfrac{1}{8}H_R$.

In spite of the marked difference in the absolute values of the estimates in the two sowings the relative values of the two estimates are remarkably consistent: the additive variation accounting for 35 % and 33 % (narrow heritability) of the total variation in the two sowings, respectively.

Since the parents of the biparental progenies were scored for flowering-time in the previous season, we can calculate the parent–offspring covariances. These covariances between the mid-parent values and the progeny family means are 10·2748 and 6·8190 for sowings 1 and 2, respectively. Since there are no common environmental effects between parents and offspring the covariances are direct estimates of $\tfrac{1}{4} D_R$. Bringing in these covariances thus allows us to estimate all three parameters with the following result.

	Sowing 1	Sowing 2
$\hat{D}_R =$	41·08	27·28
$\hat{H}_R =$	7·68	− 38·40
$\hat{E}_w =$	37·92	23·07

Allowing H_R to take a value other than zero has had no effect on the relative values of D_R and E_w and an effect on their absolute values only in sowing 2. Furthermore, it is quite clear from the relatively small value of H_R in sowing 1 and the negative value in sowing 2 that $H_R = 0$ is a reasonable assumption. However, the large negative value of H_R in sowing 2 requires further comment. It is estimated as

$$16(\sigma_b^2 - W_{POR})$$

It is, therefore, very sensitive to small differences in the relative values of σ_b^2 and W_{POR} and hence it has a correspondingly large sampling error. Unfortunately, we have no estimate of the sampling error because the differences between the two sowings are too large and too consistent to be regarded as the result of sampling variation. It is not unlikely, however, that the large value of H_R in sowing 2 would be non-significant against its sampling error.

It should further be observed that the differences between the two sowings show flowering to be very sensitive to time of sowing. Indeed a difference of 28 days between the times of sowing has produced a two-fold difference in the magnitudes of the variances and covariances. The parents were, of course, grown in the previous season at yet another time of sowing which, however, was closer to the time of sowing 1 than of sowing 2. In view of the marked effect of time of sowing, we must, therefore, expect genotype × environment interactions to influence the magnitude of the parent–offspring covariances. Furthermore, we would expect this influence to be greater for sowing 2 than for sowing 1 because the parents were grown under conditions more comparable with the latter. Since we have not allowed for such interactions in the model this in itself could account for the large negative value of H_R.

This example underlines the weakness of the biparental design, in that, to achieve a solution on the simplest model we require an organism in which single individual randomization is possible so that $E_b = 0$, and one in which genotype × environment interactions are unlikely to have too much influence on the parent–offspring covariance. The other designs, to a lesser or greater degree, overcome these difficulties (see also Section 40).

(ii) In the biparental mating design we are limited by the fact that only two degrees of relationship exist among the progeny: they are either full-sibs or unrelated. To produce other degrees of relationships, for example half-sibs, it is necessary to produce families some of which have only one parent in common. The simplest method of random mating which achieves this leads to our second design.

A random sample of males, or in the case of hermaphrodite species, individuals designated as males, are chosen and each male is mated to a number of females, or individuals so designated, taken at random. From the analytical viewpoint it is preferable for each male to be allotted the same number of females but this is not essential. Thus each male is involved in a number of matings but each female is mated once only so that the males, but not the females, are common parents to a number of progeny families.

One can, of course, carry out the reciprocal of this design, i.e. use each female in a number of successive matings and each male once only so that the common parent is the mother rather than the father. But while this design might be useful, for example, in assessing the effect of a common maternal environment on the performance of the half-sib progenies, practical difficulties would rule it out in most cases except the trivial one in which the designation of hermaphrodite organisms as mothers or fathers is in any case an arbitrary distinction.

There are, on the other hand, practical advantages in using the males as the common parent. Thus there is usually no problem in using the same individual male repeatedly as a parent over a relatively short period although the maximum number of times that it can be used successfully will vary with the species and the length of the period over which the matings can be spread and yet the progenies still be regarded as being raised simultaneously in the same environment. Another advantage is that it is common practice, particularly with larger mammals, to maintain populations in which considerably fewer males are allowed to reach sexual maturity than females. In such populations the relative frequencies of the two sexes at maturity will automatically lead to each male being the common parent in many matings.

If we again approach the analysis through the analysis of variance we have the same two main items as before, namely, between families and within families, but we can now partition the former into two parts by grouping families with a common parent. We can then distinguish the variation between groups, that is between common parents, from that between non-common parents within the groups. The analysis for a situation in which c common parents are each crossed to n non-common parents to give nc progenies, each of size m is as follows:

Item	df	Expected MS
Between common parents	$c-1$	$\sigma_w^2 + m\sigma_{nc}^2 + nm\sigma_c^2$
Within common parents (between non-common parents)	$c(n-1)$	$\sigma_w^2 + m\sigma_{nc}^2$
Within families	$cn(m-1)$	σ_w^2

where
$$\sigma_w^2 = \tfrac{1}{4}D_R + \tfrac{3}{16}H_R + E_w$$
$$\sigma_c^2 = \tfrac{1}{8}D_R = \text{covariance of half-sibs} \left.\begin{array}{c} \\ \\ \end{array}\right\} = \text{covariance of full-sibs.}$$
$$\sigma_{nc}^2 = \tfrac{1}{8}D_R + \tfrac{1}{16}H_R + E_b$$

The advantages of this design are now apparent. The analysis of variance provides a test of significance for D_R as $\tfrac{1}{8}nmD_R$ and the variance components allow us to estimate D_R, H_R and E_w when E_b is zero as

$$D_R = 8\sigma_c^2$$
$$H_R = 16(\sigma_{nc}^2 - \sigma_c^2)$$

and
$$E_w = \sigma_w^2 - 2\sigma_c^2 - \tfrac{3}{16}(\sigma_{nc}^2 - \sigma_c^2)$$

and D_R, E_w and E_b when H_R is zero as

$$D_R = 8\sigma_c^2$$
$$E_b = \sigma_{nc}^2 - \sigma_c^2$$
$$E_w = \sigma_w^2 - 2\sigma_c^2.$$

When E_b is not zero, the first solution will lead to an inflated value for H_R and an underestimate of E_w. When H_R is not zero, the second solution will lead to inflated values for both E_w and E_b. We can, therefore, achieve a satisfactory solution with organisms in which single individual randomization of the progenies is practicable. We can also achieve a satisfactory solution in one other circumstance, namely where an estimate of E_b can be obtained by splitting each progeny between two or more replicate sites within the total environment available. The other possibility, namely, where an estimate of the parent–offspring covariance can be obtained free of common-environmental effects and genotype × environment interactions does not help in this instance, except in so far as it provides a further estimate of D_R. It does not allow us to estimate all four parameters even though it provides a fourth statistic. On the other hand, under circumstances where common environmental effects are expected between parent and offspring or where genotype × environment interactions are expected between the different generations these can be detected by comparing the estimate of D_R obtained from the parent–offspring covariance with that obtained from the half-sib covariance.

We may illustrate the analysis by referring once more to the data of Kearsey (1965). These consist of 44 progeny families produced by crossing each of 4 male parents chosen from the experimental population of *Papaver dubium* to an independent random sample of 11 female parents. Each progeny was represented by 5 plants in each of 2 blocks in each of the 2 sowings. They were scored for flowering-time. The appropriate analysis of variance is summarized in Table 90.

Since the plants were individually randomized within the blocks, the within family mean square is used to test the significance of the block interactions. Only one of these interactions, namely, that within the progeny means of

TABLE 90
Analysis of variance of design (ii) for *Papaver dubium*

Item	Sowing 1			Sowing 2		
	df	MS	P	df	MS	P
Between fathers	3	761·43	<0.01	3	214·09	<0·05
Between mothers (within fathers)	40	138·63	<0·001	40	69·95	<0·001
Block × fathers	3	41·74	NS	3	28·02	NS
Blocks × mothers	40	33·46	NS	40	28·16	<0·02
Within families	336	36·42		336	18·76	

mothers crossed to the same father in the second sowing is significant, although this interaction is clearly no larger than the non-significant interaction with the progeny means of fathers in the same sowing. Block interactions can arise only if there are genotype × environment interactions resulting from the environmental differences between blocks. They are, therefore, comparable with, though smaller in magnitude than, the differences between the two sowings.

In the absence of block interactions, as in sowing 1, the mean square between mothers is tested against the mean square within families. Where the block interaction is significant, as in sowing 2, the mean square between mothers is tested against its block interaction. Where the mean square between mothers is significant, as it is in both sowings, it is the appropriate mean square for testing the mean square between fathers. In both sowings the latter mean square is significant, hence D_R is significantly different from zero (page 239). We may now estimate the significant variance components in the manner described earlier. In view of the significance of one of the block interactions in the second sowing we shall not pool these interactions with the mean square within families in sowing 2 to estimate the mean within family variance.

The estimates of the variance components for the two sowings and their expectations in terms of the D_R, H_R and E_w model are as follows:

		Sowing 1	Sowing 2	Expectations
$\hat{\sigma}_c^2$ (fathers)	=	5·66	1·31	$\frac{1}{8}D_R$
$\hat{\sigma}_{nc}^2$ (mothers)	=	10·39	4·18	$\frac{1}{8}D_R + \frac{1}{16}H_R$
$\hat{\sigma}_w^2$	=	36·42	18·76	$\frac{1}{4}D_R + \frac{3}{16}H_R + E_w$

from which it follows that

$\hat{D}_R$	=	45·28	10·48
$\hat{H}_R$	=	75·68	45·92
$\hat{E}_w$	=	10·91	7·53

The estimates differ between the sowings in the same direction and to the same extent as those obtained from the biparental progenies. But while the estimates of D_R are in good agreement with those obtained from the biparental progenies the estimates of H_R and E_w appear to be quite different. We will, however, defer discussing this further until we have examined the outcome of the third method of ensuring random mating.

Although the inclusion of parent–offspring covariances does not allow us to fit a model including more than the three parameters D_R, H_R and E_w, it increases the precision with which these parameters can be estimated. Using a least squares solution on the four statistics we obtain the following estimates.

	Sowing 1	Sowing 2
$\hat{D}_R$	41·92	23·92
$\hat{H}_R$	82·42	17·30
$\hat{E}_w$	10·49	10·86

A typical example of the problems encountered in applying this mating design to animal species is provided by the data of Dawson (1965) on random mating in the flour beetles, *Tribolium castaneum* and *T. confusum*. In each species 30 males were chosen at random and mated to three different females each, 10 eggs being collected from each of the 90 matings. Two males which failed to produce 5 or more progeny from 2 or more mates were excluded from the analysis.

Since the members of the same family, i.e. full-sibs, were raised in a common environment, we cannot justifiably assume that $E_b = 0$. Furthermore, since the crosses were not replicated we have no internal estimate of E_b. The character recorded was time required for development, measured as the number of days taken to reach pupation, which is known to be maternally determined in part. The analysis of variance gave the following results:

| | | *T. castaneum* | | *T. confusum* |
Item	df	MS	df	MS
Between fathers	29	5·9491	27	3·2900
Between mothers (within fathers)	56	3·9247	49	2·6045
Within families	692	1·3140	607	0·7482

Because of the variable number of females (2 or 3) mated to each male and number of individuals in the progenies (5 to 10) it is necessary to compute the coefficients of the variance components by the method of Snedecor (1956). This allows a test of significance. The resulting estimates are

$$\hat{\sigma}_c^2 = \qquad 0·0755 \qquad 0·0250$$
$$\hat{\sigma}_{nc}^2 = \qquad 0·2907 \qquad 0·2121$$
$$\hat{\sigma}_w^2 = \qquad 1·3140 \qquad 0·7482$$

which lead to the following solution

$$\hat{D}_R \qquad 0·6040 \qquad 0·2000$$
$$\hat{H}_R \qquad 3·4432 \qquad 2·9936$$
$$\hat{E}_w \qquad 0·5174 \qquad 0·1369$$

The most striking feature of these estimates is the large value of H_R relative to those of D_R and E_w. However, in fitting the model we have made no allowance for either common environmental effects, E_b, or maternal effects, M. Both appear in the expectation for σ_{nc}^2 but not in the other expectations (see Sections 51 and 52). When present they introduce the following biases in the estimates of D_R, H_R and E_w

$$D_R \text{ estimates } D_R \text{ only}$$
$$H_R \text{ estimates } H_R + 16(E_b + M)$$
$$E_w \text{ estimates } E_w - 3(E_b + M).$$

Thus we expect the estimate of D_R to be unbiased even when maternal and common family environmental effects are present but the estimate of H_R is seriously inflated and that of E_w deflated by these effects. Within the context of this experiment we cannot estimate the magnitude of these biases. To do so requires a random mating experiment in which half-sib families have a common mother as well as a common father, that is, random mating carried out by design 3.

The second design for random mating is the most complex that is practicable for a majority of animals. Given an adequate experimental design which either ensures that E_b is zero or provides an estimate of E_b this crossing programme is sufficient for fitting an additive-dominance model for the heritable variation irrespective of whether the parent–offspring covariance is available or not.

39. EXPERIMENTAL POPULATIONS: NORTH CAROLINA DESIGN 2 AND DIALLELS

(iii) The third design extends further the restraints on the choice of parents for random mating so that every progeny family has half-sib relationships both through a common mother and through a common father. This is most readily achieved by a systematic crossing programme in which n_1 males and n_2 females are chosen at random from the population and each male is crossed to each of the females in turn to yield $n_1 \times n_2$ progenies. In the case of hermaphrodite plants, the two samples are arbitrarily designated male and female parents although in such circumstances every cross may be carried out reciprocally if we wish to test for differences between reciprocal crosses.

The practical limitation of this design is the number of times that the same female parent can be successfully mated since this determines the magnitude of n_1, the number of male parents. In most animals this is likely to be small. Furthermore, successive matings would be separated in time and could lead to systematic differences between progenies due to effects of parity and age of mother as well as increasing the random effects of environmental differences because of the long period over which the matings must be spread (Broadhurst and Jinks, 1965). It is, therefore, necessary either to estimate the systematic effects in a control experiment in which the same mating is repeated a number of times using the same male and female parents, or to carry out the successive matings with each female with the male parents arranged in a random order. In the latter case the systematic effects of parity and age would be randomized among the effects of different male parents on the progenies of the same female parent and would be confounded with the other random environmental effects. They would hence contribute to the value of E_b.

With hermaphrodite plants these problems do not arise to the same extent. Different flowers on the same plant can usually be pollinated by different male parents during a short period in the same season. The only limitation is imposed by the number of flowers per plant and the number of seeds produced by each pollinated flower. Only rarely are one or both of these so small as to impose serious limitations on the size of n_1, the number of male parents.

The total variation among the individuals belonging to the $n_1 n_2$ families can be partitioned into four items. The first subdivision is into the variation between families and the variation within full-sib families which is common to all analyses of random mating designs. The variation between families can then be partitioned into that due to differences between fathers, that due to differences between mothers and the remainder which is the interaction between mothers and fathers. For families of constant size m, the analysis of variance takes the form

Item	df	Expected MS
Between fathers	$n_1 - 1$	$\sigma_w^2 + m\sigma_{fm}^2 + n_2 m\,\sigma_f^2$
Between mothers	$n_2 - 1$	$\sigma_w^2 + m\sigma_{fm}^2 + n_1 m\,\sigma_m^2$
Fathers × mothers	$(n_1 - 1)(n_2 - 1)$	$\sigma_w^2 + m\sigma_{fm}^2$
Within families	$n_1 n_2 (m - 1)$	σ_w^2

where

$$\left.\begin{aligned}\sigma_f^2 &= \tfrac{1}{8}D_R\\\sigma_m^2 &= \tfrac{1}{8}D_R\end{aligned}\right\} \begin{array}{l}\text{equals half-sib}\\ \text{covariance}\end{array} \left.\begin{array}{l}\\ \\ \end{array}\right\} \begin{array}{l}\text{equals full-sib}\\ \text{covariance}\end{array}$$

$$\sigma_{fm}^2 = \tfrac{1}{16}H_R + E_b$$
$$\sigma_w^2 = \tfrac{1}{4}D_R + \tfrac{3}{16}H_R + E_w$$

The mean squares between mothers and between fathers provide independent tests of the significance of D_R and independent estimates of D_R Where single individual randomization has been practised so that $E_b = 0$ the interaction mean square provides a test of the significance of H_R and an independent estimate of this parameter. Where the experimental design does not ensure that $E_b = 0$ we can test the significance of and estimate H_R only if we have an estimate of the variation between replicated families.

Although design three provides four statistics two, σ_f^2 and σ_m^2, are identical on the simple additive-dominance model. On the other hand, they offer a test for differences between reciprocal crosses arising from either maternal or paternal effects. With animals this provides the only method of detecting differences between reciprocal crosses in a randomly mating population. With hermaphrodite plants we always have the alternative possibility of making every cross reciprocally to detect such differences whichever mating design is used (Section 52).

We may illustrate the analysis in its simplest form by reference once more to the population of poppies investigated by Kearsey (1965). The relevant experiment consisted of all possible crosses among seven plants designated as male parents and seven designated as female parents making 49 crosses in all. Five progeny were grown from each cross as single randomized plants in each of 2 blocks in each of 2 sowings. The analysis of variance of the flowering-time is as follows.

Item	df	Sowing 1 MS	Sowing 2 MS
Between fathers	6	404·13	233·03
Between mothers	6	413·35	103·53
Mothers × fathers	36	146·53	49·46
Blocks × fathers	6	11·44	24·41
Blocks × mothers	6	51·36	17.50
Blocks × father × mothers	36	42·24	21·74
Within families	377	48·47	23·13

Since the block interactions are all non-significant when tested against the within-family mean square, they may be pooled to provide the estimate of the within-family component of variation. The interaction between fathers and mothers is tested against the mean square within families and it is significant in both sowings. There is, therefore, a significant H_R component of variation. Since the interaction is significant it must be used to test the significance of the father's and mother's mean squares. In the first sowing both of these mean squares are significant and equal in magnitude. In the second sowing only the mean square attributable to the fathers is individually significant, but there is no significant difference between the father's and mother's mean squares. Thus there is a significant D_R component in both sowings and no evidence of significant differences between reciprocal crosses.

The estimates of the components of variation are

	Sowing 1	Sowing 2
$\hat{\sigma}_f^2 =$	3·68	2·62
$\hat{\sigma}_m^2 =$	3·81	0·77
$\hat{\sigma}_{fm}^2 =$	9·92	2·65
$\hat{\sigma}_w^2 =$	47·38	22·96

which leads to the following estimates

$\hat{D}_R =$	29·96	13·56
$\hat{H}_R =$	158·72	42·40
$\hat{E}_w =$	10·13	11·62

Although the inclusion of the parent–offspring covariance does not allow us to fit a more complex model in this instance, it again increases the precision with which we can estimate the parameters. The least squares estimates obtained by including the covariance are

$\hat{D}_R$	37·37	22·71
$\hat{H}_R$	158·72	42·40
$\hat{E}_w$	8·28	9·33

Both sets of estimates suggest that $\hat{H}_R$ is large relative to $\hat{D}_R$, thus supporting the conclusions from design (ii).

The particular advantages of design (iii) over design (ii), however, are best illustrated by reference to the experiments on random mating in *Tribolium* species (Dawson, 1965). It is possible to mate any one female with two different males because if one male is replaced by a second, after a few days, all the eggs laid by the female are fertilized by the second male. Even so this limits the number of male parents, n_1, to two. To overcome this limitation, independent sets of 2×2 matings were set up, the pairs of males and females in any one set being chosen at random. Rate of development was recorded for the progenies of 43 such 2×2 sets in *T. castaneum* and of 28 sets in *T. confusum*.

Within any 2×2 set we can carry out the following analysis of variance for constant family size, m.

	df
Between fathers	1
Between mothers	1
Fathers × mothers	1
Within families	$4(m-1)$

By summing the corresponding items over the independent sets we arrive at the following analyses for the two species.

Item	*T. castaneum* df	MS	*T. confusum* df	MS
Between fathers	43	6·5821	28	1·0474
Between mothers	43	9·2764	28	1·7125
Fathers × mothers	43	5·3030	28	0·8333
Within families	1217	1·8599	832	0·4372

In the investigation of the same populations using design (ii) (page 242) we saw that the estimates of H_R and E_w were biased by maternal and common family environmental effects. Although the present design (iii) experiment as carried out by Dawson does not provide an estimate of E_b it allows us to estimate M as the difference between the half-sib covariance based on common fathers (σ_f^2) and the same covariance based on common mothers (σ_m^2). The estimates of these and of the other components of variation are as follows.

	T. castaneum	*T. confusum*
$\hat{\sigma}_f^2 =$	0·0897	0·0204
$\hat{\sigma}_m^2 =$	0·2785	0·0834
$\hat{\sigma}_{fm}^2 =$	0·4566	0·0908
$\hat{\sigma}_w^2 =$	1·8599	0·4372

which leads to the following estimates of the parameters in the model

	T. castaneum	*T. confusum*
$\hat{D}_R$	0·7176	0·1632
$\hat{M}$	0·1888	0·0630
$\hat{H}_R + 16\hat{E}_b$	7·3055	1·4528
$\hat{E}_w - 3\hat{E}_b$	0·3107	0·1240

The estimates of D_R are virtually the same as those obtained from design (ii) and the estimates of E_w are very similar. Again the estimates of H_R are extremely large even though we have now allowed for the effects of M, thus suggesting that E_b rather than M may be the major source of the inflation. There is, however, one further source of inflation that is not allowed for in the model, namely, that arising from each progeny reacting differently to the maternal influence of the mother. An interaction between progeny genotype and the mother would modify the magnitude of the maternal effect from one half-sib family to another (see Section 52). Clearly where such a possibility exists design (iii) is inadequate as a method of obtaining an unbiased estimate of H_R.

Additional evidence of a maternal effect in $T.$ $castaneum$ is provided by the parent–offspring regression obtained from random biparental progenies produced from the same base population according to design (i). This regression was $2\frac{1}{2}$ times greater when estimated for offspring on mothers than for offspring on fathers. Unfortunately, there is no confirmatory evidence about the significance and magnitude of E_b. We are thus unable to estimate or to allow for this component in fitting a model to the data.

We have now considered three mating programmes that will ensure that a population is randomly breeding and illustrated their practical usefulness by applying them to the analysis of poppy and $Tribolium$ populations. This allows us to compare the three designs for their ability to detect and estimate the contributions of additive, dominance and environmental effects to the variation in the populations. We can, however, compare them from a different viewpoint, namely, the relative accuracy with which they allow us to estimate the different parameters and the correlations between these estimates. For comparative purposes we shall confine our attention to situations where a D_R, H_R and E_w model is adequate and where there are sufficient statistics to allow us to fit such a model, that is design (i) plus the parent–offspring covariance and designs (ii) and (iii) with and without the covariance. So as to compare the experimental designs on a standard basis we shall suppose that all observed statistics have unit error variance, and are based on infinitely sized samples. Table 91 contains the variance–correlation matrices of the estimates of the three parameters for each of the five experimental situations. The error variances lie on the diagonal and the error correlations off the diagonals of each matrix.

It is quite clear from Table 91 that all the experiments provide considerably less information about H_R than about D_R or E_w. It is also clear that an appropriate estimate of the parent–offspring covariance has a marked effect on the amount of information about D_R although this also improves as we go from design (i) through (ii) to design (iii). There is also a marked reduction in the correlations as we go from design (i) to (iii), although this reduction is confined to the correlations involving D_R. Indeed, the correlation between H_R and E_w remains relatively large irrespective of the experimental design.

On balance, therefore, there is an advantage in using design (iii) in that it provides more information about each of the parameters than the other designs and it also yields the lowest correlations. However, even design (iii) is

TABLE 91

Error variances (diagonal terms) and correlations (off diagonals) of the estimators of D_R, H_R and E_w in the five experimental situations on the assumption of unit error variances of the observed statistics

Design (i) Biparental progenies plus parent–offspring covariance

	D_R	H_R	E_w
D_R	16·00	−0·71	0·53
H_R		512·00	−0·95
E_w			14·00

Design (ii) Common fathers only (or common mothers only)

	Without po covariance			With po covariance		
	D_R	H_R	E_w	D_R	H_R	E_w
D_R	64·00	−0·71	0·30	12·80	−0·41	0·14
H_R		512·00	−0·86		307·20	−0·91
E_w			11·00			10·20

Design (iii) Common fathers and common mothers (also applicable to design (iv) diallels)

	Without po covariance			With po covariance		
	D_R	H_R	E_w	D_R	H_R	E_w
D_R	32·00	0·00	−0·25	10·66	0·00	−0·25
H_R		256·00	−0·87		256·00	−0·92
E_w			12·00			10·66

far from satisfactory as far as H_R and its correlation with E_w are concerned. To this, of course, we must add the further difficulty of separating E_b from H_R when the design fails either to ensure that $E_b = 0$ or to provide an estimate of E_b from replication of the families. While design (ii) also has the advantage over design (i) it is worth noting that design (i) including the parent–offspring covariance, is as good as, if not better than, design (ii) without the covariance.

If we are mainly concerned with partitioning the total variation within a randomly mating population among its principal or significant causes rather than in estimating the relative sizes of D_R, H_R, E_w, E_b, M, etc., the relative amounts of information about these contributors is somewhat different from that in Table 91. For example, the contribution of additive, dominance and general environmental sources to the variation in the population are $\frac{1}{2}D_R$, $\frac{1}{4}H_R$ and E_w, respectively. Hence the relative amounts of information will be 4, 16 and 1 times the reciprocal of the diagonal terms in the matrices for each of the designs for $\frac{1}{2}D_R$, $\frac{1}{4}H_R$ and E_w, respectively. This considerably increases the information for the additive and dominance contributions to the total variation relative to that for the environmental contributions, without, of course, affecting the correlations. However, it still leaves the dominance contribution less reliably estimated than the other contributions.

Although the results in Table 91 serve to illustrate the relative advantages and disadvantages of the alternative, random mating, designs, they are artificial in that they are based on the assumption of infinite numbers of families each of infinite size. To illustrate the consequences of relaxing this assumption the error variances and correlations have been recalculated for the two extreme designs for the special case of families of size 5 and for samples of 10 male and 10 female parents (Table 92). Imposing these small sample sizes on design (i) has minimal effect on the variances and correlations; on the other hand, they reduce the advantage of design (iii) over all other designs in that the correlations between D_R and E_w are increased and the estimates of D_R and H_R are no longer independent. Sample size, however, has more important consequences on the efficiencies of these designs which will be discussed in Section 61.

TABLE 92
Error variances (diagonal terms) and correlations (off diagonals) of the estimators of D_R, H_R and E_w for design (i) and design (iii) (see Table 91) for families of size 5 and samples of 10 male and 10 female parents

Design (i) Biparental progenies plus parent–offspring covariance

	D_R	H_R	E_w
D_R	*16·0*	−0·70	0·51
H_R		*522·30*	−0·94
E_w			*15·56*

Design (iii) Common fathers and common mothers without parent–offspring covariance

	D_R	H_R	E_w
D_R	*32·25*	0·08	−0·45
H_R		*269·82*	−0·89
E_w			*14·57*

If we now reconsider the estimates of the parameters obtained by applying the analyses appropriate for the designs (i), (ii) and (iii) to the poppy and *Tribolium* populations, we find, as expected, that the estimates of D_R are much more consistent from one design to another than the estimates of the other parameters. The estimates of H_R, on the other hand, vary from being almost zero to being considerably larger than D_R (see Tables on pp. 237–246). Furthermore, the changes in H_R are inversely related to changes in the estimates of E_w as would be expected from the high correlation between them. Thus while the analyses leave little doubt that there is dominance variation, we can place little reliance on the estimates of its magnitude even from the more sophisticated experimental designs.

Table 93 summarizes the estimates of the three components D_R, H_R and E_w and the total variation which equals $\frac{1}{2}D_R + \frac{1}{4}H_R + E_w$ for the poppy population from the two designs, NC 1 and NC 2, which allow their separation. There are differences between the estimates from the two designs that we have already noted but these are less marked and less consistent than those between the estimates from the two sowings which are brought out particularly clearly by averaging them over the designs. The total variation is over twice as high after sowing 1 than after sowing 2, but while both of its genetical components

TABLE 93
Variation in flowering-time of a population of poppies (Kearsey, 1965)

		Experiment			
	Sowing	NC1	NC2	Mean	Raio 1/2
D_R	1	45	30	37·5	3·13
	2	10	14	12.0	
H_R	1	76	159	117·5	2·67
	2	46	42	44·0	
E_w	1	11	10	10·5	1·05
	2	8	12	10·0	
V	1	89	65	77·0	2·13
	2	43	29	36·0	

are about three times as high after sowing 1 the environmental component hardly changes between sowings. The expression of the genes mediating the variation in flowering-time are changing markedly with time of sowing but the microenvironmental variation within sowings is uninfluenced. The difference in the variation between environments is thus unlikely to be one that can be scaled out by transforming the metric on which the character has been measured.

(iv)   The fourth design, a diallel set of crosses, is limited in its application to hermaphrodite species. This design places the ultimate restraint on the choice of parents in that a random sample of n individuals is chosen and each individual is crossed to every other in the sample. If every individual is crossed to every other including itself, both as male and as female parents, it leads to a full n^2 diallel set of crosses. If the selfs are omitted, this is equivalent to design (iii) in which the n_1 male parents and the n_2 female parents are identical and indeed it may be analysed in the same way as data from a design (iii) experiment. However, we have the additional expectation that the two halves of the diallel table of results will be mirror images in the absence of any effects leading to differences between reciprocal crosses. Hence alternative analyses are possible which recognize this expectation. An additional consequence of this expectation is that where differences between reciprocal crosses seem unlikely or are known to be absent it is necessary to carry out only the $\frac{1}{2}n(n-1)$ crosses required to complete half of the diallel table by making only one of the two possible crosses between any pair of individuals. The practical limitation of this design among hermaphrodite species is, as in design (iii), the number of crosses that can be made onto a single plant used as the female parent.

Many alternative analyses of variance are available for diallel sets of crosses which are more appropriate than that used for design (iii). These are discussed for the special case of diallel sets of crosses among a fixed sample of inbred lines

(Section 43). The many advantages which this design holds when applied to inbred lines do not extend to diallel sets of crosses among random samples of individuals from a randomly mating population.

Nevertheless, the Hayman (1954a) analysis of variance (Section 43) of a complete diallel set of crosses including selfs is probably the most sensitive means available of detecting non-additive variation and maternal sources of reciprocal differences and dominance, if the model is adequate, in a randomly mating population; although, of course, the selfs which constitute an essential part of the diallel set of crosses for this analysis are not a part of the randomly mating population produced.

40. THE TRIPLE TEST CROSS

The most efficient analysis of randomly mating populations is possible where two contrasting inbred lines are available using the extension of the design (3) experiment of Comstock and Robinson (1952) devised by Kearsey and Jinks (1968) and described in Sections 27 and 32. In this design each individual (i) in the population sample is crossed to both inbred lines and their F_1 to produce three progeny families L_{1i}, L_{2i} and L_{3i} which are raised in a replicated experiment. The contribution of a single gene difference to the progeny family means is then as shown in Table 94.

TABLE 94
Backcrosses of a population to contrasting inbred lines and their F_1

Contrasting inbred lines and their F_1	Frequency Genotype	Population			Mean
		u_a^2 AA	$2u_a v_a$ Aa	v_a^2 aa	
$\bar{L}_1$ AA	d_a	$\frac{1}{2}d_a + \frac{1}{2}h_a$	h_a	$u_a d_a + v_a h_a$	
$\bar{L}_2$ aa	h_a	$-\frac{1}{2}d_a + \frac{1}{2}h_a$	$-d_a$	$-v_a d_a + u_a h_a$	
$\bar{L}_3$ Aa	$\frac{1}{2}d_a + \frac{1}{2}h_a$	$\frac{1}{2}h_a$	$-\frac{1}{2}d_a + \frac{1}{2}h_a$	$\frac{1}{2}(u_a - v_a)d_a + \frac{1}{2}h_a$	
$\bar{L}_{1i} + \bar{L}_{2i} + \bar{L}_{3i}$	$\frac{3}{2}(d_a + h_a)$	$\frac{3}{2}h_a$	$\frac{3}{2}(-d_a + h_a)$	$\frac{3}{2}[(u_a - v_a)d_a + h_a]$	
$\bar{L}_{1i} - \bar{L}_{2i}$	$d_a - h_a$	d_a	$d_a + h_a$	$d_a + (u_a - v_a)h_a$	
$\bar{L}_{1i} + \bar{L}_{2i} - 2\bar{L}_{3i}$	0	0	0	0	

If we carry out an analysis of variance of the three orthogonal comparisons between the family means for each individual i in the original population sample, i.e. $\bar{L}_{1i} + \bar{L}_{2i} + \bar{L}_{3i}$, $\bar{L}_{1i} - \bar{L}_{2i}$ and $\bar{L}_{1i} + \bar{L}_{2i} - 2\bar{L}_{3i}$ this automatically leads to a separation of the additive (d_a^2) and dominance (h_a^2) effects of the genes, to tests of their significance and a test for the presence of non-allelic interactions (see Sections 27 and 32 and Table 54 where the analysis is described for the special case of equal gene frequencies).

In extending the model to any number of genes a problem arises which is not encountered in the earlier application where the population was an F_2 and the testers the parents and F_1 from which the F_2 was derived. This problem is the choice of a contrasting pair of inbred lines. If there are k loci segregating in the population, the inbred lines must differ at all k loci in order that the estimate of the additive variation will have any real meaning. Thus if the two inbred lines differ at k' loci where $k' < k$ the expected variance of the sum and differences are

$$\tfrac{1}{2} \underset{k}{S}\, uvd^2 + \tfrac{1}{2} \underset{k-k'}{S}\, uvh^2 - \underset{k-k'}{S}\, uvdh$$

and $\tfrac{1}{2} \underset{k'}{S}\, uvh^2$, respectively.

Hence, the estimate of the additive component contains the dominance contributions of the loci which are segregating in the population but at which the inbred parents do not differ, while the estimate of the dominance component contains the dominance contribution of only the loci for which the parents differ. It follows, therefore, that if $k' < k$ the analysis will indicate whether dominance is present but it will not provide estimates of the magnitudes of either the additive or the dominance component of variation. To exploit the full advantages of this breeding programme and its analysis it is, therefore, necessary to ensure that k' approaches k in magnitude. In the case of randomly mating populations, this can probably be best accomplished by using two inbred lines, P_1 and P_2, selected from the population for high and low manifestation of the character.

However, even if P_1 and P_2 do not differ at all k loci, if the variance of $\bar{L}_{1i} - \bar{L}_{2i}$ is not significant, i.e. there is no dominance, the variance of $\bar{L}_{1i} + \bar{L}_{2i} + \bar{L}_{3i}$ will provide an unbiased estimate of the total additive variance, D.

A further advantage of this mating programme is that it provides a test for non-allelic interaction that is valid for the k' genes for which the inbreds P_1 and P_2 differ, irrespective of whether k' is equal to or less than k. Indeed, the procedure for detecting epistasis in Chapters 6 and 7 is directly applicable to random mating populations without modification.

41. RANDOM MATING OF A PARTIALLY INBRED POPULATION

So far we have considered the analysis of randomly mating populations and in the next chapter we will describe the results of randomly mating a collection of completely inbred lines (Chapter 9). As a preliminary, we will consider an intermediate possibility, namely, the consequences of imposing random mating on a partially inbred population.

In order to specify completely the genotypic frequencies in such a population for a single gene difference, A–a, two independent parameters are needed. Of the various methods of specifying these two parameters, we will use that in which we retain u_a and v_a, the allele frequencies, in combination with f, the coefficient of inbreeding. Thus

Genotype	Frequency
AA	$u_a^2 + u_a v_a f$
Aa	$2u_a v_a (1-f)$
aa	$v_a^2 + u_a v_a f$

The contribution of this locus to the variation in the population is then, following Dickinson and Jinks (1956), and Jinks and Broadhurst (1965)

$$2u_a v_a (1+f) d_a^2 + 2u_a v_a (1-f) h_a^2 - 4u_a^2 v_a^2 (1-f)^2 h_a^2 - 4u_a v_a (1-f)(u_a - v_a) d_a h_a$$

which for any number of independent genes equals

$$\tfrac{1}{2}(1+f)D + \tfrac{1}{2}(1-f)H_1 - \tfrac{1}{4}(1-f)^2 H_2 - \tfrac{1}{2}(1-f)F,$$

D, H_1, H_2 and F being the diallel forms of the parameters (Chapter 9).

If this population is now randomly mated using any of the mating designs described in Sections 38 and 39 the total genetic variation among the progenies will become (because f now equals zero)

$$\tfrac{1}{2}D_R + \tfrac{1}{4}H_R \quad \text{or} \quad \tfrac{1}{2}D + \tfrac{1}{2}H_1 - \tfrac{1}{4}H_2 - \tfrac{1}{2}F,$$

in the random mating and the diallel forms of the parameters, respectively.

Similarly, the parent–offspring covariance is

$$\tfrac{1}{4}(1+f)D + \tfrac{1}{4}(1-f)H_1 - \tfrac{1}{4}(1-f)H_2 - \tfrac{1}{4}F$$

which can be written only in the diallel form.

With all organisms which will produce more than one offspring from a pair of parents, use of any one of the random mating programmes enables us to partition the total variation into that between family means (rank 1) and that within families (rank 2). The contribution of a single gene difference to these two variances is:

variation between family means

$$= u_a v_a (1+f) d_a^2 + u_a v_a (1+f) h_a^2 + u_a^2 v_a^2 (3-f)(1+f) h_a^2 - 2u_a v_a (1+f) d_a h_a$$

mean variation within families

$$= u_a v_a (1-f) d_a^2 + u_a v_a (1-f) h_a^2 - u_a^2 v_a^2 (1-f)^2 h_a^2 - 2u_a v_a (1-f) d_a h_a$$

which for any number of independent genes and allowing for the environmental components of variation become

$$\tfrac{1}{4}(1+f)D_R + \tfrac{1}{16}(1+f)^2 H_R + E_b$$

or

$$\tfrac{1}{4}(1+f)D + \tfrac{1}{4}(1+f)H_1 - \tfrac{1}{16}(3-f)(1+f)H_2 - \tfrac{1}{4}(1+f)F + E_b$$

and

$$\tfrac{1}{4}(1-f)D_R + \tfrac{1}{16}(1-f)(3+f)H_R + E_w$$

or

$$\tfrac{1}{4}(1-f)D + \tfrac{1}{4}(1-f)H_1 - \tfrac{1}{16}(1-f)^2 H_2 - \tfrac{1}{4}(1-f)F + E_w,$$

respectively, in the random mating and the diallel forms of the parameters.

Unless f is known, there are six or nine parameters respectively that must be estimated from the four statistics available to describe the contribution of the genetic and environmental components to the total variation. This is because f and f^2 cannot be separated from D_R and H_R or D, H_1, H_2, and F. By carrying out the random mating in a way that generates half-sib relationships (Section 34) the variation between family means can be partitioned. Thus the variance of half-sib group means (the covariance of half-sibs) equals

$$\tfrac{1}{8}(1+f)D_R$$

or $\qquad \tfrac{1}{8}(1+f)D + \tfrac{1}{8}(1+f)H_1 - \tfrac{1}{8}(1+f)H_2 - \tfrac{1}{8}(1+f)F$

and the mean variance between families belonging to the same half-sib group is,

$$\tfrac{1}{8}(1+f)D_R + \tfrac{1}{16}(1+f)^2 H_R + E_b$$

or

$$\tfrac{1}{8}(1+f)D + \tfrac{1}{8}(1+f)H_1 - \tfrac{1}{16}(1+f)(1-f)H_2 - \tfrac{1}{8}(1+f)F + E_b$$

which together equal the variation between family means.

Where half-sib groups can be recognized on the basis of common fathers as well as common mothers we can obtain both variances of half-sib group means, one based on family groups around common fathers, the other as groups around common mothers, both of which have the same expectation. The remaining variation between family means, namely, the interaction between common mothers and common fathers then equals

$$\tfrac{1}{16}(1+f)^2 H_R + E_b \text{ or } \tfrac{1}{16}(1+f)^2 H_2 + E_b.$$

Where, as a result of an adequate experimental design (Section 38), we have an estimate of E_b this statistic provides a test for the presence of dominance variation and an estimate of $(1+f)^2 H_2$. However, even though we now have the maximum number of statistics that can be obtained from randomly mating a population we still cannot estimate the remaining parameters without an independent estimate of f (Broadhurst and Jinks, 1965).

By a further generation of random mating we can obtain estimates of the parameters for $f = 0$ given in Section 34, and it is then possible to combine the statistics to estimate either D_R, $f D_R$, H_R, $f H_R$ and $f^2 H_R$ or D, $f D$, H_1, $f H_1$, H_2, $f H_2$, $f^2 H_2$, F and $f F$. In addition to the usual comparisons that one can make between these estimates we can in addition test the hypothesis that $D_R = f D_R$ etc., and hence test for inbreeding in the original population.

Although the prospects of successfully analysing a partially inbred population in this way are perhaps remote, the present treatment has considerable theoretical value in that the important special cases in practice, namely the analysis of randomly mating populations and of populations of inbred lines, can be obtained by substituting $f = 0$ and $f = 1$ in the expectations, respectively.

9 Diallels

42. COMPONENTS OF VARIATION IN DIALLELS

In Chapter 6 we considered the generations that can be derived from an initial cross between a pair of inbred lines by selfing, sib-mating, random mating and backcrossing. To obtain sufficient statistics to carry out a biometrical genetical analysis in such cases and to determine the adequacy of the genetical model, it is necessary to collect data from a number of successive generations or from a number of different mating systems carried out simultaneously. Where a number of inbred lines are available an alternative approach is possible which consists of crossing the lines in all possible combinations including crossing each line to itself such that n lines yield n^2 progeny families. This crossing programme allows a genetic analysis to be carried out after one generation and provides tests of the adequacy of the model.

If we consider a single gene difference A–a for which the inbred lines differ, a proportion u_a of the lines will be AA and a proportion $v_a (= 1 - u_a)$ will be aa. The frequencies and means of the different kinds of progeny families produced by crossing these inbred lines in all possible combinations will then be as shown in Table 95.

There will be three kinds of progeny families with genotypes AA, Aa and aa occurring with the Hardy–Weinberg equilibrium frequencies of $u_a^2 : 2u_a v_a : v_a^2$, respectively. The contribution of this gene difference to the total variance among the family means is, therefore,

$$u_a^2 d_a^2 + 2u_a v_a h_a^2 + v_a^2 d_a^2 - [(u_a - v_a)d_a + 2u_a v_a h_a]^2$$
$$= 2u_a v_a [d_a - (u_a - v_a)h_a]^2 + 4u_a^2 v_a^2 h_a^2$$

which is identical with the contribution of a single gene difference to the total variation in a randomly mating population (see Chapter 8).

In a diallel set of crosses the total variation among progeny family means can be attributed to three causes, differences among maternal parents, differences among paternal parents and the interaction between them. These are, of course, the standard items of an analysis of variance of a square table in which differences between row totals, column totals and the interaction

TABLE 95
The contribution of a single gene difference to the means of families produced by crossing inbred lines in all possible combinations

		Maternal parents		
	Genotype Frequency Mean	AA u_a d_a	aa v_a $-d_a$	Paternal array means (row means)
Paternal parents	AA u_a d_a	AA u_a^2 d_a	Aa $u_a v_a$ h_a	$u_a d_a + v_a h_a$
	aa v_a $-d_a$	Aa $u_a v_a$ h_a	aa v_a^2 $-d_a$	$-v_a d_a + u_a h_a$
Maternal array means (column means)		$u_a d_a + v_a h_a$	$-v_a d_a + u_a h_a$	Overall mean $(u_a - v_a)d_a + 2u_a v_a h_a$

between rows and columns can be recognized. In the special case of a diallel table, however, where on a simple model reciprocal crosses have identical expectations (see Chapter 10), the variation attributable to differences among maternal and among paternal parents should be identical apart from sampling variation. Indeed, we can go further. Where the simple model applies, apart from sampling variation, the two halves of a diallel table about the leading diagonal should be mirror images and each row in the table should be identical with its corresponding column, i.e. the first row with the first column, etc.

The expected variance due to differences among maternal (or paternal) parents can be obtained as the variance of column (or row) means of Table 95 (i.e. the array means) around the overall progeny mean, i.e.

$$V_{\bar{r}} = u_a(u_a d_a + v_a h_a)^2 + v_a(-v_a d_a + u_a h_a)^2 - [(u_a - v_a)d_a + 2u_a v_a h_a]^2$$
$$= u_a v_a [d_a - (u_a - v_a)h_a]^2.$$

Since families belonging to the same row or column (array) have one parent in common, the progeny families in the same array are related as half-sibs. Hence, the expected variance due to differences between maternal (or paternal) parents can also be obtained as the covariance of half-sib family means (see also Chapter 8).

The interaction between maternal and paternal parents can be obtained as the difference between the total variation among progeny family means and the sum of the variances due to differences among maternal and among paternal (= maternal) parents. This equals,

$$V_I = 2u_a v_a [d_a - (u_a - v_a)h_a]^2 + 4u_a^2 v_a^2 h_a^2 - 2u_a v_a [d_a - (u_a - v_a)h_a]^2$$
$$= 4u_a^2 v_a^2 h_a^2.$$

Generalizing for any number of genes these variances can be written in terms of the random mating forms of D and H, namely D_R and H_R. Thus the total variation among progeny family means equals $\frac{1}{2}D_R + \frac{1}{4}H_R$, the variance due to differences among maternal, and equally among paternal parents, equals $\frac{1}{4}D_R$ and the interaction equals $\frac{1}{4}H_R$.

These expectations can also be derived as a special case of randomly mating a population which has been partially inbred, by substituting $f = 1$ (the inbreeding coefficient) into the expectations (see page 253).

By using an appropriately replicated experimental design to provide estimates of the environmental components of variation of the progeny family means, estimates of D_R and H_R can be obtained from the analysis of a diallel table and their significances tested. Analysed in this way, however, a diallel set of crosses between inbred lines has little or no prospective advantage over any other multiple mating design applied to either samples of inbred lines or samples of individuals from a randomly mating population. But the random mating forms of D and H are neither the most useful nor the most efficient components that can be estimated from a diallel set of crosses.

The most important shortcoming of D_R and H_R is that unless $u = v = \frac{1}{2}$ at each locus, D_R does not contain only additive genetic effects and H_R contains less than the whole of the dominance effects, with the consequence that the ratio of H_R to D_R is not a measure of the degree of dominance. If we expand the expression for D_R we obtain

$$D_R = S4uvd^2 + S4uvh^2 - S16u^2v^2h^2 - S8uv(u-v)dh$$
$$= D + H_1 - H_2 - F$$

where
$$D = S4uvd^2$$
$$H_1 = S4uvh^2$$
$$H_2 = S16u^2v^2h^2 = H_R$$
and
$$F = S8uv(u-v)dh$$

which are the diallel form of the components, and which we have already used in Chapter 8.

In this form there are four genetic components, one of which, D, measures only additive effects; two, H_1 and H_2, measure only dominance effects, and one of these, H_1, has the same coefficient as D so that the ratio $\sqrt{(H_1/D)}$ $= \sqrt{(Suvh^2/Suvd^2)} = h/d$ is a measure of the average degree of dominance (weighted by uv) and the ratio $H_2/4H_1 = \overline{uv}$ measures the average value of uv over all loci. This average value of uv is weighted by uvh^2 and so will be sensitive to any correlation that may exist between the gene frequencies and the magnitude of the dominance effects. Where $u = v = \frac{1}{2}$ at all loci $D = Sd^2$, $H_1 = H_2 = Sh^2$ and $F = 0$, which are the forms appropriate to a cross between two inbred lines.

In the presence of unequal gene frequencies the sign and magnitude of F can be used to determine the relative frequencies of dominant to recessive alleles in the parental population and the variation in the dominance level over loci. Thus if the increasing alleles are in overall excess, i.e. $u > v$ at the majority of loci and the increasing allele is dominant at more loci than the decreasing allele, i.e. h is positive at most loci, then F will be positive. Similarly if the decreasing alleles are in excess, i.e. $v > u$ at the majority of loci, and the decreasing allele is dominant at more loci than the increasing allele, i.e. h is negative for most loci, then F will again be positive. Hence, F will be positive whenever the dominant alleles are more frequent than the recessive alleles, irrespective of whether or not the dominant alleles are increasers or decreasers.

For a single gene difference the ratio

$$\frac{(\frac{1}{2}F)^2}{D(H_1 - H_2)} = \frac{16u^2v^2(u-v)^2d^2h^2}{16u^2v^2(u-v)^2d^2h^2} = 1$$

since $H_1 - H_2 = 4uv(u-v)^2h^2$.

This expectation will hold good for any number of genes if $(u-v)h/d$ is constant over all loci. However, if the values of $(u-v)h$ and d, taking sign into account, vary independently over loci, the ratio will have an expected value of 0. Hence, irrespective of the level of dominance and the gene frequencies, the ratio (or its square root) can be used to measure the extent to which $(u-v)h/d$ varies from one locus to another.

To estimate these four components four statistics, whose expectations can be expressed in terms of them, are required. Although the analysis of progeny family means provides three statistics, two of these have identical expectations, leaving effectively only two for estimating the four components. The variation within the families is, of course, entirely environmental because all families are either inbred parents or the F_1s resulting from crosses between two such parents.

Two further statistics can be obtained from the variation among the parental lines and the covariation between parents and progenies. Reference to Table 95 shows that the variance of the inbred parents has the expectation,

$$V_P = u_a d_a^2 + v_a d_a^2 - [(u_a - v_a)d_a]^2 = 4u_a v_a d_a^2$$

which summed over all relevant genes equals D or D_p.

The covariance between parents and progeny family means can be derived in a number of ways. The simplest is to obtain the covariance between array (row or column) means and the common parent of the array as follows.

$$\begin{aligned} W_{\bar{r}} &= u_a d_a (u_a d_a + v_a h_a) - v_a d_a (-v_a d_a + u_a h_a) \\ &\quad - (u_a - v_a)d_a[(u_a - v_a)d_a + 2u_a v_a h_a] \\ &= 2u_a v_a d_a^2 - 2u_a v_a (u_a - v_a)d_a h_a \end{aligned}$$

which on generalizing becomes $\frac{1}{2}D - \frac{1}{4}F$ or D_w.

Thus, given an experimental design which provides estimates of the environmental components of variation, there are sufficient statistics available from a diallel table to provide estimates of the four genetic components. The environmental components of the four statistics depend on the experimental design employed. In the general case where E_P is the environmental component (error variance) of the parental family means and E_F is the comparable component of the F_1 family means, the environmental components of the four statistics are:

$$V_P = D + E_P$$

$$V_{\bar{r}} = \tfrac{1}{4}D + \tfrac{1}{4}H_1 - \tfrac{1}{4}H_2 - \tfrac{1}{4}F + \frac{(n-1)}{2n^2}E_F + \frac{1}{n^2}E_P$$

$$V_I = \tfrac{1}{4}H_2 + \frac{n-1}{2n}E_F + \frac{1}{n}E_P$$

$$W_{\bar{r}} = \tfrac{1}{2}D - \tfrac{1}{4}F + \frac{1}{n}E_P.$$

The reasoning behind these expectations is as follows. The diallel table of n^2 progeny family means consists of n parental families on the leading diagonal and $n(n-1)$ F_1 families. Since the reciprocal F_1 families have identical expectations on the simple model, reciprocal families will in general be averaged before computing these statistics – hence halving the environmental component ($\tfrac{1}{2}E_F$). It follows, therefore, that the weighted mean environmental component of each entry in the diallel table is $1/n(E_P + \tfrac{1}{2}(n-1)E_F)$. This is, therefore, the environmental contribution of the interaction variance (V_I). It follows that the variance of array means ($V_{\bar{r}}$) has an environmental component which is equal to $1/n$ of this value.

In most cases the covariance will have an environmental component of $(1/n)E_P$. This is because the leading diagonal entries in the diallel table, which constitute $1/n$ of all progeny families, are frequently used both to provide an estimate of the mean of the parental family and of the progeny family produced by crossing each line with itself. Hence $1/n$ of the items contributing to the parent–offspring covariance will be the squares of these leading diagonal terms. The covariance will, therefore, have an environmental component which is $1/n$ of that of the variance of the parents.

This general situation may be modified in many ways, the commonest being:

(i) Twice as many individuals per parental family are raised so that the environmental component per family is comparable for the diagonal and off-diagonal entries in the diallel table after averaging over reciprocal F_1 families.

(ii) Differences between reciprocal F_1 families may be known or assumed to be absent so that only one of each possible pair of reciprocal crosses is raised, again giving the same family size for parents and F_1s.

(iii) Independent families of each parental line are raised in each block, one of which is used as the parental family and the other is used as the progeny family produced by crossing a parental line to itself. The effect of this is to remove the environmental component from the parent–offspring covariance.

Further simplification is possible if in any body of data E_P is found not to differ significantly from E_F.

The source of the estimate of E_P and E_F will depend on the experimental design. If the progenies are raised in plots and each plot is independently randomized, replicate plots must be raised for each progeny and parental family to provide the estimates of E_P and E_F which will be obtained from the variances of replicate plot means. If, on the other hand, individuals belonging to the same family are randomized as single individuals along with members of all other families, E_P and E_F are estimated from the variances within families by dividing the latter by the number of individuals per family.

43. ANALYSIS OF VARIANCE OF DIALLEL TABLES

The square tables of results from a replicated diallel set of crosses lend themselves to a variety of alternative analyses of variance which can be used to test the significance of some of the genetical components of variation and also the validity of some of the subsidiary assumptions underlying the simple model, such as the absence of differences between reciprocal crosses. Analyses have been described which allow for every conceivable variation in experimental design including the presence and absence of parental means and of reciprocal crosses (e.g. Griffing, 1956; Hayman, 1954a; Jones, 1965; Yates, 1947) and differing relative degrees of replication of diagonal (parental) and off-diagonal (F_1) entries in the diallel table (Jones, 1965). There is a corresponding range of alternative methods of partitioning the total variation (Griffing, 1956; Hayman, 1954b; Henderson, 1952; Jinks, 1954; Jinks and Mather, 1955) and methods of deriving the variance components from the mean squares according to whether maternal or reciprocal effects are present or not (Griffing, 1956; Wearden, 1964) and whether the parental lines are a fixed sample or a random sample of a population of inbred lines (Griffing, 1956; Hayman, 1960b; Wearden, 1964). Since all these analyses have been described and their uses illustrated in the literature, and the relative advantages and disadvantages of the alternative analyses reviewed, no attempt will be made to describe them in detail here.

The general requirements of any analysis of variance of a diallel table are that it provides appropriate tests of significance of the principal genetic components, namely additive and non-additive effects, irrespective of whether there are reciprocal differences among the progeny families, and provides a test for the presence of the latter. The most satisfactory analysis for a complete diallel set of crosses from these points of view is the analysis of Hayman (1954a). This analysis will be illustrated for the commonest situation, namely, a diallel set of crosses among a fixed sample of inbred lines. For this purpose we shall use the flowering-time data from a 9×9 diallel set of crosses among 9 inbred lines of *Nicotiana rustica* (Jinks, 1954). Each progeny family is

TABLE 96
Plot means for a 9 × 9 diallel set of crosses among nine inbred lines of *Nicotiana rustica* for the character flowering-time

				Blocks I and II summed female parent						
	1	2	3	4	5	6	7	8	9	yr.
Male parent 1	77·8	53·4	79·6	69·6	50·2	59·6	71·4	67·6	50·6	579·8
2	47·8	54·1	50·0	46·2	43·0	52·4	46·8	41·2	40·4	421·9
3	68·8	53·2	97·6	59·1	50·0	63·0	72·2	48·8	52·0	564·7
4	72·2	47·0	62·4	68·2	46·8	58·7	54·4	44·6	50·0	504·3
5	53·0	46·4	52·0	51·0	53·2	55·0	54·4	40·4	48·4	453·8
6	56·8	48·2	60·6	63·8	48·3	54·0	55·4	44·8	49·6	481·5
7	73·8	49·4	83·6	67·8	60·2	59·6	74·0	48·8	58·2	575·4
8	53·6	38·6	55·6	44·2	38·4	37·6	45·4	30·6	43·6	387·6
9	50·6	46·6	49·8	48·0	45·0	42·6	54·8	38·0	50·8	426·2
y.s	554·4	436·9	591·2	517·9	435·1	482·5	528·8	404·8	443·6	4395·2
yr. + y.s	1134·2	858·8	1155·9	1022·2	888·9	964·0	1104·2	792·4	869·8	8790·4
yr. − y.s	25·4	−15·0	−26·5	−13·6	18·7	−1·0	46·6	−17·2	−17·4	560·3
yr. + y.s − 9yr	434·0	371·9	277·5	408·4	410·1	478·0	438·2	517·0	412·6	3747·7
1		−5·6	−10·8	2·6	2·8	−2·8	2·4	−14·0	0·0	
2	101·2		3·2	0·8	3·4	−4·2	2·6	−2·6	6·2	
3	148·4	103·2		3·3	2·0	−2·4	11·4	6·8	−2·2	
4	141·8	93·2	121·5		4·2	5·1	13·4	−0·4	−2·0	
ysr + yrs 5	103·2	89·4	102·0	97·8		−6·7	5·8	−2·0	−3·4	ysr − yrs
6	116·4	100·6	123·6	122·5	103·3		4·2	−7·2	−7·0	
7	145·2	96·2	155·8	122·2	114·6	115·0		−3·4	−3·4	
8	121·2	79·8	104·4	88·8	78·8	82·4	94·2		−5·6	
9	101·2	87·0	101·8	98·0	93·4	92·2	113·0	81·6		
yr. + y.s − 2yr	978·6	750·6	960·7	885·8	782·5	856·0	956·2	731·2	768·2	

Right-hand column notations: y.. , 2y.. , y. , 2y.. − 9y.

represented by a plot of five plants in each of two replicate blocks. The plot means summed over the two blocks are given in Table 96. The following nomenclature will be used:

y is the score of any of the n^2 families made up of n inbreds and $n(n-1)$ crosses;

yr is the score of the $r = 1$ to n inbreds which occupy the leading diagonal and $y\cdot$ is their sum $= \underset{r}{S}\, vr$;

yrs is the score of the cross between inbreds r and s; that is, the entry in the rth row and sth column, where $r = 1$ to n and $s = 1$ to n but $r \neq s$;

ysr is the score of the reciprocal cross;

$yr.$ is the sum of the n scores, consisting of 1 inbred and $n - 1$ crosses in the rth array (row);

$y\cdot s$ is the sum of the n scores in the sth array (column)

$y..$ is the sum of all n^2 entries in the diallel table $= Sy = \underset{r}{S}\, yr. = \underset{s}{S}\, y.s$

The various sums and differences required to derive the sums of squares of the analysis are shown for block sums in Table 96. Table 97 contains intermediate sums of squares computed for blocks separately and combined, from which the final items in the analysis can be derived. These are based on the assumption that the sum of squares for the b item is estimated as the total sum of squares (Sy^2 for I + II) minus the sums of squares for the a, c and d items which are simpler to compute directly. Similarly they assume that the sum of squares for the b_3 item is estimated as the sum of squares for b minus the sums of squares for b_1 and b_2 which are also simpler to compute directly. This approach also avoids the necessity of using the sums of reciprocal crosses ($yrs + ysr$). The alternative expectation for b in Table 98 also allows its direct calculation. The sum of squares for b_3 can be calculated directly from a diallel table reduced to the $n(n-1)$ crosses by omitting the n inbreds

TABLE 97
Intermediate sums of squares in the analysis of variance of diallel tables. The values in column (I + II) have been halved to allow for the summation over blocks

	Blocks		
	I	II	(I + II)
Sy^2	61 955·7169	63 036·8448	124 676·0224
$y^2 ../n^2$	58 655·4579	60 598·0277	119 245·5743
$S(yr. + y.s)^2/2n$	119 731·9300	122 834·7711	242 446·9828
$(y.. - ny.)^2/n^2(n-1)$	216·4356	115·0139	323·5002
$S(yr.+y.s - nyr)^2/n(n-2)$	5 895·3457	6 814·0446	12 671·3763
$(2y.. - ny.)^2/n^2(n-2)$	5 747·3493	6 654·8611	12 385·5867
$S(yr. - y.s)^2/2n$	80·7833	102·9422	135·5061
$\underset{r>s}{S}(ysr - yrs)^2/2$	205·9550	216·7800	290·0875

The main items in the analysis of variance derivable from the intermediate sums of squares in Table 97 are listed in Table 98

from the leading diagonal. The three terms in the sum of squares can then be obtained from the $\frac{1}{n}(n-1)$ sums of reciprocal crosses ($yrs + ysr$), the new array totals ($yr. + y.s - 2yr$) which equal the array totals of the full diallel table ($yr. + y.s$) minus the inbred contribution, $2yr$, and the total of the reduced table $y.. - y$.

The mean squares derived from the sums of squares in Table 98 and the appropriate tests of significance are given in Table 99. In general we must test each main effect (a to d) against its own interaction over blocks (B × a to B × d respectively). However, if the error variances are homogeneous they may be pooled to give a block interaction mean square as a common error variance.

Where the additive-dominance and additive-environmental model is adequate and there are no reciprocal differences, the mean squares of most of the items in the analysis of variance can be interpreted in relatively simple terms. Thus the a item tests the significance of the statistically additive effects of the genes, that is D_R and the b items the dominance effects, that is $H_R = H_2$. If the latter are non-significant, the a item is a test of the additive genetic component $D = S4uvd^2$. In the presence of dominance, however, the additive genetic effects can be tested for by substituting $Syr^2 - y.^2/n$ and its block interaction for the corresponding a items which is a regular feature of some of the alternative analyses (Jinks and Mather, 1955; Walters and Gale, 1977). The b_1 item tests the mean deviation of the F_1s from their mid-parental values. It is significant only if the dominance deviations of the genes are predominantly in one direction, that is, there is a directional dominance effect. The b_2 item tests whether the mean dominance deviation of the F_1 from their mid-parental values within each array differs over arrays. It will do so if some parents contain considerably more dominant alleles than others. The b_3 item tests that part of the dominance deviation that is unique to each F_1. This item is equivalent to the specific combining ability of Griffing (1956) for a fixed model where the inbred lines are omitted from the analysis.

On the assumption of no genotype × environment interactions and no differences between reciprocal crosses the means squares for c, d and the block interactions are all estimates of E, the environmental component of variation. If reciprocal crosses differ, c detects the average maternal effects of each parental line and d the reciprocal differences not ascribable to c (see Chapter 10). If genotype × environment interactions are present they will be detected as a difference between the block interactions for the a and b items if the additive and dominance variations are influenced to different extents by the environment (see Section 33). The presence of either differences between reciprocal crosses or genotype × environment interactions requires modifications in the tests of significance of the a and b items to which we will return later.

Applying these tests to the analysis of variance of flowering-time in *N. rustica* (Table 99) the a and b items are significant, irrespective of whether we test their significance against a pooled block interaction mean square or against their own individual block interactions. To test for additive genetical

TABLE 98
Derivation of the sums of squares in the analysis of variance of a diallel table

Item	SS	df
a	$S(yr.+y.s)^2/2n - 2y^2../n^2$ for (I+II) = 3955·836	$(n-1)$ = 8
b	$Sy^2_{yr} + y^2../n^2 - \underset{r>s}{S}(yrs-ysr)^2/2 - S(yr.+y.s)^2/2n$ for (I+II) = 1184·524	$\frac{1}{2}n(n-1)$ = 36
	or, $S^2_{yr} + \underset{r>s}{S}(yrs+ysr)^2/2 - S(yr.+y.s)^2/2n + y^2../n^2$ for (I+II)	
b_1	$(y..-ny.)^2/n^2(n-1)$ for (I+II) = 323·500	1 = 1
b_2	$S(yr.+y.s-nyr)^2/n(n-2) - (2y..-ny.)^2/n^2(n-2)$ for (I+II) = 285·790	$(n-1)$ = 8
b_3	$\underset{r>s}{S}(yrs+ysr)^2/2 - S(yr.+y.s)^2/2n - 2yr^2/n^2(n-2)$ for (I+II) = 575·235	$\frac{1}{2}n(n-3)$ = 27
c	$S(yr.-y.s)^2/2n$ for (I+II) = 135·506	$(n-1)$ = 8
d	$\underset{r>s}{S}(ysr-yrs)^2/2 - S(yr.-y.s)^2/2n$ for (I+II) = 154·581	$\frac{1}{2}(n-1)(n-2)$ = 28
BLOCKS	7·911	$(r-1)$ = 1
B×a*	103·894	$(r-1)(n-1)$ = 8
B×b	72·087	$(r-1)\frac{1}{2}n(n-1)$ = 36
B×b_1	7·949	$(r-1)$ = 1
B×b_2	21·390	$(r-1)(n-1)$ = 8
B×b_3	42·747	$(r-1)\frac{1}{2}n(n-3)$ = 27
B×c	48·219	$(r-1)(n-1)$ = 8
B×d	84·428	$\frac{1}{2}(r-1)(n-1)(n-2)$ = 28
BLOCK INTERACTIONS	308·628	(n^2-1) = 80
TOTAL	5746·897	$(2n^2-1)$ = 161

* The block interactions are obtained as the excess of the sum of the sums of squares for each item obtained from blocks I and II separately over the corresponding sums of squares from block sums (I+II)

TABLE 99
The mean squares and degrees of freedom for the analysis of variance of a 9×9 diallel

Item	MS	df	VR*	P	VR[†]	P
a	494·479	8	38·08	<0·001	128·17	<0·001
b	32·903	36	16·43	<0·001	8·53	<0·001
b_1	323·500	1	40·70	0·05–0·10	83·85	<0·001
b_2	35·724	8	13·36	<0·001	9·26	<0·001
b_3	21·305	27	13·46	<0·001	5·52	<0·001
c	16·938	8	2·81	0·05–0·10	4·39	<0·001
d	5·521	28	1·83	0·05–0·10	1·43	0·10–0·20
BLOCKS	7·911	1				
B × a	12·987	8				
B × b	2·002	36				
B × b_1	7·949	1				
B × b_2	2·674	8				
B × b_3	1·583	27				
B × c	6·027	8				
B × d	3·015	28				
BLOCK INTERACTIONS	3·858	80				

* Each item tested against its own block interactions
[†] All items tested against the pooled block interaction MS

effects, D, as opposed to the statistically additive effects of the genes, D_R, we must therefore calculate the sum of squares of the nine inbred family means. This leads to a mean square of 2386.49 for 8 df which is very highly significant against both its own block interaction and the pooled block interaction mean squares. The c and d items, however are not significant against their block interactions, although c is significant against the pooled interactions. If we accept this significance the a item must be tested against the c mean square, but this does not alter the significance of a. Little evidence of genuine maternal effects for adult characters in *N. rustica* has been found since this experiment was conducted in 1951. We shall not, therefore, fit to these data a model which allows for maternal effects.

The b_2 and b_3 portions of the b item are significant on both tests but b_1 is significant only against the pooled error. This discrepancy for b_1 is a result of its own block interaction being based on a single degree of freedom. We may conclude, therefore, that the mean flowering-times of F_1s and parents differ and reference to Table 96 shows that the F_1s have the earlier flowering-times on average. The significant b_2 implies asymmetry of gene distribution, i.e. $H_1 > H_2$, and the significant b_3 shows that there are dominance effects specific to individual crosses and not therefore attributable to b_1 and b_2.

We could, of course, now estimate D_R, H_R and E directly from the mean squares of the analysis of variance, but we have already noted that these are not the most useful parameters to estimate. Furthermore we have as yet no indication as to whether an additive-dominance, additive-environmental model is adequate for these data.

44. THE RELATIONS OF V_r AND W_r

So far we have proceeded on the assumption that an additive-dominance model with additive environmental effects and independence of the genes in action and in distribution is an adequate description of the situation found in diallel sets of crosses between inbred lines. Before fitting such a restrictive model to data we must have some indication whether or not it is a realistic desription of the data. That is, we require the equivalent of scaling tests (Section 14) to determine the validity of the assumptions of the model.

The essentials of a scaling test are predictable relationships between observed statistics that hold, within the sampling errors of the statistics, only in so far as the assumptions underlying the expected relationship apply to the data from which the statistics were obtained. In a diallel set of crosses such relationships may be found among the second degree statistics that hold only in so far as an additive-dominance model involving independently distributed genes provides an adequate description of the data.

One such relationship is that between the variance (V_r) and parent–offspring covariance (W_r) of members of the same array. In respect of a single gene difference A-a, there are two kinds of array, those with a common parent (maternal or paternal) with genotype AA, and those with a common parent with genotype aa. The variance of the progeny family means within each of these two kinds of array around their own array mean can be derived from Table 95 as follows:

$$V_{rA} = u_a d_a^2 + v_a h_a^2 - (u_a d_a + v_a h_a)^2 \quad = u_a v_a (d_a - h_a)^2$$
$$V_{ra} = v_a d_a^2 + u_a h_a^2 - (-v_a d_a + u_a h_a)^2 = u_a v_a (d_a + h_a)^2.$$

The covariance (W_r) of the progeny family means within each kind of array on the non-common parental family means is

$$W_{rA} = u_a d_a . d_a - v_a d_a . h_a - [(u_a - v_a) d_a (u_a d_a + v_a h_a)]$$
$$= 2u_a v_a d_a (d_a - h_a)$$
$$W_{ra} = v_a d_a . d_a + u_a d_a . h_a - [(u_a - v_a) d_a (-v_a d_a + u_a h_a)]$$
$$= 2u_a v_a d_a (d_a + h_a).$$

There is clearly a simple relationship between the effects of substituting AA for aa in the common parent of an array on the expected values of V_r and W_r that can be expressed in a number of ways. For example,

$$W_{rA} - W_{ra} = V_{rA} - V_{ra} = -4u_a v_a d_a h_a.$$

or

$$W_{rA} - V_{rA} = W_{ra} - V_{ra} = u_a v_a (d_a^2 - h_a^2),$$

or again,

$$b_{W_r/V_r} = \frac{W_{rA} - W_{ra}}{V_{rA} - V_{ra}} = 1$$

where b_{W_r/V_r} is the linear regression coefficient of W_r on V_r over arrays. The most useful of these relationships is the second.

Extended to an arbitrary number of independent genes, ignoring for the time being the environmental components,

$$W_{ri} - V_{ri} = \tfrac{1}{4}D - \tfrac{1}{4}H_1$$

for each array ($i = 1$ to n) of a diallel set. In the regression form

$$y = a + bx$$

this can be expressed as

$$W_{ri} = \tfrac{1}{4}(D - H_1) + bV_{ri} \text{ where } b = 1.$$

W_{ri} and V_{ri} are, of course, related to the statistics used to estimate the additive and dominance components of a set of diallel crosses. Thus the mean value of W_{ri} over all arrays ($\overline{W}_r$) equals the previous parent–offspring covariance. Returning to our single gene model

$$\overline{W}_r = u_a W_{rA} - v_a W_{ra} = 2u_a v_a d_a^2 - 2u_a v_a (u_a - v_a) d_a h_a$$

which summing over all genes

$$= \tfrac{1}{2}D - \tfrac{1}{4}F.$$

Similarly the mean value of V_r over all arrays ($\overline{V}_r$) on the single gene model equals

$$u_a V_{rA} + v_a V_{ra} = u_a v_a d_a^2 + u_a v_a h_a^2 - 2u_a v_a (u_a - v_a) d_a h_a$$

which summing over all genes gives

$$V_r = \tfrac{1}{4}D + \tfrac{1}{4}H_1 - \tfrac{1}{4}F$$

which in turn equals the sum of the variance of array means ($V_{\bar{r}} = \tfrac{1}{4}D + \tfrac{1}{4}H_1 - \tfrac{1}{4}H_2 - \tfrac{1}{4}F$) plus the interaction variance ($V_1 = \tfrac{1}{4}H_2$). Each V_r will have an environmental component identical with that of the interaction variance while each W_r will have an environmental component if the same families serve both as parents and progeny families of the inbred lines. Under these conditions the environmental component of the W_rs will always be $(1/n)$th of that appropriate for the variance of the parents, as we have already seen.

A further relationship between the statistics derivable from a diallel set of crosses which can be used to test the adequacy of the simple model is that between W_r and W_r', the covariance of members of an array with the means of arrays whose common parents are the non-common parents of the members of the array. On the single gene model of Table 95

$$W_r' \text{ for array AA} = u_a d_a (u_a d_a + v_a h_a) + v_a h_a (u_a h_a - v_a d_a)$$
$$- (u_a d_a + v_a h_a)[(u_a - v_a) d_a + 2u_a v_a h_a]$$
$$= u_a v_a d_a^2 - 2u_a^2 v_a d_a h_a + u_a v_a (u_a - v_a) h_a^2$$

and

$$W_r' \text{ for array aa} = u_a h_a (u_a d_a + v_a h_a) - v_a d_a (u_a h_a - v_a d_a)$$
$$- (u_a h_a - v_a d_a)[(u_a - v_a)d_a + 2u_a v_a h_a]$$
$$= u_a v_a d_a^2 + 2u_a^2 v_a d_a h_a - u_a v_a (u_a - v_a)h_a^2.$$

These expectations bear a simple relationship to the corresponding array variances (V_r) and covariances (W_r). For example, the regression of W_r' on W_r has the expectation

$$\frac{W_{ra}' - W_{rA}'}{W_{ra} - W_{rA}} = \frac{1}{2} - \frac{(u_a - v_a)h_a}{2d_a}.$$

This holds for any number of genes which are independent in action and in distribution. Hence, we expect a linear regression relationship with a slope greater or less than $\frac{1}{2}$ according to the sign of $(u - v)h$. However, this is a useful relationship for testing the adequacy of the simple model only when the frequencies of the alleles are $u = v = \frac{1}{2}$ at all loci, when of course we expect a linear relationship with a regression slope of $\frac{1}{2}$ which passes through the origin. This situation is encountered when an orthogonal set of homozygous substitution lines of *Drosophila* provides the parental material (Breese and Mather, 1957, 1960; Hill, 1964) and it may be assumed where the inbred lines have been derived without selection from an F_2 of a cross between two inbred lines (Jinks *et al.*, 1969).

The W_r, V_r statistics have a further useful property, namely that they provide estimates of the relative number of dominant to recessive genes present in the common parents of the arrays. Let us assume that the inbred lines differ at k loci and inbred line P_i contains k' dominant alleles and $k - k'$ recessive alleles. The expected V_{ri} and W_{ri} for the array of which P_i is the common parent are

$$V_{ri} = \underset{k'}{S} uv(d - h)^2 + \underset{k - k'}{S} uv(d + h)^2$$

$$W_{ri} = \underset{k'}{S} 2uvd(d - h) + \underset{k - k'}{S} 2uv(d + h).$$

Clearly if h is positive V_{ri} and W_{ri} have a minimum value when $k' = k$, i.e. P_i contain all the dominant alleles, their values then being,

$$V_{ri}(\text{min}) = \underset{k}{S} uv(d - h)^2$$

and

$$W_{ri}(\text{min}) = \underset{k}{S} 2uvd(d - h).$$

Similarly if h is positive their maximum values occur when $k' = 0$, i.e. P_i contains all the recessive alleles, the values of V_{ri} and W_{ri} then being

$$V_{ri}(\text{max}) = \underset{k}{S} uv(d + h)^2$$

$$W_{ri}(\text{max}) = \underset{k}{S} uvd(d + h).$$

The maximum and minimum values are, of course, reversed if h is negative. Thus if we plot W_{ri} against V_{ri} we obtain a linear regression with the points nearest the origin arising from the arrays whose common parents contains most dominant alleles whether increasers or decreasers, i.e. whether h is positive or negative, and the points further from the origin arising from the arrays whose common parents contain most recessives. The point at which the line cuts the W_r axis, that is where $V_r = 0$ is given by the regression formula

$$W_{ri} = \tfrac{1}{4}(D - H_1) + bV_{ri} = \tfrac{1}{4}(D - H_1), \text{ when } V_{ri} = 0.$$

The point of intersection thus depends on the overall level of dominance over all loci, since, when $V_{ri} = 0$,

$$W_{ri} = \underset{k}{S}\, uvd^2 - \underset{k}{S}\, uvh^2.$$

The expected graphical relationships between W_r and V_r for different relative values of $Suvd^2 : Suvh^2$ are given in Fig. 18. The statistical inequality

$$W_{ri}^2 \leqslant V_{ri} \cdot V_P \quad \text{i.e.} \quad r = \frac{W_{ri}}{\sqrt{(V_{ri} \cdot V_P)}} \leqslant 1$$

means that the points (W_{ri}, V_{ri}) can only lie on those parts of the lines of slope 1 that lie inside the limiting parabola

$$W_{ri}^2 = V_{ri} \cdot V_P.$$

The completely recessive parents correspond to points at the upper end of the regression lines where they cut the limiting parabola, and completely

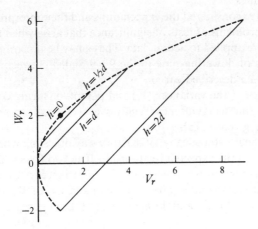

Figure 18 The theoretical regressions of W_r on V_r for various degrees of dominance, i.e. h/d ratios. The curve (broken line) joins the points of the arrays whose common parents contain all the dominant or all the recessive alleles. (Reproduced by permission from Jinks, 1954.)

dominant parents to points at the lower ends of the regression lines where they cut the limiting parabola. When there is no dominance ($H_1 = 0$) all the W_{ri}, V_{ri} points cluster at a single point where

$$V_{ri} = \tfrac{1}{4}D = \tfrac{1}{4}V_P$$

and
$$W_{ri} = \tfrac{1}{2}D = \tfrac{1}{2}V_P.$$

Thus the graph of W_r on V_r prospectively provides information on three points. First it supplies a test of the adequacy of the model; in the absence of non-allelic interaction and with independent distribution of the genes among the parents W_r is related to V_r by a straight regression line of unit slope. Second, given that the model is adequate, a measure of the average level of dominance is provided by the departure from the origin of the point where the regression line cuts the W_r axis. The distance of this point from the origin is $\tfrac{1}{4}(D - H_1)$: $D > H_1$ when the intercept is positive, $D = H_1$ where the line passes through the origin, and $D < H_1$ when the intercept is negative. Finally, the relative order of the points along the regression line indicates the distribution of dominant and recessive genes among the parents: the points nearest the origin stem from the arrays derived from parents with most dominant genes and the points furthest from the origin stem from arrays derived from parents with fewest dominant genes.

Since the first and third of these conclusions depend only on the relative positions of the W_r, V_r points on the graph these statements hold irrespective of whether or not the W_{ri}, V_{ri} values have been corrected for their environmental components. Since, however, the second conclusion depends on the position of the line in relation to the axis, the statement holds only for the genetic part of the W_{ri}, V_{ri} values.

So far we have considered the expectations and their interpretation without regard to the problems of tests of significance that arise when the models and expectations are applied to actual data. These may be illustrated by reference to the analysis of flowering-time in the 9×9 diallel between inbred lines of *Nicotiana rustica* described earlier.

The estimates of the variances (V_r) and parent–offspring covariances (W_r) for each of the nine arrays after averaging over reciprocal crosses in each of the two blocks are given in Table 100.

There is no single, statistically satisfactory way of testing whether the values of W_r and V_r have the expected relationship. The two tests which jointly give the most satisfactory answer are an analysis of variance to test the consistency of $W_r - V_r$ over arrays and a joint regression analysis of W_r on V_r to test the agreement between blocks and the agreement of the joint regression slope with unity.

Taking first the analysis of variance: the total variation among the 18 V_r and 18 W_r values can be subdivided into that among the 18 ($W_r + V_r$) values for 17 df and that among the 18 ($W_r - V_r$) values for 17 df. The variation among the 18 values of ($W_r + V_r$) and of ($W_r - V_r$) can be subdivided into the

TABLE 100
Estimates of the array variances (V_r) and array parent–offspring covariances (W_r) for the nine arrays in each of the two blocks for flowering-time in *Nicotiana rustica*

Array	Block I				Block II			
	V_r	W_r	$W_r + V_r$	$W_r - V_r$	V_r	W_r	$W_r + V_r$	$W_r - V_r$
1	34·1053	47·8333	81·9386	13·7280	29·1719	32·5907	61·7626	3·4188
2	6·7275	18·3775	25·1050	11·6500	4·2950	10·4363	14·7313	6·1413
3	75·2369	84·2917	159·5286	9·0547	55·3278	57·4336	112·7614	2·1058
4	36·4178	43·8708	80·2886	7·4530	16·2428	25·3674	41·6101	9·1246
5	7·7686	22·0458	29·8144	14·2772	7·1044	12·0231	19·1275	4·9186
6	14·6428	36·4867	51·1295	21·8439	10·5311	25·6797	36·2108	15·1486
7	35·1475	56·1150	91·2625	20·9675	28·0453	44·2576	72·3029	16·2124
8	18·0850	37·0925	55·1775	19·0075	19·4011	35·8115	55·2126	16·4104
9	6·8086	22·3433	29·1519	15·5347	5·2011	9·8922	15·0933	4·6911
Totals	234·9400	368·4567	603·3967	133·5167	175·3206	253·4921	428·8126	78·1715

differences among the 9 arrays for 8 df and the differences among the 9 pairs of values for each array in the two blocks for 9 df (Allard, 1956). The result of this partitioning of the total variation is given in Table 101.

TABLE 101
An analysis of variance of the estimates of W_r and V_r for the 9×9 diallel set of crosses in *N. rustica*

Item	df	MS	VR	P
($W_r + V_r$) Array differences	8	2735·9396	9·67	<0·001
($W_r + V_r$) Block differences	9	282·8816		
($W_r - V_r$) Array differences	8	50·8497	1·96	0·10–0·20
($W_r - V_r$) Block differences	9	25·9480		

The difference in the magnitude of ($W_r + V_r$) over arrays is highly significant when compared with the differences over replicate blocks. Hence the arrays have different ($W_r + V_r$) values. Since W_r and V_r and therefore ($W_r + V_r$) are constant over arrays in the absence of non-additive genetic variance we can conclude that there is non-additive genetic variation for flowering-time.

The difference in the magnitude of ($W_r - V_r$) over arrays, in contrast, is not significant when compared with the differences over replicate blocks. Since ($W_r - V_r$) is expected to be constant if an additive-dominance model with independent gene distribution is adequate we have no reason to doubt the adequacy of this model for the variation in flowering-time. Taking the two parts of the analysis together we can conclude that the non-additive genetic variance which leads to differences in the magnitude of ($W_r + V_r$) over arrays may be ascribed solely to the dominance effects of genes, which are independently distributed among the parental lines.

The second approach to the analysis of the W_r, V_r relationship is to carry out a joint regression analysis of W_r on V_r for the two blocks. The outcome of applying this analysis to the values for flowering-time is given in Table 102.

TABLE 102
Joint regression analysis of W_r on V_r for estimates from two blocks for a 9×9 diallel in N. rustica

Item	df	MS	VR	P
Joint regression	1	5101·20	167·5	<0·001
Heterogeneity of regression	1	1·07		
Remainder	14	30·46		

The joint regression is highly significant and the replicate blocks are in complete agreement. The joint linear regression coefficient is $b = 0.912 \pm 0.0708$ which differs significantly from zero but not from the value of one ($P = 0.30$–0.20) which is expected if non-additive genetic variation is present but as dominance only. Thus the two analyses agree in showing that the simple model is satisfactory. We can therefore proceed with the interpretation of the relative values of the statistics obtainable from a diallel table. However, it cannot be stressed too strongly that only if both analyses agree in showing the expected regression and a consistent order of the array values of W_r and V_r in all replicates, should one proceed with an interpretation as set out below.

From the W_r, V_r values themselves, we can determine the distribution of dominant and recessive genes in the parental lines. The relative values of W_r and V_r, which are consistent over blocks, show that line 2 has the lowest values and hence contains most dominant genes while line 3 has the highest values and hence the highest proportion of recessive genes. The other lines fall between these in the order 2, 9, 5, 6, 8, 4, 1, 7, 3.

From the mean values of W_r and V_r over all arrays we may determine the relative sizes of D and H_1 as $\bar{W}_r - \bar{V}_r = \frac{1}{4}D - \frac{1}{4}H_1$ and from the variance of $W_r - V_r$ over arrays and blocks we can assess its significance. Before doing so, however, the values of $\bar{W}_r$ and $\bar{V}_r$ must be corrected for their environmental components. A comparison of the block interactions for the parental families and the F_1 families of the diallel set show that while the former is the larger, there is in fact no significant difference between them (VR for 8 and 71 df $= 1.5$, $P = 0.10$–0.20). Therefore, $\hat{E}_P = \hat{E}_F$ and both equal the block interaction mean square for the 81 replicated families of the diallel. This is in fact 3·8580. The values of $W_r - V_r$ in Table 100 have been obtained after averaging over reciprocal F_1 families, $\bar{W}_r$ is therefore inflated by $\frac{1}{9}(3.8580)$ and $\bar{V}_r$ by $\frac{5}{9}(3.8580)$, i.e. $\bar{W}_r - \bar{V}_r = \frac{1}{4}D - \frac{1}{4}H_1 + \frac{4}{9}(3.8580) = 11.7588 + 1.7147 = 13.4735$. The appropriate error is the blocks $\times (W_r - V_r)$ mean square of the analysis in Table

101 divided by 18, the number of $W_r - V_r$ values included in the block interaction mean square, which equals 1·4416. The standard error of $\frac{1}{4}D$ $-\frac{1}{4}H_1 = 13\cdot4735$ is, therefore $\sqrt{1\cdot4416}$ which equals $\pm 1\cdot2006$. This leads to a value of t for nine degrees of freedom of 11·22 ($P < 0\cdot001$). Hence H_1 is smaller than D. Since the previous analyses established that H_1 is greater than zero we can conclude that dominance is incomplete.

The last remaining information that can be obtained from the W_r, V_r values is the direction of the dominance. The relative values of $W_r + V_r$ over arrays indicate the relative number of dominant to recessive alleles in the common parents of the arrays. By comparing the $W_r + V_r$ value for each array with the mean of the common parent, i.e. comparing ($W_{ri} + V_{ri}$) with $\bar{P}_i$, we can see whether the distribution of dominant to recessive alleles is correlated with the phenotype ($\bar{P}$) of the recurrent parent of the array. $W_r + V_r$ is plotted against $\bar{P}$ in Fig. 19, from which it is clear that, in general, the later flowering lines give the

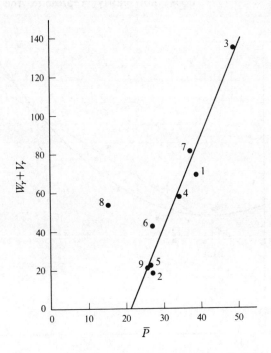

Figure 19 $W_r + V_r$ from each array of the Nicotiana rustica diallel plotted against P the mean flowering time (expressed in days after 1 July) of the parental line giving rise to that array. Note that all the points lie on a straight regression line except that from parental line 8. With the exception of line 8, the earlier the flowering of the parent, the smaller the corresponding $W_r + V_r$, showing that, in general, the alleles for earlier flowering are dominant to those for later flowering. The position of point 8, however, indicates that this dominance relation no longer holds when the parent's flowering time is earlier than mid-July. (From Mather and Jinks, 1977.)

larger values of $W_r + V_r$ and so must carry fewer dominant genes. There is, in fact, a significant correlation of $r = 0.78$ between $\bar{P}$ and $W_r + V_r$. Evidently the alleles which give early flowering tend to be dominant. The anomalous position of line 8, which while being the earliest flowering parent of all has an intermediate value of $W_r + V_r$, shows however that not all the alleles for early flowering can be dominant, and suggests an ambidirectional element in the dominance.

The occurrence of ambidirectional dominance emerges more strikingly from diallel analysis of the variation for sternopleural and coxal chaeta numbers among inbred lines derived from the 'Texas' laboratory population of *Drosophila melanogaster* (Caligari and Mather, 1980; Caligari, 1981). Figure 20 shows the relation $W_r + V_r$ bears to $\bar{P}$ found by Caligari (1981) for sterno-pleural chaeta number in a half-diallel experiment using 16 of these inbred lines. It is clear that the best fitting linear regression ($Y = 8.419 + 0.928 (x - 18.506)$ i.e. $Y = 0.928x - 8.759$) does not fairly represent the relation of

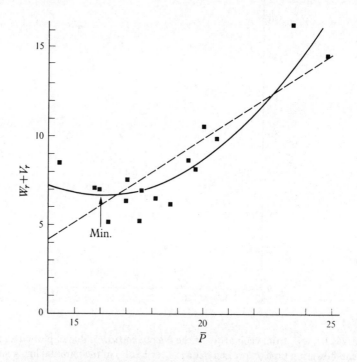

Figure 20 $W_r + V_r$, for sternopleural chaeta number, from each array of a half-diallel in *Drosophila melanogaster*, plotted against P the mean chaeta number of the parental line giving rise to that array. The broken line shows the best fitting linear regression of $W_r + V_r$ on P, and the solid line shows the best fitting quadratic regression. The arrow marks minimum point of the quadratic regression. (Data kindly supplied by Dr P. D. S. Caligari.)

$W_r + V_r (= y)$ to $\bar{P} (= x)$ because the value of $W_r + V_r$ is rising at the lower end, as well as at the higher end, of the distribution of $\bar{P}$: the phenotype associated with the maximum concentration of dominant genes is clearly at an intermediate value of $\bar{P}$, with the proportion of recessives rising with both decrease and increase from this value. The quadratic regression best fitting these results. $Y = 40.074 - 4.161x + 0.130x^2$ (found by use of the multiple regression technique) enables us to estimate the value of $\bar{P}$ at which $W_r + V_r$ is minimal. This turns out to be $\bar{P} = 16.039$. Too much weight must not, however, be given to this value as it is no more than assumption that the relationship can be represented adequately by a quadratic regression: indeed, inspection suggests that something like an inverted normal curve might fit the data better. We shall return to this matter of the estimation of the phenotype of maximum dominance in the next section, when we have considered the components of variation.

45. ESTIMATING THE COMPONENTS OF VARIATION

Provided that the relations between W_r and V_r, as described in the last section, give no reason to doubt the adequacy of the simple model the components of variation D, H_1, H_2, F and E may be estimated. The relevant statistics for the 9×9 diallel in *N. rustica* are summarized for the two blocks separately in Table 103.

Since E is derived from the block interactions of the family means (see Section 4) it has the same value for both blocks. The perfect fit solution of the parameters is given by the equations

$$E = E$$
$$D = V_P - E$$
$$H_1 = 4\bar{V}_r + V_P - 4\bar{W}_r - \frac{3n-2}{n} E$$
$$H_2 = 4\bar{V}_r - 4V_{\bar{r}} + \frac{2(n^2-1)}{n^2} E$$
$$F = 2V_P - 4\bar{W}_r - \frac{2(n-2)}{n} E.$$

These lead to the solutions in Table 104.

Although no worthwhile estimate of the errors of these components is available the consistency of their relative magnitudes, if not their absolute magnitudes, over blocks, means that we can interpret their values with some confidence. Taking first the values of D and H_1 we see that dominance is complete as suggested by the W_r, V_r analysis which showed that $0 < H_1 < D$. The component analysis allows us to go further and estimate the dominance ratio $\sqrt{(H_1/D)}$.

TABLE 103
The second degree statistics for flowering-time in the 9×9 diallel set of crosses between inbred lines of *N. rustica*

Statistic	Blocks		Mean	Model
	I	II		
V_P	111·1000	82·6195	96·8597	$D + E$
$\bar{V}_r$	26·1044	19·4801	22·7922	$\frac{1}{4}D + \frac{1}{4}H_1 - \frac{1}{4}F + \frac{5}{9}E$
$\bar{W}_r$	40·9363	28·1658	34·5510	$\frac{1}{2}D - \frac{1}{4}F + \frac{1}{9}E$
$V_{\bar{r}}$	16·7706	11·3800	14·0753	$\frac{1}{4}D + \frac{1}{4}H_1 - \frac{1}{4}H_2$
				$\quad - \frac{1}{4}F + \frac{5}{81}E$
E	3·8580	3·8580	3·8580	E

TABLE 104
Perfect fit estimates of the components

Component	Block		Mean
	I	II	
D	107·2420	78·7615	93·0017
H_1	41·0556	38·1600	39·1076
H_2	29·7144	24·7796	27·2468
F	52·4532	48·5744	49·5136
E	3·8580	3·8580	3·8580
$\sqrt{(H_1/D)}$	0·6187	0·6961	0·6407
$\overline{uv}$	0·1809	0·1620	0·1742
Heritability $\begin{cases} \text{narrow} \\ \text{broad} \end{cases}$	74·6%	68·9%	72·2%
	91·3%	88·1%	89·9%

H_2 is smaller than H_1, hence there are unequal allele frequencies, i.e. $u \neq v$ at the relevant loci. The mean value of uv over all these loci, estimated from the ratio $\frac{1}{4}H_2/H_1$, is less than its maximum value of 0·25 which arises when $u = v = 0.5$ at all loci.

The positive value of F shows that there are more dominant alleles present in the inbred lines than recessive alleles, irrespective of whether these are increasing or decreasing in their effect on flowering-time. This is, of course, equally shown by the fact that $D_R = S\{4uv[d - (u-v)h]^2\} = 55·3489$ is less than $D = S(4uvd^2) = 93·0017$. The square root of the ratio $F^2/4D(H_1 - H_2) = 0·5556$, which is 0·745, suggests that $(u-v)h/d$ is relatively consistent over all loci. Hence, the overall picture of incomplete dominance is probably the result of incomplete dominance at all loci rather than complete dominance at some loci and no dominance at others.

Since we have worked throughout on plot means within blocks with five plants per plot, the E in Table 104 is, in fact, $E_b + \frac{1}{5}E_w$ in terms of the between-

and within-family components of environmental variation. If single plant randomization had been used, E would, of course, have equalled $\frac{1}{5}E_w$. The total variation among the family means in the diallel set of crosses is, therefore,

$$\frac{1}{2}D_R + \frac{1}{4}H_R + E_b + \frac{1}{5}E_w = \frac{1}{2}D + \frac{1}{2}H_1 - \frac{1}{4}H_2 - \frac{1}{2}F + E_b + \frac{1}{5}E_w.$$

The proportion of variation which is due to any particular cause may, therefore, be estimated as its contribution to this total variation. Two such widely used proportions, namely, the narrow and broad heritabilities, can be obtained from the estimates in Table 104 as

$$\text{narrow heritability} = \frac{\frac{1}{2}D_R}{\frac{1}{2}D_R + \frac{1}{4}H_R + E} = \frac{\frac{1}{2}D + \frac{1}{2}H_1 - \frac{1}{4}H_2 - \frac{1}{2}F}{\frac{1}{2}D + \frac{1}{2}H_1 - \frac{1}{4}H_2 - \frac{1}{2}F + E};$$

$$\text{broad heritability} = \frac{\frac{1}{2}D_R + \frac{1}{4}H_R}{\frac{1}{2}D_R + \frac{1}{4}H_R + E} = \frac{\frac{1}{2}D + \frac{1}{2}H_1 - \frac{1}{4}H_2 - \frac{1}{2}F}{\frac{1}{2}D + \frac{1}{2}H_1 - \frac{1}{4}H_2 - \frac{1}{2}F + E}.$$

For the *N. rustica* data these have values of 70% and 90%, respectively (Table 104).

Equally we could relate the contribution of the various sources of variation to the total variation among family means in the fixed sample of inbred lines, the latter being,

$$D + E_b + \frac{1}{5}E_w$$

but this contains no sources of variation that cannot be estimated from the parents alone. However, unless the complete diallel set of crosses are raised and the appropriate analyses carried out, we have no way of knowing that the additive-genetic, additive-environmental model of the variation among the parental lines is adequate.

Turning to chaeta numbers in *Drosophila*, all the sternopleural and coxal chaeta characters analysed by Caligari and Mather (1980) and Caligari (1981) gave similar results. In every case $D > D_R$, so showing simultaneously that the allele frequencies are unequal and that dominant genes are preponderantly more common than their recessive alleles irrespective of the direction of dominance. F is, of course, positive also in all cases. Furthermore in no case is there any suggestion of wide variation in the value of $(u - v)h/d$ among the loci.

The values of the components of variation and of the ratios to which they lead are shown in Table 105 for sternopleural chaetae number in Caligari's (1981) 16×16 half-diallel, to which we referred in the previous section. The degree of dominance, measured by h/d, is rather lower for the genes mediating this chaeta number than for those mediating flowering-time in *N. rustica*, whereas the difference in allele frequencies, $u - v$, is rather greater; but the broad picture is much the same for these two widely different kinds of character, one in an insect and the other in a flowering plant.

Perhaps, however, the most important resemblance between the chaeta characters in *Drosophila* and flowering-time in *N. rustica* is that in all cases

TABLE 105
Components of variation for sternopleural chaeta
number from a 16×16 half-diallel in *Drosophila*
(Caligari, 1981)

D	7·577	$\frac{1}{2}F/\sqrt{[D(H_1 - H_2)]} = 0·793$
H_1	1·637	$\sqrt{(H_1/D)} = \frac{h}{d} = 0·465$
H_2	0·678	$\frac{1}{4}H_2/H_1 = \overline{uv} = 0·104$
F	4·274	$\therefore u, v = 0.118, 0·882$

$D = D_P > D_w > D_R$, or to put it the other way that F is positive, which means that in all of them the dominant alleles are preponderantly the common alleles. Now, where this is the case, Caligari and Mather (1980) have shown that the data from a diallel or half-diallel, can be used in another way to estimate the phenotype associated with maximum dominance. We take an array, i.e. a set of crosses which have one parent in common, and find the excess in the expression of the character in each cross over the average of its two parents, i.e. find $F_1 - \frac{1}{2}(\bar{P}_C + \bar{P}_N)$ where $\bar{P}_C$ is the expression in the recurrent parent of the array and $\bar{P}_N$ that in the non-recurrent parent. Within certain limitations, this difference is expected to give a straight, or nearly straight, line of negative slope when regressed on $\bar{P}_N$. Furthermore the regression line cuts the abscissa as far to the right (or left) of P_M, the expression associated with maximum dominance, as $\bar{P}_C$ lies to the left (or right) of it. Thus knowing $\bar{P}_C$ and $\bar{P}_1$, the intersection of regression line and abscissa, we can find P_M as $\frac{1}{2}(\bar{P}_C + \bar{P}_1)$, as illustrated in Fig. 21.

With a diallel size of n there are n arrays and hence n such estimates of P_M. These will not be independent of each other since in a diallel each parental value is used n times, once as P_C and $n - 1$ times as a P_N. Nonetheless, the mean of the n estimates of P_M will be better than the individual estimates themselves even though its error variance must reflect the correlations arising from multiple usage of the parental values, not to mention the double usage that must be made of the F_1 values in a half-diallel.

The estimate of P_M obtained in the case of the sternopleural chaetae is 17·70, which does not depart significantly from the overall mean of the 16 parental lines, $18·48 \pm 0·70$, (Caligari, 1981). P_M for flowering-time in *N. rustica*, though still falling within the range of phenotypes shown by the nine parental lines, was however found to lie markedly below their mean – as indeed would be expected since only one of the nine lines showed signs of carrying a preponderance of dominant genes for early flowering. Caligari and Mather (1980) follow Mather (1960b) and Breese and Mather (1960) in relating ambidirectional dominance to stabilizing selection, and they give reasons for regarding P_M, the phenotype of maximum dominance, as the optimal phenotype towards which the forces of selection must have been acting to build up the genetical architecture that we observe for the character.

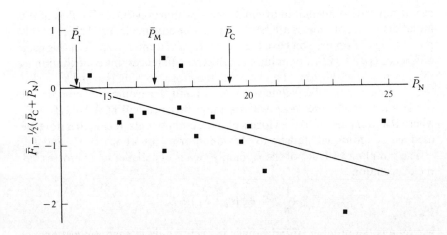

Figure 21 Estimating the phenotype of maximum dominance (P_M) in respect of sternopleural chaeta number from an array taken from a half-diallel in *Drosophila melanogaster*. P_C is the phenotype of the constant or recurrent parent of the array, and P_N is the non-recurrent parent in each cross. $F_1 - \frac{1}{2}(P_C + P_N)$ is regressed on P_N and the point (P_I) at which the regression line intersects the abscissa is expected to be as far away from P_M as is P_C, but on the opposite side of it. Thus P_M is found as $\frac{1}{2}(P_C + P_I)$. (Data kindly supplied by Dr P. D. S. Caligari.)

46. DERIVED DIALLELS

The analyses described in this chapter may, with appropriate modifications, be applied to the generations derivable from a diallel set of crosses by selfing, sib-mating and backcrossing (Jinks, 1956). Indeed where, as in many hermaphroditic plants, it is easier to raise large F_2s than to produce F_1 seed in large quantities, it may be advantageous to carry out the analyses on the progenies produced by selfing the parental and F_1 families of the diallel set of crosses.

Selfing or sib-mating a diallel set of crosses produces a table of progeny means in which the leading diagonal entries are parental families and the off-diagonal entries, F_2 families. All the analyses and expectations for the original diallel still apply, except that the coefficient of the dominance contributions to means and variances must be changed to allow for the halving of the frequency of heterozygotes in each of the off-diagonal entries. The coefficients of all h terms in the expectations must thus be halved. The contribution of the dominance component to family and generation means is correspondingly halved, the coefficients of H_1 and H_2 (terms in h^2) must be quartered and the coefficient of F (term in h) halved. The resulting loss of information about the contribution of dominance to means and variances can be compensated by raising correspondingly larger F_2 progenies.

One further consequence is that the variation within families will no longer be entirely environmental in origin. Since a proportion $2uv$ of families in the initial diallel set of crosses are heterozygous for any one locus (Table 95) this proportion of families will have the typical F_2 variation in respect of this gene difference ($\frac{1}{2}d_a^2 + \frac{1}{4}h_a^2$). Summing over all gene differences this contribution to the mean variance within families will therefore be $S2uv(\frac{1}{2}d^2 + \frac{1}{4}h^2) = \frac{1}{4}D + \frac{1}{8}H_1$. Since all families will have a within-family environmental component of variation of E_w the mean within-family variance will be $\frac{1}{4}D + \frac{1}{8}H_1 + E_w$ where D and H_1 are the diallel forms of these components. It can, therefore, be used as an additional statistic to provide information about D, H_1 and E_w.

The sampling variance of the F_2 family means of a diallel set of crosses has the expectation

$$\frac{1}{m}(\tfrac{1}{4}D + \tfrac{1}{8}H_1 + E_w) + E_b$$

where m is the number of individuals in each family in each replication.

Where the parents, F_1s and F_2s, of a diallel set of crosses are grown simultaneously the information from the two analyses can be combined in a number of ways (Hayman, 1958; Jinks, 1956). To test the adequacy of the additive-dominance model the W_r, V_r analysis can be carried out on both sets of data independently. In addition we have the expectation that the regression line for W_r on V_r for the F_2 will fall half-way between that for the F_1 and the point for no dominance, since the W_r, V_r relationship for the F_2 is comparable to that of an F_1 showing half the dominance contribution (see Fig. 18) of the actual F_1 diallel. In addition, each cell of the diallel set of crosses should satisfy the C scaling test criterion (Chapter 4) if the model is adequate.

The process of estimating the components of variation also provides tests of the adequacy of the model, since on combining the statistics of an F_1 and an F_2 diallel set of crosses there are more statistics than parameters. Hence, least squares estimates of the components can be obtained, followed by a test of the goodness of fit of the model.

The raising of backcross families from an initial diallel set of crosses produces no new problems of analysis, since averaging over the two possible backcrosses in respect of each F_1 gives a table of results, whose expectations are identical with those of the corresponding F_2s if the model is adequate ($\bar{F}_2 = \frac{1}{2}\bar{B}_1 + \frac{1}{2}\bar{B}_2$, see Chapter 4), and these can be analysed as for an F_2 diallel. The backcrosses, however, provide additional tests of the adequacy of the model in the form of A and B scaling tests. Ample illustrations of these analyses are given by Jinks (1956), Hayman (1957, 1958), Jinks and Jones (1958) and Jinks and Stevens (1959).

47. NON-ALLELIC INTERACTION IN DIALLELS

The effects of interaction on the results of a set of diallel crosses involving two gene pairs has been given by Coughtrey and Mather (1970). Assuming that the genes are distributed independently of each other among the parents, which are all homozygous, the frequencies of the different types of parents and the means of their progenies are shown in Table 106. For the sake of brevity i_{ab} is written as i, $j_{a/b}$ as j_a, $j_{b/a}$ as j_b and l_{ab} as l in this table. The expressions for V_r and W_r in the four arrays of the diallel table are set out in the upper part of Table 107 and for V_P, $\bar{V}_r$, $V_{\bar{r}}$, $\bar{W}_r$ and $W_{\bar{r}}$ in the lower part of that table.

TABLE 106
Phenotypes and array means of parents and progeny of a diallel cross involving two gene pairs with interaction

Parental line	AABB	AAbb	aaBB	aabb
Phenotype	$d_a + d_b + i$	$d_a - d_b - i$	$-d_a + d_b - i$	$-d_a - d_b + i$
Frequency	$u_a u_b$	$u_a v_b$	$v_a u_b$	$v_a v_b$
AABB $(u_a u_b)$	$d_a + d_b + i$	$d_a + h_b + j_a$	$h_a + d_b + j_b$	$h_a + h_b + l$
AAbb $(u_a v_b)$	$d_a + h_b + j_a$	$d_a - d_b - i$	$h_a + h_b + l$	$h_a - d_b - j_b$
aaBB $(v_a u_b)$	$h_a + d_b + j_b$	$h_a + h_b + l$	$-d_a + d_b - i$	$-d_a + h_b - j_a$
aabb $(v_a v_b)$	$h_a + h_b + l$	$h_a - d_b - j_b$	$-d_a + h_b - j_a$	$-d_a - d_b + i$
Array means	$u_a d_a + u_b d_b$ $+ v_a h_a + v_b h_b$ $+ u_a u_b i + u_a v_b j_a$ $+ u_b v_a j_b + v_a v_b l$	$u_a d_a - v_b d_b$ $+ v_a h_a + u_b h_b$ $- u_a v_b i + u_a u_b j_a$ $- v_a v_b j_b + v_a u_b l$	$- v_a d_a + u_b d_b$ $+ u_a h_a + v_b h_b$ $- v_a u_b i - v_a v_b j_a$ $+ u_a u_b j_b + u_a v_b l$	$- v_a d_a - v_b d_b$ $+ u_a h_a + u_b h_b$ $+ v_a v_b i - u_a v_b j_a$ $- u_a v_b j_b + u_a u_b l$

Mean of parents $= (u_a - v_a)d_a + (u_b - v_b)d_b + (u_a - v_a)(u_b - v_b)i$

Mean of all progeny $= (u_a - v_a)d_a + (u_b - v_b)d_b + 2u_a v_a h_a + 2u_b v_b h_b$
$\qquad + (u_a - v_a)(u_b - v_b)i + 2u_b v_b(u_a - v_a)j_a + 2u_a v_a(u_b - v_b)j_b$
$\qquad + 4u_a v_a u_b v_b l.$

Three basic categories of components of heritable variation can be recognized in these variances and covariances. First, the components of which $V_{\bar{r}}$ and $\bar{V}_r$ are compounded have the same definitions as we found for D_R, H_R, I_R, J_R and L_R in randomly breeding populations (Section 35) and the results from the two-gene case here can be generalized in the same way as in that section.

Secondly only d and i contribute to the components of the variance of the parents, V_P, as is to be expected since the parents are all homozygous. Two components of variation can therefore be recognized, which can be generalized as

$$D_P = S4u_a v_a [d_a + S_b i_b (u_b - v_b)]^2 \text{ and } I_P^2 = 4S_{ab} u_a v_a u_b v_b i_{ab}.$$

Third, the covariances, $\bar{W}_r = W_{\bar{r}}$ again include only D and I type terms, but

TABLE 107

Variances and covariances of the diallel table for two gene pairs with interaction

Array	Parent	V_r	W_r
1	AABB	$u_a v_a[(d_a + v_b j_a + u_b i)] - (h_a + u_b j_b + v_b l)]^2 + u_b v_b[(d_b + v_a j_b + u_a i)] - (h_b + u_a j_a + v_a l)]^2 + u_a u_b v_a v_b (i - j_a - j_b + l)^2$	$2u_a v_a[d_a + (u_b - v_b)i][(d_a + v_b j_a + u_b i) - (h_a + u_b j_b + v_b l)] + 2u_b v_b[d_b + (u_a - v_a)i][(d_b + v_a j_b + u_a i) - (h_b + u_a j_a + v_a l)] + 4u_a u_b v_a v_b i(i - j_a - j_b + l)$
2	AAbb	$u_a v_a[(d_a + u_b j_a - v_b i)] - (h_a - v_b j_b + u_b l)]^2 + u_b v_b[(d_b + v_a j_b + u_a l)] - (h_b + u_a j_a + v_a l)]^2 + u_a u_b v_a v_b (i + j_a - j_b - l)^2$	$2u_a v_a[d_a + (u_b - v_b)i][(d_a + u_b j_a - v_b i) - (h_a - v_b j_b + u_b l)] + 2u_b v_b[d_b + (u_a - v_a)i][(d_b + v_a j_b + u_a i) - (h_b + u_a j_a + v_a l)] + 4u_a u_b v_a v_b i(i + j_a - j_b - l)$
3	aaBB	$u_a v_a[(d_a + v_b j_a + u_b i)] + (h_a + u_b j_b + v_b l)]^2 + u_b v_b[(d_b - v_a j_a + u_a l)] - (h_b - v_a j_a + u_a i)]^2 + u_a u_b v_a v_b (i - j_a - j_b + l)^2$	$2u_a v_b[d_a + (u_b - v_b)i][(d_a + v_b j_a + u_b i) + (h_a + u_b j_b + v_b l)] + 2u_b v_b[d_b + (u_a - v_a)i][(d_b + u_a j_b - v_a i) - (h_b - v_a j_a + u_a l)] + 4u_a u_b v_a v_b i(i - j_a - j_b + l)$
4	aabb	$u_a v_a[(d_a + u_b j_a - v_b i)] + (h_a - v_b j_b + u_b l)]^2 + u_b v_b[(d_b - v_a j_a + u_a l)] + (h_b - v_a j_a + u_a l)]^2 + u_a u_b v_a v_b (i + j_a + j_b + l)^2$	$2u_a v_a[d_a + (u_b - v_b)i][(d_a + u_b j_a - v_b i) + (h_a - v_b j_b + u_b l)] + 2u_b v_b[d_b + (u_a - v_a)i][(d_b + u_a j_b - v_a i) + (h_b - v_a j_a + u_a l)] + 4u_a u_b v_a v_b i(i + j_a + j_b + l)$

Variance of array means $= u_a v_a\{[d_a + 2u_b v_b j_a + (u_b - v_b)i] - (u_a - v_a)[h_a + (u_b - v_b)j_b + (u_a - v_a)l]\}^2 + u_b^2 v_b^2[(d_b + 2u_a v_a j_b + (u_a - v_a)i] - (u_b - v_b)[h_b + (u_a - v_a)j_a + (u_b - v_b)l]\}^2 + u_a u_b v_a v_b[i - (u_b - v_b)(u_a - v_a) - (u_a - v_a)(u_b - v_b)l]^2$

Mean variance of arrays $= V_{\bar{r}}$ as above $+ 2u_a u_b v_a v_b[i - (u_b - v_b)j_a - (u_a - v_a)][h_a + (u_b - v_b)j_b + (u_b - v_b)j_b]^2 + 4u_b^2 v_b^2[h_b + 2u_a v_a l]^2 + (u_a - v_a)j_a]^2 + 8u_a u_b v_a^2 v_b^2[j_a - (u_a - v_a)l]^2 + 8u_a^2 v_a^2 u_b v_b[j_b - (u_b - v_b)l]^2 + 16u_a^2 u_b v_a^2 v_b^2 l^2$

Covariance of array means $=$ Mean covariance of arrays
$= 2u_a v_a[d_a + (u_b - v_b)i]\{[d_a + 2u_b v_b j_a + (u_b - v_b)i] - (u_a - v_a)[h_a + (u_b - v_b)j_b + 2u_b v_b l]\} + 2u_b v_b[d_b + (u_a - v_a)i]\{[d_b + 2u_a v_a j_b + (u_a - v_a)i] - (u_b - v_b)[h_b + (u_a - v_a)j_a + 2u_a v_a l]\} + 4u_a u_b v_a v_b i[i - (u_b - v_b)j_a - (u_a - v_a)j_b + (u_a - v_a)(u_b - v_b)l]$

Variance of parents $= 4u_a v_a[d_a + (u_b - v_b)i]^2 + 4u_b v_b[d_b + (u_a - v_a)i]^2 + 16u_a u_b v_a v_b i^2$

the generalized definitions of the two components of variation now become

$$D_W = 4S_a u_a v_a [d_a + S_b i_b (u_b - v_b)] [d_a + 2S_b (u_b v_b j_{ab}) + S_b (u_b - v_b) i_{ab}]$$
$$I_W = 16 S_{ab} u_a v_a u_b v_b i_{ab} [i_{ab} - (u_b - v_b) j_{ab} - (u_a v_a) j_{ba}$$
$$+ (u_a - v_a)(u_b - v_b) l_{ab}].$$

We can thus write

$$V_{\bar{r}} = \tfrac{1}{4} D_R + \tfrac{1}{16} I_R$$
$$\bar{V}_r = \tfrac{1}{4} D_R + \tfrac{1}{4} H_R + \tfrac{3}{16} I_R + \tfrac{1}{8} J_R + \tfrac{1}{16} L_R$$
$$\bar{W}_r = W_{\bar{r}} = \tfrac{1}{2} D_W + \tfrac{1}{4} I_W$$

and

$$V_P = D_P.$$

It will be noted that

$$V_{\bar{r}} + \bar{V}_r = \tfrac{1}{2} D_R + \tfrac{1}{4} H_R + \tfrac{1}{4} I_R + \tfrac{1}{8} J_R + \tfrac{1}{16} L_R$$

which equals (as it should) V_R, the heritable variance of the corresponding randomly breeding population (see Table 87 in Section 35).

It will be seen also that items of which D_W is composed are the geometric means of the corresponding items in D_P and D_R, and similarly for the items in I_W. Thus $W_{\bar{r}}$ can never exceed $\sqrt{(V_P V_{\bar{r}})}$, just as in the case where interaction is absent, and it will fall short of $\sqrt{(V_P V_{\bar{r}})}$ because of interaction only to the extent that the ratio $\{[2u_b v_b j_{ab} + (u_b - v_b) i_{ab}](u_b - v_b) i_{ab} + [(u_b - v_b) j_{ab} + (u_a - v_b) j_{ba} - (u_{ab} - v_a)(u - v_b) l_{ab}]\}/i_{ab}$ varies among the pairs of loci. The comparison of W^2 and $V_P \cdot V_{\bar{r}}$ thus still provides a test of homogeneity of the properties of the relevant loci in the presence of interaction, as it does of $(u-v)h/d$ where interaction is absent.

We can still formally equate D_R with $D + H_1 - H_2 - F$ but the interpretation of the results will now be more complex since all these four analytical components will be affected by interaction as well as by d and h. The use of $\sqrt{(H_1/D)}$ as a measure of h/d and of $H_2/4H_1$ as a measure of $\overline{uv}$ thus becomes suspect. The values of these ratios will reflect the effects of interaction as well as of dominance and inequality of the gene frequencies respectively, and further investigation of their relations to interaction are required before the implications can be assessed.

Diallel analysis, however, provides us with a test for the presence of interaction, as well as a means of detecting and measuring dominance. In the absence of interaction, the points on the W_r/V_r graph, one from each array, coincide, apart from sampling variation, when only additive variation is present ($h = 0$). With dominance ($h \neq 0$) these points fall, again within the limits of sampling variation, on a straight line of slope 1, which intersects the ordinate at a point $\tfrac{1}{4}(D - H_1)$ above the origin. It is clear, however, that these simple relations fail when interaction is present, and they may fail in either or both of two ways as we can see by considering the case of two gene pairs (Table 107).

In the first place, the points from the four arrays no longer coincide in the absence of dominance. The precise way in which the points then scatter depends on the type of interaction and on the gene frequencies. With j type interactions the scattering is always irregular, but with i type interaction alone or l type interaction alone the points scatter along a line of unit slope when $u_a = v_a = u_b = v_b = \frac{1}{2}$, even in the absence of dominance. A line of unit slope is, therefore, not a completely unequivocal indication of dominance, though the limitations in the circumstances under which it can spring from interaction leave little uncertainty in its attribution to dominance. It might be observed that even with equal gene frequencies the rectilinear relation is lost if i and h are present together, though it is retained when both l and h are present. When i interaction is present by itself but the gene frequencies are not equal a curvilinear relation is found, the curves being concave upwards when i is positive and concave downwards when it is negative. As a special case, with $u_a = u_b$ and $d_a = d_b$ a straight line is obtained but with a slope other than 1 except when $u_a = u_b = \frac{1}{2}$. These special cases are, however, of little more than academic interest in view of the unlikelihood that one type of interaction would occur without any of the others and of the restricted gene frequencies under which they arise. The most interesting of the results is that i type interaction can cause the points to scatter along a curve and even give a straight line when the gene frequencies are equal, when $h = j = l = 0$ for, in the absence of dominance, it would seem very likely that j and l types of interaction would be absent too.

The second way in which interactions cause the simple relation of V_r and W_r to fail is the departure from linearity which they bring about when dominance is present: the array points will still scatter when $h \neq 0$, but they will not generally scatter along a straight line of unit slope. Again, the general case is too complex for detailed consideration, but a class of special cases is of interest. As we have already had occasion to observe, the classical complementary interaction is shown by a pair of genes when the js are positive and i and l have the same sign as the hs, and the duplicate interaction when the js are negative and i and l have the opposite sign to the hs (Mather, 1967b). We will consider the case of complementary action arising when the hs, i, js and l are all positive and of duplicate action arising when the hs are positive and i, js and l all negative. In effect this is a complete treatment since change in sign of the hs reorders the points on the diallel graph and so while changing the relation of the arrays to the points on the graph makes no alteration in the overall pattern of the points in relation to one another. We will follow Mather (1967b) in considering partial as well as full complementary and duplicate interaction by setting $d_a = d_b = h_a = h_b$ and $i = j_a = j_b = l$ but with $i = \theta d_a$ where θ specifies the type and strength of interaction. It is positive for complementary and negative for duplicate interaction, and ranges from 1 for full interaction to 0 for no interaction.

Table 108 sets out the expressions for V_r and W_r in terms of d and θ where

TABLE 108
Variances and covariances in the diallel table with complementary and duplicate interaction

Array	Recurrent parent	V_r	W_r
1	AABB	0	0
2	AAbb	$4u_b v_b d^2 (1+\theta)^2$	$4u_b v_b d^2 [1-(v_a-u_a)\theta][1+\theta]$
3	aaBB	$4u_a v_a d^2 (1+\theta)^2$	$4u_a v_a d^2 [1-(v_b-u_b)\theta][1+\theta]$
4	aabb	$4u_a v_a d^2 [1-(v_b-u_b)\theta]^2$	$4u_a v_a d^2 [1-(v_b-u_b)\theta]^2$
		$+ 4u_b v_b d^2 [1-(v_a-u_a)\theta]^2$	$+ 4u_b v_b d^2 [1-(v_a-u_a)\theta]^2$

$d = d_a = d_b = h_a = h_b$ and $\theta d = i = j_a = j_b = l$, as defined above. The table also gives the excesses of the values of V_r and W_r for arrays 2, 3 and 4 over those for array 1. It will be seen that when $\theta = 0$ these expressions reduce to those for the uncomplicated diallel as derived in the previous section. Also, no matter what the value of θ, V_r and W_r for array 1 are both 0, so in other words the point arising from the array always falls on the origin of the graph. Now W_r for array 4 exceeds W_r for array 1 by exactly the same amount as does V_r. The points for the extreme arrays, 1 and 4, will therefore always lie on a line of unit slope passing through the origin. When we turn, however, to the intermediate arrays, 2 and 3, the situation is different since the excess over array 1 in W_r is no longer equal to the excess in V_r. For array 2 minus array 1,

$$\Delta V_r = 4u_b v_b d^2 (1+\theta)^2$$

and

$$\Delta W_r = 4u_b v_b d^2 [1-(v_a-u_a)\theta][1+\theta]$$

giving

$$\Delta V_r - \Delta W_r = 4u_b v_b d^2 . 2v_a \theta (1+\theta).$$

Similarly for array 3 minus array 1,

$$\Delta V_r - \Delta W_r = 4u_a v_a d^2 . 2v_b \theta (1+\theta).$$

Then when interaction is absent and $\theta = 0$ these differences vanish, indicating that V_r and W_r change by the same amount and all the array points fall on the straight line of unit slope. With complementary action, however, θ is positive and $\Delta V_r - \Delta W_r$ is positive for both arrays 2 and 3. The change in W_r is thus less than that in V_r, and the points for array 2 and 3 must lie to the right of the straight line of unit slope joining the points from the extreme arrays 1 and 4. Complementary action thus characteristically gives a curve which is concave upwards.

The effect of duplicate interaction is a little more complicated. When the interaction is complete $\theta = -1$ and $(1+\theta) = 0$. Then $\theta(1+\theta) = 0$ and $\Delta V_r - \Delta W_r$ is 0 for both arrays 2 and 3. A straight line relation thus appears to be obtained with complete duplicate interaction. The reason for this becomes clear when we look at Table 108. When $\theta = -1$, $V_r = W_r = 0$ for arrays 2 and 3 as well as for array 1. Arrays 1, 2 and 3 then give points coinciding at the origin and since the only other point is from array 4, a straight line of unit slope ensues.

When, however, $-1 < \theta < 0$, $\theta(1 + \theta)$ lies between 0 and $-\frac{1}{4}$ and the points from arrays 2 and 3 lie to the left of the line which joins the extreme points from arrays 1 and 4. This gives a curve which is concave downwards, i.e. in the direction opposite to that with complementary interaction. If we can imagine a super-duplicate interaction defined by $\theta < -1$, $\theta(1 + \theta)$ becomes positive again and the points for arrays 2 and 3 fall to the right of the line of unit slope, to give a curve concave upwards. This is the same type of curvature as with complementary interaction, but the two are hardly likely to be confused with each other because with $\theta < -1$, W_r becomes negative for both the

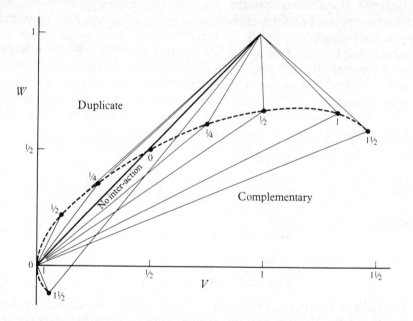

Figure 22 The effect of complementary and duplicate interactions between two gene pairs on the W_r/V_r graph from a diallel cross. With no interaction ($\theta = 0$) the graph is a straight line with the middle point halfway between the two extreme points. Complementary interaction results in this middle point moving to the right and upwards, and duplicate interaction results in it moving to the left and downwards. The W_r/V_r graph thus ceases to be a straight line where interaction supervenes, being concave upwards with complementary and concave downwards with duplicate interaction. The dotted curve shows the path of the middle point as it moves under the influence of interaction, the numbers indicating the values of θ to which the points correspond. (Reproduced by permission from Mather, 1967b.)

intermediate arrays, which thus give points falling below the abscissa – a result which can never arise from the complementary relation. This is brought out in Fig. 22, taken from Mather (1967b). For ease of representation in the figure it is assumed that $u_a = u_b = v_a = v_b = \frac{1}{2}$, and as a consequence the points for arrays 2 and 3 coincide giving only three, rather than the more general four, points on the line or curve. Also, for clarity of representation, the value of d assumed in each case has been adjusted so that the extreme point from array 4 would be constant in position on the graph. The figure also brings out a further point, that for corresponding values of θ the departure of the centre point to the left of the line for duplicate interaction is less than the departure to the right for complementary action. Thus, in general, duplicate interaction will be harder to detect than complementary interaction by the curvature it causes. By the same token, however, duplicate interaction leads to a greater tendency for the points from arrays 2 and 3 to cluster with the point from array 1 than does complementary interaction to bring about clustering at the other end of the line.

One final point remains to be made about the diallel graph in the important special case of complementary and duplicate interaction. As we have seen, the line through the points from arrays 1 and 4 is of unit slope. If, therefore, we can ascertain which are the extreme points on such a graph, we can determine the basic line of unit slope even where complementary or duplicate interaction is present.

48. GENE ASSOCIATION AND DISPERSION

Linkage as such has no effect on the results of a set of diallel crosses, since we are dealing only with family means, and indeed in the most common diallel analysis we are dealing only with the means of true breeding lines and their F_1s. The results of diallel analysis are, however, affected by a phenomenon somewhat similar to linkage, and one which may at times arise from the genes being linked, though it does not necessarily do so. This phenomenon is the departure from independence of the genes in their distribution among the parental lines. Thus with independence we should expect AABB, AAbb, aaBB and aabb to occur with frequencies $u_a u_b$, $u_a v_b$, $v_a u_b$, $v_a v_b$ among the parents, but they may not in fact do so. Departure from this expectation may be accommodated, and indeed any set of frequencies specified completely, by the introduction of a third parameter, c, such that the frequencies of the parental types are respectively, $u_a u_b + c$, $u_a v_b - c$, $v_a u_b - c$, $v_a v_b + c$ (Coughtrey and Mather, 1970). When c is positive the parents with reinforcing combinations of genes are in excess and the genes may be said to be predominantly associated. When c is negative the parents with balancing combinations of genes are in excess and the genes may be said to be predominantly dispersed.

With the four parental genotypes present in these proportions and in the absence of non-allelic interactions, the variances and covariances of the arrays are as shown in Table 109. When the distributions of the genes in the parents are independent of one another $c = 0$ and the expressions for V_r and W_r reduce to those already discussed earlier in the chapter, giving the rectilinear relation, with unit slope, between W_r and V_r. When $c \neq 0$, however, this rectilinear relation is lost. Table 110 sets out the increments by which V_r and W_r in arrays 2, 3 and 4 exceed V_r and W_r in array 1. It will be seen that for array 4,

$$\Delta V_r = \Delta W_r = 4u_a v_a d_a h_a + 4u_b v_b d_b h_b + 4c(d_a h_b + d_b h_a).$$

TABLE 109
Variances and covariances in a diallel where the genes show predominant association (c positive) or dispersion (c negative) in the parental lines

	Recurrent parent and frequency	V_r	W_r
1	AABB $u_a u_b + c$	$u_a v_a(d_a - h_a)^2$ $+ u_b v_b(d_b - h_b)^2$ $+ 2c(d_a - h_a)(d_b - h_b)$	$2u_a v_a d_a(d_a - h_a)$ $+ 2u_b v_b d_b(d_b - h_b)$ $+ 2c[d_a(d_b - h_b) + d_b(d_a - h_a)]$
2	AAbb $u_a v_b - c$	$u_a v_a(d_a - h_a)^2$ $+ u_b v_b(d_b + h_b)^2$ $+ 2c(d_a - h_a)(d_b + h_b)$	$2u_a v_a d_a(d_a - h_a)$ $+ 2u_b v_b d_b(d_b + h_b)$ $+ 2c[d_a(d_b + h_b) + d_b(d_a - h_a)]$
3	aaBB $v_a u_b - c$	$u_a v_a(d_a + h_a)^2$ $+ u_b v_b(d_b - h_b)^2$ $+ 2c(d_a + h_a)(d_b - h_b)$	$2u_a v_a d_a(d_a + h_a)$ $+ 2u_b v_b d_b(d_b - h_b)$ $+ 2c[d_a(d_b - h_b) + d_b(d_a + h_a)]$
4	aabb $v_a v_b + c$	$u_a v_a(d_a + h_a)^2$ $+ u_b v_b(d_b + h_b)^2$ $+ 2c(d_a + h_a)(d_b + h_b)$	$2u_a v_a d_a(d_a + h_a)$ $+ 2u_b v_b d_b(d_b + h_b)$ $+ 2c[d_a(d_b + h_b) + d_b(d_a + h_a)]$
	$(2+3)-(4+1)$	$-8ch_a h_b$	0

Variance of parents $V_P = 4u_a v_a d_a^2 + 4u_b v_b d_b^2 + 8cd_a d_b$
Variance of array means $V_{\bar{r}} = u_a v_a[d_a + (v_a - u_a)h_a]^2 + u_b v_b[d_b + (v_b - u_b)h_b]^2$
$\qquad + 2c[d_a + (v_a - u_a)h_a][d_b + (v_b - u_b)h_b]$
Mean variance of arrays $\bar{V}_r = V_{\bar{r}} + 4u_a^2 v_a^2 h_a^2 + 4u_b^2 v_b^2 h_b^2 + 8c^2 h_a h_b$
Covariances $W_{\bar{r}} = \bar{W}_r = 2u_a v_a d_a[d_a + (v_a - u_a)h_a]$
$\qquad + 2u_b v_b d_b[d_b + (v_b - u_b)h_b]$
$\qquad + 2cd_a[d_b + (v_b - u_b)h_b] + 2cd_b[d_a + (v_a - u_a)h_a]$

Thus in the diallel graph the line giving the points derived from arrays 1 and 4 is of unit slope. As will be seen from the table, however, this is not true of the relations between array 1 on the one hand and arrays 2 and 3 on the other. With both arrays 2 and 3, ΔW_r exceeds ΔV_r in value by $\Delta W_r - \Delta V_r = 4ch_a h_b$. Then with association c is positive and the points for the intermediate arrays 2 and 3 lie to the left of the line joining the extreme points 1 and 4, thus giving a curvilinear relation with the curve concave downwards. With dispersion c is negative and the points for the intermediate arrays 2 and 3 will lie to the right

TABLE 110

Changes between arrays of variances and covariances in a diallel table with association and dispersion

Arrays	ΔV_r	ΔW_r
$2-1$	$4u_b v_b d_b h_b + 4c(d_a h_b - h_a h_b)$	$4u_b v_b d_b h_b + 4c d_a h_b$
$3-1$	$4u_a v_a d_a h_a + 4c(d_b h_a - h_a h_b)$	$4u_a v_a d_a h_a + 4c d_b h_a$
$4-1$	$4u_a v_a d_a h_a + 4u_b v_b d_b h_b$ $+ 4c(d_a h_b + d_b h_a)$	$4u_a v_a d_a h_a + 4u_b v_b d_b h_b + 4c(d_a h_b$ $+ d_b h_a)$
$3-2$	$4u_a v_a d_a h_a - 4u_b v_b d_b h_b$ $+ 4c(d_b h_a - d_a h_b)$	$4u_a v_a d_a h_a - 4u_b v_b d_b h_b + 4c(d_b h_a$ $- d_a h_b)$

of the line joining the extreme points from arrays 1 and 4. The curve is then concave upwards. We might observe in passing that the difference in both V_r and W_r between arrays 2 and 3 is

$$\Delta V_r = \Delta W_r = 4u_a v_a d_a h_a - 4u_b v_b d_b h_b + 4c(d_b h_a - d_a h_b),$$

and the line joining the points derived from these arrays is also of unit slope. It will thus be parallel to the line points 1 and 4, with which it coincides when $c = 0$.

Before proceeding, we should note that since, for points 2 and 3, $\Delta W_r - \Delta V_r = 4c h_a h_b$, a change in sign of one of the hs would appear to change the sign of the difference and hence the direction of curvature of the W_r/V_r line. But should one of the hs change sign it has the effect of removing the points deriving from arrays 1 and 4 from their extreme positions and putting them in the intermediate position on the W_r/V_r graph, the points from arrays 2 and 3 taking up the extreme positions. Thus change of sign in h alters the relation of the arrays to the points on the graph, but it does not alter the overall pattern of the points and so has no effect on the relation between the sign of c and the direction of curvature.

Since association gives a curvature concave downwards its effect is at least superficially similar to that of a duplicate type interaction, and dispersion is correspondingly at least superficially similar to complementary type interaction. A difference between the effects of non-independent distribution of the genes and interaction should, however, be noted. In general the curves stemming from duplicate interaction are expected to depart by lesser amounts from the straight line than do those stemming from complementary interaction (see Fig. 22): curves concave downwards will be shallower than curves concave upwards. With non-independent gene distribution the opposite should in general be true. Where only two gene pairs are involved the effects of association and dispersion will be equal and opposite, but whereas with more than two loci all of them can show association with one another, they cannot all show dispersion relative to one another, just as with linkage any number of genes can be coupled but not all of them can be in repulsion with one another if more than two loci are involved (Section 31). Thus with more than two loci a mutually reinforcing system of departures from rectilinearity of the W_r/V_r relation can arise from association, but not from dispersion. The maximum curvature which is concave downwards is then greater than the maximum curvature which is concave upwards.

One last point remains to be made about the W_r/V_r line. Since in respect of the extreme arrays 1 and 4

$$\Delta V_r = \Delta W_r = 4u_a v_a d_a h_a + 4u_b v_b d_b h_b + 4c(d_a h_b + d_b h_a),$$

both differences are increased when c is positive and both reduced when c is negative. In other words, association lengthens the W_r/V_r line and dispersion shortens it, though the slope of the line through the extreme points is always one.

The expressions for V_P, $V_{\bar{r}}$, $\bar{V}_r$, and the covariances are also given in Table 109. The hs are, of course, not involved in V_P, and they enter into $V_{\bar{r}}$ and the covariances only in the way in which they contribute to the D component of variation in a randomly breeding population. By contrast $\bar{V}_r$ includes a true H type component of variation. All the statistics, however, include terms in c. In V_P this term is necessarily positive if c is positive and in the rest it will be positive except where $h(v-u) > d$. This would require h to be greater than d and also to be negative if $v > u$. In general, therefore, we might expect $d + (v - u)h$ and with it the terms in c, to be positive. Association will be expected, therefore, generally to inflate the variances and covariances and dispersion to reduce them, while V_P will always be inflated by association and reduced by dispersion.

In these variances and covariances the coefficients of the terms in c are formally independent of the gene frequencies, but we should observe that the possible range of values of c itself is not, since the limits for c are set by the contributions of the individual gene frequencies to the overall genotypic frequencies. When the gene frequencies are all equal, $u_a = v_a = u_b = v_b = 0.5$ and c must lie between 0.25 and -0.25. If $c = 0.25$ there is complete association of the genes and only the extreme genotypes are present. Equally when $c = -0.25$ only the intermediate genotypes are present. With unequal gene frequencies the range of values of c is correspondingly smaller. When positive, c cannot exceed in value $u_a v_b$ or $v_a u_b$, whichever is the lesser; and when negative, c cannot exceed in value $u_a u_b$ or $v_a v_b$, whichever is the lesser. Thus the more unequal the gene frequencies the smaller c must be and the less marked are the effects of association and dispersion in distorting the W_r/V_r graph, though of course association may still have important effects if a large number of genes are involved.

Other parameters than c can be devised, and indeed have been used (Hill, 1964; Jinks and Stevens, 1959) to represent and quantify association and dispersion. The special value of c is that it brings out particularly clearly the effects of non-independent gene distribution on both the variances and covariances derivable for the diallel table and the W_r/V_r relationship. As it stands, however, it cannot provide an unambiguous measure of the degree of association or dispersion because its range of values is limited by the gene frequencies.

10 *Departures from simple disomic inheritance*

49. SEX-LINKAGE: CROSSES BETWEEN INBRED LINES

In the models and analyses discussed so far we have assumed, or shown, that reciprocal crosses are identical. There are two situations where this is not so. These are where we have sex-linked genes or maternal influences affecting the phenotype of the offspring. The contributions of these to the differences between reciprocal crosses will be discussed and methods for estimating their individual contributions and for distinguishing between them will be described.

Let us consider first the consequences of sex-linkage. In the homogametic sex (XX) there are three possible genotypes in respect of a gene pair, A–a, namely XAXA, XAXa and XaXa with contributions to the mean phenotype of d_a, h_a and $-d_a$, respectively. In the heterogametic sex (XY), on the other hand, only two genotypes are possible, XAY and XaY, with contributions to the genotype which we will designate d'_a and $-d'_a$, respectively.

The contribution that gene A–a makes to the generations that can be derived from an initial cross between two inbred lines are summarized in Table 111. The larger inbred line has the genotypes XAXA and XAY for the two sexes and the smaller, P_2, the genotypes XaXa and XaY. On the basis of these expectations, which apply also of course to the summed effects of sex-linked genes, the diagnostic properties of sex-linkage are:

(i) A difference between reciprocal crosses in the F_1 generation which is confined to the heterogametic sex;

(ii) A difference between reciprocal crosses in the F_2 generation which is confined to the homogametic sex;

(iii) A difference between reciprocal crosses in the two backcrosses B_1 and B_2, in both sexes, but a difference between the two reciprocal crosses using the inbred line as the homogametic parent which is confined to the individuals of that sex.

If the genes in the homozygous state make the same contributions to the phenotype as in the hemizygous state, i.e. if $d_a = d'_a$, there are simple relationships between the magnitudes of the reciprocal differences in the

TABLE 111
The contribution of a single gene difference, A–a, to the means of the generations derivable from an initial cross between two inbred lines

Generation	Mating		Progeny mean	
	Homogametic parent	Heterogametic parent	Homogametic sex	Heterogametic sex
P_1	P_1	$\times P_1$	d_a	d'_a
P_2	P_2	$\times P_2$	$-d_a$	$-d'_a$
F_1	$\begin{cases} P_1 \\ P_2 \end{cases}$	$\times P_2$ $\times P_1$	h_a h_a	d'_a $-d'_a$
F_2	$\begin{cases} F_1(P_1 \times P_2) \\ F_1(P_2 \times P_1) \end{cases}$	$\times F_1(P_1 \times P_2)$ $\times F_1(P_2 \times P_1)$	$\frac{1}{2}d_a+\frac{1}{2}h_a$ $-\frac{1}{2}d_a+\frac{1}{2}h_a$	0 0
B_1	$\begin{cases} P_1 \\ P_1 \\ F_1(P_1 \times P_2) \\ F_1(P_2 \times P_1) \end{cases}$	$\times F_1(P_1 \times P_2)$ $\times F_1(P_2 \times P_1)$ $\times P_1$ $\times P_1$	d_a h_a $\frac{1}{2}d_a+\frac{1}{2}h_a$ $\frac{1}{2}d_a+\frac{1}{2}h_a$	d'_a d'_a 0 0
B_2	$\begin{cases} P_2 \\ P_2 \\ F_1(P_1 \times P_2) \\ F_1(P_2 \times P_1) \end{cases}$	$\times F_1(P_1 \times P_2)$ $\times F_1(P_2 \times P_1)$ $\times P_2$ $\times P_2$	h_a $-d_a$ $-\frac{1}{2}d_a+\frac{1}{2}h_a$ $-\frac{1}{2}d_a+\frac{1}{2}h_a$	$-d'_a$ $-d'_a$ 0 0

various generations. For example, the magnitude of the reciprocal difference among homogametic progeny in the F_2 should be half that among heterogametic progeny in the F_1 for any number of independent genes.

In deriving the relationships between generation means that provide the scaling tests (Section 14) the effects of sex-linkage were ignored. However, none of the relationships or conclusions reached are affected by the presence of sex-linkage providing that reciprocal crosses are made and the generation means obtained by averaging over all reciprocal crosses. Thus examination of the expectations in Table 111 shows that the A, B and C scaling tests will hold on the means of reciprocal crosses if an additive-dominance model is adequate. These expectations, however, do not hold if carried out on each reciprocal cross separately.

We may illustrate the detection and estimation of the effects of sex-linkage by reference to the unpublished data of Dr R. Killick. These consist of the parental F_1, F_2 and backcross generations of a cross between two inbred lines of *Drosophila melanogaster*, namely 6Cl(P_1) and the Birmingham strain of Oregon (P_2) scored for number of sternopleural chaetae. The data, which are the means of twenty flies from each of three replicate matings, are summarized in Table 112.

An analysis of variance of the reciprocal differences within the F_1, F_2 and backcross generations shows that all except the F_1 reciprocal difference in the females and the F_2 difference in the males are significant. This pattern of

TABLE 112

The mean sternopleural chaetae number of generations derived from a cross of inbred lines, 6Cl and Oregon of *Drosophila melanogaster*

Generation	Mating		Progeny			
	Female parent	Male parent	Female Observed	Female Model	Male Observed	Male Model
P_1	6Cl	×6Cl	34·88	$m+[d]+[d_x]$	33·15	$m+[d]+[d'_x]$
P_2	O	×O	22·80	$m-[d]-[d_x]$	22·63	$m-[d]-[d'_x]$
F_1	⎰ 6Cl	×O	28·98	$m+[h]+[h_x]$	29·58	$m+[h]+[d_x]$
	⎱ O	×6Cl	28·70	$m+[h]+[h_x]$	27·22	$m+[h]-[d_x]$
F_2	⎰ F_1(6Cl×O)	×F_1(6Cl×O)	30·32	$m+\frac{1}{2}[h]+\frac{1}{2}[d_x]+\frac{1}{2}[h_x]$	27·32	$m+\frac{1}{2}[h]$
	⎱ F_1(O×6Cl)	×F_1(O×6Cl)	27·88	$m+\frac{1}{2}[h]-\frac{1}{2}[d_x]+\frac{1}{2}[h_x]$	27·35	$m+\frac{1}{2}[h]$
B_1	⎧ 6Cl	×F_1(6Cl×O)	30·98	$m+\frac{1}{2}[d]+\frac{1}{2}[h]+[d_x]$	32·42	$m+\frac{1}{2}[d]+\frac{1}{2}[h]+[d'_x]$
	⎪ 6Cl	×F_1(O×6Cl)	32·85	$m+\frac{1}{2}[d]+\frac{1}{2}[h]+[h_x]$	31·42	$m+\frac{1}{2}[d]+\frac{1}{2}[h]+[d'_x]$
	⎨ F_1(6Cl×O)	×6Cl	32·50	$m+\frac{1}{2}[d]+\frac{1}{2}[h]+\frac{1}{2}[d_x]+\frac{1}{2}[h_x]$	29·75	$m+\frac{1}{2}[d]+\frac{1}{2}[h]$
	⎩ F_1(O×6Cl)	×6Cl	30·78	$m+\frac{1}{2}[d]+\frac{1}{2}[h]+\frac{1}{2}[d_x]+\frac{1}{2}[h_x]$	30·02	$m+\frac{1}{2}[d]+\frac{1}{2}[h]$
B_2	⎧ O	×F_1(6Cl×O)	25·33	$m-\frac{1}{2}[d]+\frac{1}{2}[h]+[h_x]$	24·65	$m-\frac{1}{2}[d]+\frac{1}{2}[h]-[d'_x]$
	⎪ O	×F_1(O×6Cl)	26·73	$m-\frac{1}{2}[d]+\frac{1}{2}[h]-[d_x]$	24·58	$m-\frac{1}{2}[d]+\frac{1}{2}[h]-[d'_x]$
	⎨ F_1(6Cl×O)	×O	25·80	$m-\frac{1}{2}[d]+\frac{1}{2}[h]-\frac{1}{2}[d_x]+\frac{1}{2}[h_x]$	26·70	$m-\frac{1}{2}[d]+\frac{1}{2}[h]$
	⎩ F_1(O×6Cl)	×O	25·43	$m-\frac{1}{2}[d]+\frac{1}{2}[h]-\frac{1}{2}[d_x]+\frac{1}{2}[h_x]$	25·62	$m-\frac{1}{2}[d]+\frac{1}{2}[h]$

Averaged over reciprocals

Generation	Female Observed	Female Model	Male Observed	Male Model
F_1	28·84	$m+[h]+[h_x]$	28·40	$m+[h]$
F_2	29·10	$m+\frac{1}{2}[h]+\frac{1}{2}[h_x]$	27·33	$m+\frac{1}{2}[h]$
B_1	31·78	$m+\frac{1}{2}[d]+\frac{1}{2}[h]+\frac{1}{2}[d_x]+\frac{1}{2}[h_x]$	30·90	$m+\frac{1}{2}[d]+\frac{1}{2}[h]+\frac{1}{2}[d'_x]+\frac{1}{2}[h_x]$
B_2	25·82	$m-\frac{1}{2}[d]+\frac{1}{2}[h]-\frac{1}{2}[d_x]+\frac{1}{2}[h_x]$	25·39	$m-\frac{1}{2}[d]+\frac{1}{2}[h]-\frac{1}{2}[d'_x]+\frac{1}{2}[h_x]$

Significance of reciprocal differences

Generation	Female	Male
F_1	$P>0·05$	$P<0·001$
F_2	$P<0·001$	$P>0·05$
B_1	$P<0·001$	$P<0·001$
B_2	$P=0·05–0·01$	$P<0·001$

significant reciprocal differences is exactly that which is expected if they are due to sex-linked genes (Table 111).

After averaging over reciprocal crosses within each generation the scaling test relationships between generation means should hold if an additive-dominance model is adequate. The results of the A, B and C scaling tests are summarized in Table 113. All the deviations are smaller than their standard errors apart from that of the C scaling test when applied to males. This deviation is significantly greater than zero with a probability of 0·05–0·02. Thus an additive-dominance model is adequate for the females, but there is some suggestion from one of the three tests that it may be inadequate for the males. We shall, however, proceed on the assumption that the simple model is satisfactory for both sexes.

TABLE 113
The scaling tests on the averages of reciprocal crosses

	Females	Males
A	$-0·16 \pm 0.51$	$0·25 \pm 0.48$
B	$0·00 \pm 0.45$	$-0·25 \pm 0.46$
C	$1·04 \pm 1.28$	$-3·26 \pm 1.63$

Earlier we noted that the scaling tests relationships do not hold if reciprocal crosses are tested independently. In the present data four C tests, two for each sex, can be made by considering each reciprocal series in turn. Both tests in the females show a significant deviation and one, that initiated by using P_2 as the female parent of the F_1, is significant in the males.

The additive-dominance model including autosomal, $[d]$ and $[h]$, and sex-linked, $[d_x]$ and $[h_x]$, effects given in Table 112 can now be fitted to the female and male data separately by the method of weighted least squares (Section 14). In the males four parameters must be fitted to 14 statistics leaving 10 df for testing the adequacy of the model. The estimates of these 4 parameters, all of which are significant, are given in Table 114 along with the χ^2 test of the goodness of fit of the model which is not significant. Hence the model is adequate.

In the females there are five parameters to fit to 14 statistics but 2 of these, $[h]$ and $[h_x]$, i.e. the dominance contributions to the generation means of autosomal and sex-linked genes, proved to be non-significant. A reduced model omitting these parameters was, therefore, fitted to the 14 statistics leaving 11 df for testing the goodness of fit. The estimates of the remaining 3 statistics which are all significant are given in Table 114 along with the χ^2 testing the goodness of fit. The purely additive model is clearly adequate for the females.

The estimates of the parameters, apart from $[h]$ are very similar in both

TABLE 114
Estimates of the autosomal and sex-linked effects ascribable to the additive and dominance components of the generation means

	Females		Males	
Parameter	Estimate	Parameter	Estimate
m	$28 \cdot 88 \pm 0 \cdot 083$	m	$27 \cdot 86 \pm 0 \cdot 140$
$[d]$	$4 \cdot 22 \pm 0 \cdot 309$	$[d]$	$4 \cdot 01 \pm 0 \cdot 199$
		$[h]$	$0 \cdot 49 \pm 0 \cdot 229$
$[d_x]$	$1 \cdot 81 \pm 0 \cdot 265$	$[d'_x]$	$1 \cdot 34 \pm 0 \cdot 136$
χ^2	$15 \cdot 07$		$17 \cdot 85$
df	11		10
P	$0 \cdot 20 - 0 \cdot 10$		$0 \cdot 10 - 0 \cdot 05$

sexes. Indeed, the relative proportions of the total additive difference borne on the autosomal and sex chromosomes are almost identical in the two sexes. It is worth noting that the additive effects of the sex-linked genes in the hemizygous males and the homozygous females do not differ significantly ($P = 0 \cdot 20 - 0 \cdot 10$) hence there is no evidence that $[d'_x]$ differs in magnitude from $[d_x]$ in these flies.

The procedures outlined above have allowed us not only to diagnose the presence of sex-linked gene effects, to determine the correct model for describing these effects and to estimate their magnitudes, but we have also carried out a simple chromosome assay for the X chromosome (see Chapter 1).

The additive-dominance model may be extended to include interactions between non-allelic genes for situations where the non-interactive model is inadequate. For the homogametic sex the specification of the interactions follows exactly that outlined in Section 16. In the heterogametic sex only interaction between hemizygous genes can occur, i.e. between d'_a and d'_b which we may designate as i'_{ab}. The specification then follows exactly that for haploid systems described in Section 53.

The expectations of the second degree statistics are, of course, changed in the presence of sex-linkage. Those for the F_2 and backcross generations are given in Table 115 (Cooke and Mather, 1962). By summing over statistics with identical expectations and over expectations that differ solely in the sign of F, the $S(dh)$ term, they reduce to the three expectations given at the foot of the table. These provide perfect fit solutions of the additive, dominance and environmental components of variation ascribable to sex-linked and auto-somal genes. They do not, however, allow a separation of the contributions of D_x and H_x because these occur throughout the expectations with the same coefficient and signs.

TABLE 115

Expected variance for autosomal and sex-linked genes in generations derived from a cross between two inbred lines

Generation	Mating — Homogametic parent	Heterogametic parent	Homogametic progeny	Heterogametic progeny
F_2	$F_1(P_1 \times P_2)$	$\times\ F_1(P_1 \times P_2)$	$\frac{1}{2}D + \frac{1}{4}H + \frac{1}{4}D_x + \frac{1}{4}H_x - \frac{1}{2}F_x + E_w$	$\frac{1}{2}D + \frac{1}{4}H + D'_x + E_w$
	$F_1(P_2 \times P_1)$	$\times\ F_1(P_2 \times P_1)$	$\frac{1}{2}D + \frac{1}{4}H + \frac{1}{4}D_x + \frac{1}{4}H_x + \frac{1}{2}F_x + E_w$	$\frac{1}{2}D + \frac{1}{4}H + D'_x + E_w$
B_1	P_1	$\times\ F_1(P_1 \times P_2)$	$\frac{1}{4}D + \frac{1}{4}H - \frac{1}{2}F + E_w$	$\frac{1}{4}D + \frac{1}{4}H - \frac{1}{2}F + E_w$
	P_1	$\times\ F_1(P_2 \times P_1)$	$\frac{1}{4}D + \frac{1}{4}H - \frac{1}{2}F + E_w$	$\frac{1}{4}D + \frac{1}{4}H - \frac{1}{2}F + E_w$
	$F_1(P_1 \times P_2)$	$\times\ P_1$	$\frac{1}{4}D + \frac{1}{4}H - \frac{1}{2}F + \frac{1}{4}D_x + \frac{1}{4}H_x - \frac{1}{2}F_x + E_w$	$\frac{1}{4}D + \frac{1}{4}H - \frac{1}{2}F + D'_x + E_w$
	$F_1(P_2 \times P_1)$	$\times\ P_1$	$\frac{1}{4}D + \frac{1}{4}H - \frac{1}{2}F + \frac{1}{4}D_x + \frac{1}{4}H_x - \frac{1}{2}F_x + E_w$	$\frac{1}{4}D + \frac{1}{4}H - \frac{1}{2}F + D'_x + E_w$
B_2	P_2	$\times\ F_1(P_1 \times P_2)$	$\frac{1}{4}D + \frac{1}{4}H + \frac{1}{2}F + E_w$	$\frac{1}{4}D + \frac{1}{4}H + \frac{1}{2}F + E_w$
	P_2	$\times\ F_1(P_2 \times P_1)$	$\frac{1}{4}D + \frac{1}{4}H + \frac{1}{2}F + E_w$	$\frac{1}{4}D + \frac{1}{4}H + \frac{1}{2}F + E_w$
	$F_1(P_1 \times P_2)$	$\times\ P_2$	$\frac{1}{4}D + \frac{1}{4}H + \frac{1}{2}F + \frac{1}{4}D_x + \frac{1}{4}H_x + \frac{1}{2}F_x + E_w$	$\frac{1}{4}D + \frac{1}{4}H + \frac{1}{2}F + D'_x + E_w$
	$F_1(P_2 \times P_1)$	$\times\ P_2$	$\frac{1}{4}D + \frac{1}{4}H + \frac{1}{2}F + \frac{1}{4}D_x + \frac{1}{4}H_x + \frac{1}{2}F_x + E_w$	$\frac{1}{4}D + \frac{1}{4}H + \frac{1}{2}F + D'_x + E_w$

Summed statistics

			Homogametic progeny	Heterogametic progeny
F_2 sum of reciprocal variances			$D + \frac{1}{2}H + \frac{1}{2}D_x + \frac{1}{2}H_x + 2E_w$	$D + \frac{1}{2}H + 2D'_x + 2E_w$
$B_1 + B_2$ sum of variances using inbred lines as homogametic parent			$\frac{1}{2}D + \frac{1}{2}H + 2E_w$	$\frac{1}{2}D + \frac{1}{2}H + 2E_w$
$B_1 + B_2$ sum of variances using F_1 as homogametic parent			$\frac{1}{2}D + \frac{1}{2}H + \frac{1}{2}D_x + \frac{1}{2}H_x + 2E_w$	$\frac{1}{2}D + \frac{1}{2}H + 2D'_x + 2E_w$

50. SEX-LINKAGE: DIALLELS AND RANDOMLY BREEDING POPULATIONS

In the multiple matings between many inbred lines such as a diallel set of crosses (Chapter 9) sex-linkage may be detected by comparing reciprocal crosses. Thus individuals of the homogametic sex in the F_1s will show no differences between reciprocals in the presence of sex-linkage while in the heterogametic sex a difference is expected. So if an analysis of variance is carried out on a diallel table containing the results of homogametic and heterogametic sexes separately, sex-linkage will be indicated by a significant interaction between sex and reciprocal differences, which can be confirmed by finding no significant difference between reciprocal crosses in the homogametic sex and a significant difference in the heterogametic sex. Alternatively, the diallel can be treated as a special case of random mating and tests applied of the kind described below.

The contributions of sex-linked genes to the variances and covariances that can be derived from a randomly mating population (Chapter 8) have been given by Mather and Jinks (1963). These are summarized in Table 116.

TABLE 116
The contribution of sex-linked genes to the variances and covariances derivable from a randomly mating population

Statistic	Progeny	
	Homogametic sex	Heterogametic sex
Total variation	$\frac{1}{2}D_{Rx} + \frac{1}{4}H_{Rx} + E_w + E_b$	$D'_x + E_w + E_b$
Variation between family means (covariance of full-sibs)	$\frac{3}{8}D_{Rx} + \frac{1}{8}H_{Rx} + E_b$	$\frac{1}{2}D'_x + E_b$
Mean variation within families	$\frac{1}{8}D_{Rx} + \frac{1}{8}H_{Rx} + E_w$	$\frac{1}{2}D'_x + E_w$
Parent/offspring covariances		
On homogametic parent	$\frac{1}{4}D_{Rx}$	$\frac{1}{2}D'_{xx} - \frac{1}{4}F'_{xx}$
On heterogametic parent	$\frac{1}{2}D'_{xx} - \frac{1}{4}F'_{xx}$	0
Covariance homogametic/heterogametic full-sibs	$\frac{1}{4}D'_{xx} - \frac{1}{8}F'_{xx}$	

Where:

$D_{Rx} = S4uv[d + (v-u)h_x]^2$ } Random mating forms of D and H for genes
$H_{Rx} = S16u^2v^2h_x^2$ } showing sex-linkage

$D'_x = S4uvd'^2_x$ — Diallel form of D for genes showing sex-linkage

$D'_{xx} = S4uvd_xd'_x$ } Modified diallel forms of D and F for genes
$F'_{xx} = S8uvd'_x(u-v)h_x$ } showing sex-linkage

The contributions of autosomal genes, as derived in Chapter 8, must of course be added to the sex-linked contribution to complete the expectations. The statistics listed in the table will therefore be inadequate to estimate all the parameters of an additive-dominance model although a partial separation can be achieved using the North Carolina 2 and diallel mating designs described in Section 39 (Killick, 1971a). Nevertheless it is possible to devise a number of tests which can diagnose the presence of contributions from sex-linked genes.

The most diagnostic effects of sex-linked genes are on the parent/offspring and full-sib covariances. In the absence of any contribution of sex-linked genes the parent/offspring covariances are expected to be the same and equal to $\frac{1}{4}D_R$, irrespective of the sex of the parent and of the offspring being considered. If, on the other hand, sex-linked genes are making a significant contribution the covariance between heterogametic parent and heterogametic offspring is expected to be smaller than the covariances involving the homogametic parent or homogametic offspring, or both. Furthermore, for all gene frequencies and all values of h from $h = 0$ to $h = d$ the relative magnitudes of the four covariances will be, XX parent/XY offspring = XY parent/XX offspring $\geqslant$ XX parent/XX offspring > XY parent/XY offspring, provided that the additive gene effects are more or less equal in magnitude in the homozygous and hemizygous states.

Similarly in the absence of any contribution from sex-linked genes the full-sib correlations are expected to be the same and equal to

$$\frac{\frac{1}{4}D_R + \frac{1}{16}H_R}{\frac{1}{2}D_R + \frac{1}{4}H_R + E_w + E_b},$$

irrespective of the sex of the sib being considered. The maximum value the correlation can take is 0·5, when $H_R = 0$ and $E_w = E_b = 0$. Where only sex-linked differences are involved the corresponding maxima are: correlation between XX sibs = 0·75, correlation between XY sibs = 0·50, and correlation between XX and XY sibs = 0·35, assuming that $d_x = d'_x$ for all loci.

The first diagnostic feature of these correlations is, therefore, that in the presence of sex-linkage the correlation between homogametic sibs may exceed the normal maximum value of 0·5 for autosomal genes.

If $h > 0$ the values of the correlations involving one or more homogametic sibs will depend on the values of u and v. With $h = d$ at all loci the value of the correlation between homogametic sibs ranges from 0·75 to 0·50 and that between homogametic and heterogametic sibs ranges from 0·35 to 0 as $u (= 1 - v)$ varies from 0 to 1·0. But for all values of u the relative magnitudes of the three correlations in the presence of sex-linkage are: XX sibs > XY sibs > XX sib and XY sib.

The actual values of the correlations will, of course, vary with the effects of autosomal genes and environmental agencies but assuming these to be the same in both sexes this sequence of relative values will hold and can be used to detect sex-linked inheritance.

The presence of twins in a randomly mating population does not directly assist the analysis of sex-linkage but their progeny families do. The families of monozygotic twins are genetically, although not biologically, half-sibs. Since the twins may be homogametic or heterogametic and each of these may produce homogametic and heterogametic progeny there are potentially six half-sib covariances whose expectations can be given in terms of the same components as the full-sib and parent/offspring covariances. Since all three types of covariances can be obtained from random samples of monozygotic twins and their progenies these parents and families alone provide sufficient statistics to estimate all the components of variation for the autosomal and sex-linked loci simultaneously (Haley *et al.*, 1981).

51. MATERNAL EFFECTS: CROSSES BETWEEN INBRED LINES

Maternal effects arise where the mother makes a contribution to the phenotype of her progeny over and above that which results from the genes she contributes to the zygote. These contributions may take one or more of the following forms:

(i) Cytoplasmic inheritance.
(ii) Maternal nutrition either via the egg or via pre- and post-natal supplies of food.
(iii) Transmission of pathogens and antibodies through the pre-natal blood supply or by post-natal feeding.
(iv) Imitative behaviour
(v) Interaction between sibs either directly with one another or through the mother.

Notwithstanding the diversity of underlying causes the results of these maternal influences are much the same, namely, to make the phenotype of the progeny more like that of the maternal than of the paternal strain. Where, as in some species, the father plays the more active role in rearing the offspring, paternal effects may arise which can be detected by the greater similarity of the offspring to the phenotype of the paternal strain to which they lead. We will, however, confine ourselves to maternal effects although the methods used for their specification, detection and estimation can easily be applied to the less common situation.

The net result of maternal effects is to produce differences between reciprocal crosses which, unlike those produced by sex-linkage, are shared by the offspring of both sexes in all the generations where they occur. Furthermore, unlike sex-linkage, maternal effects lead to a greater resemblance of the progeny to the maternal parent irrespective of whether this happens to be the homogametic sex in the species under consideration.

The maternal component contributed to their progenies by the three genotypes AA, Aa and aa will be specified by dm_a, hm_a and $-dm_a$, respectively. The means of the generations derivable from an initial cross between two inbred lines that differ for many pairs of independent genes are given in Table 117.

TABLE 117
The expected generation means on an additive-dominance model in the presence of maternal effects

Generation	Mating		Expected contribution to generation mean		Observed mean yield
	Maternal parent	Paternal parent	Progeny genotype	Maternal contribution	
P_1	P_1 ×	P_1	$m + [d]$	$+[dm]$	120·17
P_2	P_2 ×	P_2	$m - [d]$	$-[dm]$	114·17
F_1	$\begin{cases} P_1 \times \\ P_2 \times \end{cases}$	$\begin{cases} P_2 \\ P_1 \end{cases}$	$m + [h]$ $m + [h]$	$+[dm]$ $-[dm]$	159·67 147·33
F_2	F_1 ×	F_1	$m + \frac{1}{2}[h]$	$+[hm]$	239·33
B_1	$\begin{cases} P_1 \times \\ F_1 \times \end{cases}$	$\begin{cases} F_1 \\ P_1 \end{cases}$	$m + \frac{1}{2}[d] + \frac{1}{2}[h]$ $m + \frac{1}{2}[d] + \frac{1}{2}[h]$	$+[dm]$ $+[hm]$	117·66 259·66
B_2	$\begin{cases} P_2 \times \\ F_1 \times \end{cases}$	$\begin{cases} F_1 \\ P_2 \end{cases}$	$m - \frac{1}{2}[d] + \frac{1}{2}[h]$ $m - \frac{1}{2}[d] + \frac{1}{2}[h]$	$-[dm]$ $+[hm]$	123·75 238·83

The application of this model will be illustrated by the data of Barnes (1968) concerning the yield of adult flies in the F_1, F_2 and first backcross generations derived from a cross between two inbred lines of *D. melanogaster*, namely, Oregon (P_1) and Samarkand (P_2). The yields measured by the mean number of flies per mating in each generation averaged over three indepedently randomized replicates of each mating are also given in Table 117. The mean number of flies is treated as a character determined by the genotype of the offspring with a maternal component dependent on the mother's genotype.

Tests of the reciprocal crosses in the three generations, F_1, B_1 and B_2, in which differences arising from maternal effects might be expected, shows that only those in the B_1 and B_2 generations are significant. Reference to the model in the table suggests that since the reciprocal F_1s do not differ $[dm]$ must be zero, while the greater yields of the reciprocal backcrosses in which the F_1 was used as the mother suggests that $[hm]$ is greater than zero. A model can, therefore, be fitted based on the four parameters m, $[d]$, $[h]$ and $[hm]$. The weighted least squares estimates of these parameters obtained by the procedures described in Section 14 are,

$$\hat{m} = 109·47 \pm 10·97$$
$$[\hat{d}] = 5·17 \pm 8·95$$
$$[\hat{h}] = 36·21 \pm 17·91$$
$$[\hat{hm}] = 115.87 \pm 8·95$$

All except for $[\hat{d}]$ are significant at $P = 0·05$. This model is satisfactory ($\chi^2_{(5)} = 3·22$, $P = 0·70 - 0·50$) and clearly a model which omitted $[\hat{d}]$ would be equally satisfactory. Oregon and Samarkand do not differ either for this character or for their maternal effects. There is, however, both dominance [h] for high yield of genes dispersed between the two parents and a maternal effect of F_1 mothers [hm], which increases their yield of progeny.

The simple situation revealed by this analysis could be complicated in one of three ways, namely:

(i) There could be interaction between non-allelic genes in the progeny genotype.

(ii) There could be interaction between non-allelic genes responsible for the maternal effect.

(iii) There could be interaction between the progeny genotype and the maternal effect.

The specification and estimation of the complication described under (i) follow the usual form for non-allelic interactions described in Section 16. The specification and estimation of the effects described under (ii) also follow a modification of this scheme. For example, for interactions between the additive maternal effects we introduce the symbol [im] by analogy with [i] and for the interactions between the dominance maternal effects we can use [lm] by analogy with [l]. The interactions described under (iii) can also be specified in a similar way. Thus for an additive gene effect in the progeny which interacts with an additive maternal effect we may write [$d \cdot dm$] and for comparable dominance effects in progeny and maternal contribution we may specify the interaction by [$h.hm$]. The contributions of the interactions described under (ii) and (iii) to the generation means are given in Table 118.

Because of the variety of ways in which the simple additive-dominance model may fail and the large number of additional parameters that are required to

TABLE 118
The specification of the contribution of interactions between non-allelic genes which control the maternal effect and of interactions between progeny genotype and maternal effects to the generation means (D. W. Fulker, unpublished)

Generation	Mating	Contribution to generation mean		
		Progeny genotype and additive maternal effect	Maternal interaction	Interaction progeny genotype/maternal effect
P_1	$P_1 \times P_1$	$m + [d] + [dm]$	$+[im]$	$+[d.dm]$
P_2	$P_2 \times P_2$	$m - [d] - [dm]$	$+[im]$	$+[d.dm]$
F_1	$\begin{cases} P_1 \times P_2 \\ P_2 \times P_1 \end{cases}$	$m + [h] + [dm]$ $m + [h] - [dm]$	$+[im]$ $+[im]$	$+[h.dm]$ $-[h.dm]$
F_2	$F_1 \times F_1$	$m + \frac{1}{2}[h] + [hm]$	$+[lm]$	$+\frac{1}{2}[h.hm]$
B_1	$\begin{cases} P_1 \times F_1 \\ F_1 \times P_1 \end{cases}$	$m + \frac{1}{2}[d] + \frac{1}{2}[h] + [dm]$ $m + \frac{1}{2}[d] + \frac{1}{2}[h] + [hm]$	$+[im]$ $+[lm]$	$+\frac{1}{2}[d.dm] + \frac{1}{2}[h.dm]$ $+\frac{1}{2}[d.hm] + \frac{1}{2}[h.hm]$
B_2	$\begin{cases} P_2 \times F_1 \\ F_1 \times P_2 \end{cases}$	$m - \frac{1}{2}[d] + \frac{1}{2}[h] - [dm]$ $m - \frac{1}{2}[d] + \frac{1}{2}[h] + [hm]$	$+[im]$ $+[lm]$	$+\frac{1}{2}[d.dm] - \frac{1}{2}[h.dm]$ $-\frac{1}{2}[d.hm] + \frac{1}{2}[h.hm]$

specify all the possible types of interaction that could lead to such a failure, it is necessary to apply tests which will distinguish between them wherever this is possible. The standard scaling tests are, of course, inappropriate as these will fail in the presence of maternal effects even if a simple additive-dominance model is adequate and the tests are carried out on the means of reciprocal crosses. Nevertheless, tests can be devised which can distinguish between some of the types of interactions and which can, therefore, be used to determine the most appropriate interaction parameters to include in the model prior to fitting it to the data. For example, the three generations P_1, F_1 and B_1 can all be obtained with a constant parent, P_1, as mother, similarly, P_2, F_1 and B_2 can be obtained with P_2 as mother. The A and B scaling test relationships will then hold within these two sets of generation means provided that there are no non-allelic interactions among the genes of the progeny genotypes. Hence, failure of the A and B scaling tests when applied to these generations is still diagnostic of the presence of this type of interaction even in the presence of interactions between progeny genotype and maternal effects and of a non-allelic interaction component of the maternal effects.

A similar test can be carried out on the F_2, B_1 and B_2 generations all of which can be obtained using an F_1 mother when we expect $2\bar{F}_2 - \bar{B}_1 - \bar{B}_2 = 0$ in the absence of a non-allelic interaction contribution to the phenotype by the progeny genotype.

Provided that the simple additive-dominance model is adequate, the maternal effects will make no contribution to the variation within families of the P_1, P_2, F_1, F_2 and first backcross generations because all the members of any one family will have the same mother. This equally applies to the variances within families of any generation: for example, F_3 families obtained by selfing F_2 individuals or S_3 families raised by sib-mating pairs of F_2individuals, etc. However, in the latter generations maternal effects still contribute to the variance between families. This may be illustrated by reference to the F_3 generation. The three types of F_2 mothers in respect of a single gene difference, namely, AA, Aa and aa, will contribute dm_a, hm_a and $-dm_a$ to the mean phenotype of their progenies, respectively, while the progenies' own genotypes will contribute d_a, $\frac{1}{2}h_a$ and $-d_a$. The contribution of this gene to the mean variance within families will be $\frac{1}{4}d_a^2 + \frac{1}{8}h_a^2$ but its contribution to the variance of family means is

$$\frac{1}{2}(d_a + dm_a)^2 + \frac{1}{4}(\frac{1}{2}h_a + hm_a)^2.$$

Hence, the maternal effects will always inflate the variation between families while having no effect on the variances within families. This, however, ceases to be true once the simple model fails because of the interaction between progeny genotype and maternal effect. In these circumstances the variance within families is also inflated as the following single gene illustration shows. In an F_2, the F_1 mother will contribute a maternal effect hm_a to the progeny, but this will interact with the three progeny genotypes in different ways so that the individual mean phenotypes will be

Frequency	Progeny genotype	Progeny mean
$\frac{1}{4}$	AA	$m + d_a + hm_a + d_a hm_a$
$\frac{1}{2}$	Aa	$m + h_a + hm_a + h_a hm_a$
$\frac{1}{4}$	aa	$m - d_a + hm_a - d_a hm_a$
F_2 mean		$m + \frac{1}{2}h_a + hm_a + \frac{1}{2}h_a hm_a$.

The contribution of this gene difference to the variance within the F_2 will then be

$$\tfrac{1}{2}(d_a + d_a hm_a)^2 + \tfrac{1}{4}(h_a + h_a hm_a)^2.$$

Where, as for example in the backcross, different maternal contributions to the backcross progeny can be produced by carrying out the crosses reciprocally, it is possible to detect an interaction between progeny genotype and maternal effects as a difference in the within family variances of reciprocal crosses. Thus for a single gene the variances of the B_1 families using P_1 as mother and F_1 as mother are

$$\tfrac{1}{2}(d_a + d_a dm_a - h_a - h_a dm_a)^2$$

and

$$\tfrac{1}{2}(d_a + d_a hm_a - h_a - h_a hm_a)^2$$

respectively.

These will not necessarily differ in the presence of interactions but they will differ when the four interaction parameters are related such that $d_a dm_a - h_a dm_a \neq d_a hm_a - h_a hm_a$, i.e. the interaction between additive genetic and additive maternal effect differs in magnitude from the interaction between dominance genetic and dominance maternal effect. By raising sufficient generations from an initial cross between two inbred lines, particularly those generations that can be obtained reciprocally, it is clearly possible to obtain sufficient statistics of an appropriate kind to estimate the four interaction parameters.

52. MATERNAL EFFECTS: MULTIPLE CROSSES

Where inbred lines are available maternal effects can be detected and analysed easily and unambiguously. They are not, however, essential for this purpose in all circumstances. Thus where, as in hermaphroditic species, each individual can be used both as a maternal and as a paternal parent, maternal effects can be detected by making crosses reciprocally between pairs of individuals irrespective of whether they are homozygous or heterozygous. However, unless the breeding history of the individuals is known, no worthwhile biometrical genetical analysis or interpretation is possible. The kind of analysis that is possible when the source of the material is known may be illustrated by considering a diallel set of crosses among a random sample of individuals drawn from a randomly mating population. The contribution of a single gene difference to the maternal effect is given in Table 119.

Reference to the table shows that the maternal effect will inflate the total variation among the progenies of the diallel set of crosses and the variance of maternal array means. It does not, however, make any contribution to the variance of paternal array means. Hence, in the presence of maternal effects the variances of maternal array means will be greater than that of the paternal array means. Or, to put it another way, a maternal effect will inflate the half-sib covariance, where the common parent is the mother, relative to that where the common parent is the father. To remove the contribution of the offspring genotype to these two variances the variance of the difference between maternal and paternal array means can be computed and its magnitude depends solely on the maternal effects. Thus,

variance (maternal array mean − paternal array mean)

$$= 2u_a v_a dm_a^2 + 2u_a v_a hm_a^2 - 4u_a^2 v_a^2 hm_a^2 - 4u_a v_a (u_a - v_a) dm_a hm_a$$

which summing over all relevant genes becomes $\frac{1}{2}Dm_R + \frac{1}{4}Hm_R$ (by analogy with D_R and H_R).

The total variation due to maternal effects can similarly be computed separately from the contribution of the offspring genotype to the total variation by using the differences between the pairs of reciprocal crosses taking sign into account. The expected values of these differences is shown in Table 119 and their variance is

$$2u_a v_a dm_a^2 + 2u_a v_a hm_a^2 - 4u_a^2 v_a^2 hm_a^2 - 4u_a v_a (u_a - v_a) dm_a hm_a$$

This, of course, equals the variance of the differences between maternal and paternal array means. In other words, the latter accounts for all the variance attributable to maternal effects on the simple additive-dominance model. If

TABLE 119
The contribution of a single gene difference to the maternal effects in a diallel set of crosses in a randomly mating population and the differences between reciprocal crosses

Maternal parent		Paternal parent				
Frequency	Genotype	Frequency Genotype	u_a^2 AA	$2u_a v_a$ Aa	v_a^2 aa	Maternal array means
u_a^2	AA	Maternal effect Recip. diff.	dm_a	dm_a $dm_a - hm_a$	dm_a $2dm_a$	dm_a
$2u_a v_a$	Aa		hm_a $-dm_a + hm_a$	hm_a ·	hm_a $dm_a + hm_a$	hm_a
v_a^2	aa		$-dm_a$ $-2dm_a$	$-dm_a$ $-dm_a - hm_a$	$-dm_a$ ·	$-dm_a$
Paternal array means			$(u_a - v_a)dm_a$ $+ 2u_a v_a hm_a$	$(u_a - v_a)dm_a$ $+ 2u_a v_a hm_a$	$(u_a - v_a)dm_a$ $+ 2u_a v_a hm_a$	$(u_a - v_a)dm_a$ $+ 2u_a v_a hm_a$

this should prove not to be the case for any set of data we can take it as evidence that this simple model of the maternal effects is inadequate. But of the three ways in which the simple model can fail, as described earlier, only one will lead to a failure on this test, namely where there is an interaction between the progeny genotype and the maternal effect. Hence, by partitioning the total variation due to maternal effects into that part attributable to the variance of differences between maternal and paternal array means and that which is not attributable to this source we can detect and estimate the contribution of maternal effects and their interaction with the progeny genotype. This partitioning is the basis of the analysis of maternal effects for the special case of a diallel set of crosses between a number of inbred lines described by Hayman (1954a) and Wearden (1964). Further elaborations of the analysis to investigate the nature of the gene action underlying the maternal effects have been described by Durrant (1965).

One further crossing programme, namely, that referred to as North Carolina type 2 (Section 39), will also permit the detection of maternal effects. This provides estimates of the components of variance of maternal and paternal array means. Hence, if the former is the larger the presence of maternal effects is indicated. It is not, however, possible to test the adequacy of an additive-dominance model in this type of experiment since any interaction between maternal effect and progeny genotype will be confounded with the interaction between maternal and paternal parents which in this design provides the estimate of H_R (see Section 39).

53. HAPLOIDS

Sex-linkage and maternal effects may of course arise in diploid species and we have discussed them in relation to disomic inheritance. Many organisms, for example, most fungi, algae, protozoans and bacteria, are however haploid, while many higher plants are polyploid. In this and the next section we turn to consider the modifications in the biometrical models and analytical procedures developed for diploids that are necessary when applying them to the study of haploids and polyploids.

In a haploid only two genotypes are possible in respect of a single gene difference A–a, namely, A and a. Although we are now dealing with haploids rather than homozygous diploids the contribution of this gene pair to the mean phenotypes will be symbolized by d_a and $-d_a$, respectively. For two lines P_1 and P_2, differing by any number of independent genes, the mean phenotype is then expected to be

$$\bar{P}_1 = m + [d]$$
$$\bar{P}_2 = m - [d].$$

Hence
$$m = \tfrac{1}{2}(\bar{P}_1 + \bar{P}_2)$$

and
$$[d] = \tfrac{1}{2}(\bar{P}_1 - \bar{P}_2).$$

If we cross these two lines the haploid (F_1) progeny will segregate 1:1 for the genes at each locus for which the parents differ. Hence, the additive contributions at each locus to the progeny mean will cancel out leaving an expected progeny mean of m.

Further generations can be raised by randomly mating or backcrossing the F_1 progeny. The mean of the progeny produced by randomly mating $(\bar{R}_2)$ by any of the designs previously described (Chapter 8) will be m. A backcross to parent P_1 will have a mean $\bar{B}_1 = m + \frac{1}{2}[d]$ while that to parent P_2 will have a mean $\bar{B}_2 = m - \frac{1}{2}[d]$.

On the simple additive model there are clearly many relationships between the means of the generations derivable from an initial cross betwen two lines. For example,

$$\frac{1}{2}(\bar{P}_1 + \bar{P}_2) = \bar{F}_1 = \bar{R}_2 = \frac{1}{2}(\bar{B}_1 + \bar{B}_2) = m$$
$$\frac{1}{2}\bar{P}_1 + \frac{1}{2}\bar{F}_1 = \bar{B}_1$$
$$\frac{1}{2}\bar{P}_2 + \frac{1}{2}\bar{F}_1 = \bar{B}_2$$
$$\frac{1}{2}(\bar{P}_1 - \bar{P}_2) = (\bar{B}_1 - \bar{B}_2) = [d].$$

The relationships provide scaling tests of the adequacy of the simple model comparable to those used with diploids (Chapter 4). They will fail in the presence of non-allelic interactions which in haploids must be of the additive/additive kind only and which by analogy can be symbolized by i. The means of the four possible genotypes in respects of two gene differences A–a, B–b in the presence of interaction are

Genotype	Mean
AB	$m + d_a + d_b + i_{ab}$
Ab	$m + d_a - d_b - i_{ab}$
aB	$m - d_a + d_b - i_{ab}$
ab	$m - d_a - d_b + i_{ab}$

The two kinds of crosses in respect of these two genes, namely the association cross AB × ab and the dispersion cross Ab × aB will produce these four genotypes in equal frequencies in the absence of linkage. Hence the progeny means of both crosses will again be m irrespective of the phase of the cross or the presence of interactions. The progeny mean, however, will now no longer equal the parental mean. In fact, the difference allows us to estimate i_{ab}. Thus summing over all interacting genes

$$\bar{F}_1 - \frac{1}{2}(\bar{P}_1 + \bar{P}_2) = -[i].$$

Given, therefore, the means of the parental lines and of their F_1 progeny it is possible to estimate m, $[d]$ and $[i]$ which are all the parameters required to specify generation means in a haploid.

$$m = \bar{F}_1 \qquad\qquad V_m = V_{\bar{F}_1}$$
$$[d] = \tfrac{1}{2}(\bar{P}_1 - \bar{P}_2) \qquad V_{[d]} = \tfrac{1}{4}V_{\bar{P}_1} + \tfrac{1}{4}V_{\bar{P}_2}$$
$$[i] = \tfrac{1}{2}(\bar{P}_1 + \bar{P}_2) - \bar{F}_1 \qquad V_{[i]} = V_{\bar{F}_1} + \tfrac{1}{4}V_{\bar{P}_1} + \tfrac{1}{4}V_{\bar{P}_2}$$

The analytical procedures may be illustrated by reference to the progenies of crosses between two high (H_1 and H_2) and two low (L_1 and L_2) selections for rate of growth in the monokaryotic phase of *Schizophyllum commune* (Simchen, 1966). The mean rates of growth of the selections and of the F_1 progenies of crosses between the high and low selections are given in Table 120A. The estimates of the parameters show that the simple additive model is adequate, the estimates of $[i]$ all being smaller than their standard errors.

A further example is provided by crosses between wild isolates of *Aspergillus nidulans* (Jinks *et al.*, 1966), the character again being rate of growth (Table 120B). Analysis of the two crosses shows that the simple additive model is inadequate, significant interactions being in fact present. The estimates of $[i]$ are positive because the parental means are greater than the progeny means. Hence, the poorer growth of the progeny can be interpreted as the breakdown of interactions present in the wild isolates following an outcross between them. This is in agreement with the observation that the wild isolates appear to be genetically isolated in nature (Butcher, 1969; Jinks *et al.*, 1966).

The contribution of a single gene difference A–a to the variation among the F_1 progeny of a cross between two lines is d_a^2 and hence taking into account all relevant genes it equals $D = S d^2$. Similarly, for a pair of interacting genes A–a, B–b the contribution to the variance is $d_a^2 + d_b^2 + i_{ab}^2$, in the absence of linkage, and over all pairs of genes it equals $D + I$.

The total variation among the progeny will, of course, have an environmental component which can be estimated as the variation among replicates of the parental lines or by replication of the progeny. Since most haploid organisms may be replicated by clonal propagation or by asexually produced spores to any extent required the estimation of this component seldom presents difficulties.

We may illustrate the analysis by reference once more to the *S. commune* crosses of Simchen (1966). In this experiment two replicates of each F_1 individual were grown, one replicate in each of two blocks. The analysis of variance for the first cross (Table 120) $L_1 \times H_1$ is, therefore:

Item	df	MS
Progenies (P)	99	85·60
Blocks (B)	1	0·72
P × B	99	5·72

There are highly significant differences between the individuals in the F_1.

TABLE 120

A. Parental and F_1 means of crosses between high and low selections for rate of growth of *Schizophyllum commune*

Cross	Number of progeny	Generation means			Estimates of components		
		$\bar{P}_1$	$\bar{P}_2$	$\bar{F}_1$	m	$[d]$	$[i]$
$L_1 \times H_1$	100	97·00 ± 2·20	56·00 ± 2·20	77·23 ± 0·22	77·23	20·50	−0·73 ± 1·56
$L_1 \times H_2$	99	104·50 ± 2·19	54·75 ± 2·19	81·12 ± 0·22	81·12	24·89	−1·49 ± 1·55
$L_2 \times H_1$	100	101·00 ± 3·50	53·50 ± 3·50	75·78 ± 0·35	75·78	23·75	+1·47 ± 2·50
$L_2 \times H_2$	99	106·50 ± 3·88	44·50 ± 3·88	77·32 ± 0·39	77·32	29·00	−0·18 ± 2·76

B. Parental and F_1 means of crosses between pairs of wild type lines of *Aspergillus nidulans*

Cross	Number of progeny	Generation means			Estimates of components		
		$\bar{P}_1$	$\bar{P}_2$	$\bar{F}_1$	m	$[d]$	$[i]$
9 × 43	99	6·70 ± 0·2323	5·80 ± 0·2323	5·28 ± 0·0363	5·28	0·45	0·97 ± 0·1682
37 × 114	82	4·80 ± 0·2291	4·20 ± 0·2291	4·04 ± 0·0317	4·04	0·30	0·46 ± 0·1652

The estimate of E_w is 5·72. Since there is no evidence of non-allelic interactions $\hat{D} = \frac{1}{2}(85·60 - 5·72) = 39·94$. The estimates of D and E for all four crosses are given in Table 121.

TABLE 121
Components of variation for the crosses between selection for high and low rate of growth in *S. commune*

Cross	Components	
	D	E_w
$L_1 \times H_1$	39·94	5·72
$L_1 \times H_2$	61·70	4·91
$L_2 \times H_1$	119·94	12·33
$L_2 \times H_2$	164·28	13·85

Where the simple additive model is inadequate, as for example in the *A. nidulans* crosses (Table 120B), only $(D + I)$ and E can be estimated from the F_1 progeny. However, D and I may be separated by raising further generations such as backcrosses. The contribution of two interacting pairs of genes to the variation within the backcross to a parent with the increasing alleles in association AB is

$$\tfrac{1}{4}d_a^2 + \tfrac{1}{4}d_b^2 + \tfrac{3}{16}i_{ab}^2 + \tfrac{1}{4}(d_a + d_b)i_{ab},$$

while that to the corresponding backcross to the other parent ab is

$$\tfrac{1}{4}d_a^2 + \tfrac{1}{4}d_b^2 + \tfrac{3}{16}i_{ab}^2 - \tfrac{1}{4}(d_a + d_b)i_{ab}.$$

Summing over the two backcrosses we obtain,

$$\tfrac{1}{2}d_a^2 + \tfrac{1}{2}d_b^2 + \tfrac{3}{8}i_{ab}^2$$

which summing over all pairs of interacting genes becomes

$$\tfrac{1}{2}D + \tfrac{3}{8}I.$$

The same result is obtained on summing the variances of the two backcrosses when the genes are dispersed in the parental lines, i.e. Ab and aB. The total variance will have an environmental contribution of $2E_w$. Thus by raising backcrosses D, I and E_w can be estimated. Furthermore, the clonal nature of most haploid organisms makes it possible to raise the progenies of all the necessary generations simultaneously and randomized in the same experiment. Examples of how these advantages have been exploited are contained in reviews by Caten and Jinks (1976) and Caten (1979).

In some fungi, particularly the heterothallic basidiomycetes, there is a free living phase in the life cycle in which pairs of unlike haploid nuclei (usually of opposite mating type) are associated in the cells, sharing the same cytoplasm without fusing. In this dikaryotic condition the unlike nuclei interact to produce effects comparable with dominance and non-allelic interactions. In the dikaryotic phase of the life cycle, therefore, the expectations and models appropriate to diploid organisms may be applied to the study of continuously varying characters (Simchen and Jinks, 1964). Since the same organisms also have a free living monokaryotic haploid phase in their life cycles they offer the unique possibility of comparing the actions of the same genes in the monokaryotic (haploid) and dikaryotic (diploid) conditions. Extensive applications of these analyses to variation in *Schizophyllum commune* are contained in the papers of Simchen and Jinks (1964), Connolly and Simchen (1968) and Connolly and Jinks (1975).

54. POLYPLOIDS

A general treatment of polysomic inheritance has not been attempted because of its inherent complexity (see Fisher and Mather, 1943; Mather, 1936) and the lack of practical applications. We shall confine our attention to a few special cases involving a single locus or an arbitrary number of independent loci among which there is neither non-allelic interactions nor linkage. We shall further consider only the case in which segregation is not complicated by the occurrence of double reduction. A fuller treatment which nevertheless retains most of these limitations, is given by Kempthorne (1957) and the effect of double reduction on our expectations is described by Killick (1971b).

In a tetraploid where only one gene difference, A–a, is considered, two homozygous, AAAA and aaaa, and three heterozygous, AAAa, AAaa and Aaaa, genotypes can occur. Using the notation developed for diploid inheritance, the contribution of this gene difference to the mean phenotypes of the two homozygotes will be defined as d_a for AAAA and $-d_a$ for aaaa around a mid-homozygote value of m. In the general case the three heterozygotes will be defined as having different mean phenotypic deviations from m as follows: Aaaa, h_1; AAaa, h_2; and AAAa, h_3.

Special cases can then be derived by assuming various relationships between the relative magnitudes of h_1, h_2 and h_3.

The expected means of the generations that can be derived from an initial cross between the two homozygotes AAAA and aaaa are given in Table 122. Reference to these expectations shows that the A, B and C scaling test relationships do not hold even though the expectations contain only additive and dominance effects. They hold, however, in certain special cases. The first of these is the relatively trivial case of no dominance, i.e. $h_{a1} = h_{a2} = h_{a3} = 0$. The second depends on the assumption that the contributions to the mean phenotypes of the heterozygous genotypes AAAa and Aaaa fall midway betwen those of the duplex heterozygote, AAaa, and the homozygotes, AAAA and aaaa, respectively. The expected contributions to the mean phenotypes of the three heterozygotes are then,

Aaaa	$-\frac{1}{2}d_a + \frac{1}{2}h_a$
AAaa	h_a
AAAa	$\frac{1}{2}d_a + \frac{1}{2}h_a.$

Substituting these values for h_{a1}, h_{a2} and h_{a3} in the expectation in Table 122 we find that the A and B scaling test relationships now hold but not the C test. Killick (1971b) has shown further that the A and B scaling tests are satisfactory even with double reduction as are also the relationships

$$\bar{B}_{11} = \tfrac{3}{4}\bar{P}_1 + \tfrac{1}{4}\bar{F}_1$$
$$\bar{B}_{12} = \tfrac{1}{4}\bar{P}_2 + \tfrac{3}{4}\bar{F}_1$$
$$\bar{B}_{22} = \tfrac{3}{4}\bar{P}_2 + \tfrac{1}{4}\bar{F}_1$$
$$\bar{B}_{21} = \tfrac{1}{4}\bar{P}_1 + \tfrac{3}{4}\bar{F}_1 \qquad \text{(see Table 12)}$$

The C test, however, is satisfied only if $h_{a1} = -h_{a3}$. Thus no relative magnitudes of h_{a1}, h_{a2} and h_{a3} will satisfy the A, B and C scaling criteria simultaneously.

Two further scaling tests, however, hold for any relative values of h_1, h_2 and h_3 and are unaffected by double reduction. These are (Killick, 1971b),

$$\bar{P}_1 - \bar{F}_1 + 2\bar{B}_{21} - 2\bar{B}_{11} = 0$$
$$\bar{P}_2 - \bar{F}_1 + 2\bar{B}_{12} - 2\bar{B}_{22} = 0$$

which are derived from four relationships given above as $\bar{B}_{21} - \bar{B}_{11}$ and $\bar{B}_{12} - \bar{B}_{22}$.

TABLE 122
The means of the generations derived from an initial cross between the two homozygotes AAAA and aaaa

Generation	Expected mean
P_1	$m + d_a$
P_2	$m - d_a$
F_1	$m + h_{a2}$
F_2	$m + \tfrac{2}{9}h_{a1} + \tfrac{1}{2}h_{a2} + \tfrac{2}{9}h_{a3}$
B_1	$m + \tfrac{1}{6}d_a + \tfrac{1}{6}h_{a2} + \tfrac{2}{3}h_{a3}$
B_2	$m - \tfrac{1}{6}d_a + \tfrac{2}{3}h_{a1} + \tfrac{1}{6}h_{a2}$
F_3	$m + \tfrac{2}{9}h_{a1} + \tfrac{13}{36}h_{a2} + \tfrac{2}{9}h_{a3}$

Further inspection of these and other generations would no doubt yield additional relationships that could be used as scaling criteria.

The contribution of a single gene difference to the variances within the F_2 and backcross generations of an initial cross between parental lines AAAA and aaaa is given in Table 123: additional generations are given by Killick (1971b). In the general form, i.e. with h_{a1}, h_{a2} and h_{a3} taking arbitrary values, the additive and combined dominance contributions cannot be separated using

TABLE 123
The expected variances in the F_2 and first backcross generations following a cross between two inbred lines with genotypes AAAA and aaaa

Generation	Expected variance
F_2	$\frac{1}{18}d_a^2 + \frac{14}{81}h_{a1}^2 + \frac{1}{4}h_{a2}^2 + \frac{14}{81}h_{a3}^2 - \frac{2}{9}h_{a1}h_{a2} - \frac{8}{81}h_{a1}h_{a3} - \frac{2}{9}h_{a2}h_{a3}$
B_1	$\frac{5}{36}d_a^2 + \frac{5}{36}h_{a2}^2 + \frac{2}{9}h_{a3}^2 - \frac{1}{18}d_a h_{a2} - \frac{2}{9}d_a h_{a3} - \frac{2}{9}h_{a2}h_{a3}$
B_2	$\frac{5}{36}d_a^2 + \frac{2}{9}h_{a1}^2 + \frac{5}{36}h_{a2}^2 + \frac{2}{9}d_a h_{a1} + \frac{1}{18}d_a h_{a2} - \frac{2}{9}h_{a1}h_{a2}$

the F_2 and backcross variances. Indeed, even if these are supplemented by the variances of further selfing generations this separation can still not be made. If, however, we assume that $h_{a1} = -\frac{1}{2}d_a + \frac{1}{2}h_a$, $h_{a2} = h_a$ and $h_{a3} = \frac{1}{2}d_a + \frac{1}{2}h_a$ the expectations of the F_2 and backcross variances reduce to $\frac{1}{6}d_a^2 + \frac{1}{12}h_a^2$, $\frac{1}{12}d_a^2 + \frac{1}{12}h_a^2 - \frac{1}{6}d_a h_a$ and $\frac{1}{12}d_a^2 + \frac{1}{12}h_a^2 + \frac{1}{6}d_a h_a$, respectively, and the additive and dominance contributions can be separated.

11   *Genes, effective factors and progress under selection*

55. THE SOURCES OF ESTIMATES

The partition of variation into its different components shows us how much of it is (a) heritable and fixable in the form of differences between homozygotes, (b) heritable but unfixable in that it depends on the special properties of heterozygotes, and (c) non-heritable and hence merely serving to obscure the genetical situation. We can recognize at any rate certain kinds of genotype × environment interaction and show how the genetical differences will vary with the environment in which they are displayed. We also recognize the effects of genic interaction and linkage and so foresee up to a point the likelihood, the extent and the direction of change in the main heritable components of variation in the succeeding generations. One further piece of information is, however, required before the full import of the heritable components of variation can be assessed, i.e. the number of genes or, to be more precise, of units of inheritance contributing to the main components of variation.

If the D component depends on the difference produced by one gene only, its distribution among F_3 families, for example, will be simple. Half the families will be homozygous and hence will segregate no further. The other half will be heterozygous and will repeat the behaviour of F_2. Even in limited groups of F_3s the full potentialities of segregation will be realized. If, on the other hand, D is composed of items contributed by 10 unlinked genes, only rarely will a true breeding family be obtained in F_3, and many grades of segregation will be encountered among the various families. To this question of determining the number of genes involved we must now turn, and in answering the question we shall see that the gene is in fact no longer the unit of inheritance that we must use.

Where all the + alleles of the k genes, whose differences are involved in the cross, are concentrated in one of the true breeding parent lines, and all the − alleles in the other, half the difference between the two parental means, i.e. the deviation of either parent from the mid-parent value, will supply an estimate of $S(d_a) = [d]$ in the absence of interaction. If all these genes give equal increments, i.e. $d_a = d_b = \ldots, = d$, $S(d_a) = kd$. Furthermore, when this is true, $d_a^2 = d_b^2 = \ldots, = d^2$ and $D = S(d_a^2) = kd^2$ in the absence of interaction and linkage. Thus the ratio which the square of half the parental difference bears to D is

$$\frac{S^2(d_a)}{S(d_a^2)} = \frac{(kd)^2}{kd^2} = k.$$

This method of estimating k is the one used by Wright (1934b), except that in the absence of F_3 data he could not separate H from D, and so was forced to use an inflated value of the denominator. Charles and Goodwin (1943) and Goodwin (1944) have also used this method of estimation for a number of cases, and have indeed given a formula (though without showing the derivation) based on the same assumptions, for estimating the minimal number of genes common to two polygenic segregations. The effect of dominance in lowering the estimate of k has been discussed by Serebrovsky (1928).

Leaving the question of interaction and linkage aside, this estimate of k may be distorted in either or both of two ways even when D has been freed from H. The genes may not all give equal d increments, and the $+$ and $-$ alleles may not be distributed isodirectionally between the parents. Both inequality of the d increments and incomplete concentration of like alleles in the parents must lead to a spuriously low estimate of k. Let us consider first the effect of inequality. We may set

$$d_a = d(1 + \alpha_a), d_b = d(1 + \alpha_b) \ldots, d_k = d(1 + \alpha_k);$$

where d is the mean increment. Then

$$S(\alpha_a) = 0, S(d_a) = kd \text{ and } S^2(d_a) = k^2 d^2;$$

but
$$D = S(d_a^2) = S[d^2(1 + \alpha_a)^2] = d^2 S(1 + \alpha_a)^2$$
$$= d^2[k + 2S(\alpha_a) + S(\alpha_a^2)].$$

Now since $S(\alpha_a) = 0$, $S(\alpha_a^2) = kV_\alpha$, where V_α stands for the variance of α. Thus $D = kd^2(1 + V_\alpha)$, and K_1, the estimate of k found as the ratio borne by the square of half the parental difference to D, is therefore

$$K_1 = \frac{k^2 d^2}{kd^2(1 + V_\alpha)} = \frac{k}{1 + V_\alpha}.$$

Since V_α cannot be negative, $K_1 < k$ except in the special case of equality of all the d increments. With such equality

$$V_\alpha = 0 \text{ and } K_1 = k.$$

Where the genes are isodirectionally distributed, $\frac{1}{2}(\bar{P}_1 - \bar{P}_2) = S(d)$, but if the association of the $+$ genes in one parent and of the $-$ genes in the other is incomplete $\frac{1}{2}(\bar{P}_1 - \bar{P}_2) < S(d)$ and K_1 must be an underestimate of k. We can explore this further by noting that in general

$$\tfrac{1}{2}(\bar{P}_1 - \bar{P}_2) = S(d_+) - S(d_-) = S(d) - 2S(d_-).$$

Now $S(d) = S[\bar{d}(1 + \alpha)] = k\bar{d} + dS(\alpha) = k\bar{d}$ since $S(\alpha) = 0$ by definition; but $S(d_-) = S'[\bar{d}(1 + \alpha)] = k'\bar{d} + \bar{d}S'(\alpha)$ where S' indicates summation over the d_- group of genes and, as before (Section 13), k' is the number of these genes. Since in general S' does not involve summation over all the genes, $S'(\alpha)$ does not necessarily equal 0 and may indeed be either positive or negative. Then

$$\tfrac{1}{2}(\bar{P}_1 - \bar{P}_2) = k\bar{d} - 2[k'\bar{d} + \bar{d}S'(\alpha)] = \bar{d}[k - 2k' - 2S'(\alpha)].$$

We saw in Section 13 that the degree of gene association can be measured in the general case by

$$r_d = \frac{S(d) - 2S(d_-)}{S(d)}$$

which can now be written as

$$r_d = \frac{k - 2k' - 2S'(\alpha)}{k},$$

a factor of $\bar{d}$ cancelling out in numerator and denominator. Thus

$$[\tfrac{1}{2}(\bar{P}_1 - \bar{P}_2)]^2 = k^2 \bar{d}^2 r^2$$

and

$$K_1 = \frac{\tfrac{1}{4}(\bar{P}_1 - \bar{P}_2)^2}{D} = \frac{k^2 \bar{d}^2 r^2}{k\bar{d}^2(1 + V_\alpha)} = \frac{kr^2}{1 + V_\alpha}.$$

K_1 must therefore be an underestimate unless $r = \pm 1$ and $V_\alpha = 0$, that is, unless association is complete and all ds are equal.

An estimate of k may be reached in a similar way, but making use of the h increments in place of the ds. If all the genes have reinforcing h increments, the departure of the F_1 mean from the mid-parent value will be $S(h_a) = [h]$, which when $h_a = h_b = \ldots, = h$, is kh. Then the square of this departure divided by H, which in the absence of linkage is defined as $S(h_a^2)$, will give an estimate of k. Just as with the estimate based on the d increments, this new estimate will be reduced by any inequality of the hs and also by any opposition of the hs.

These two methods of estimating k are similar in principle, and in consequence they have the same disadvantages. A further method of estimation has, however, been proposed by Panse (1940a, b), which overcomes the difficulties of incomplete concentration and incomplete reinforcement. Half the individuals in F_2 are expected to be homozygous for one or other allele of any gene which is segregating. This gene will, therefore, contribute nothing to the heritable variance of the F_3s obtained from such individuals. The remaining F_2 individuals will be heterozygous for the gene, and it must therefore contribute $\tfrac{1}{2}d^2 + \tfrac{1}{4}h^2$ to the heritable variance of their F_3s. Let us denote $\tfrac{1}{2}d_a^2 + \tfrac{1}{4}h_a^2$ by x_a. Then the contributions of A–a to the mean variance of F_3 is, as we have already seen, $\tfrac{1}{2}(\tfrac{1}{2}d_a^2 + \tfrac{1}{4}h_a^2)$ or $\tfrac{1}{2}x_a$. But the variance of the F_3 variances (V_{VF3}) will, so far as this gene is concerned, be $\tfrac{1}{2}x_a^2 - (\tfrac{1}{2}x_a)^2$ or $\tfrac{1}{4}x_a^2$. Now with independent inheritance and independent action, the contributions of the various genes will be additive and hence $_H V_{2F3}$ will be $\frac{1}{2}S(x_a)$ and $_H V_{VF3}$ will be $\frac{1}{4}S(x_a^2)$. Then where $x_a = x_b \ldots, = x$, $_H V_{2F3} = \frac{1}{2}kx$ and $_H V_{VF3} = \frac{1}{4}kx^2$, so that

$$k = \frac{_H V_{2F3}^2}{_H V_{VF3}},$$

the subscript H in front of V indicating that it is the heritable portion of the variance which is under consideration. Where $x_a \neq x_b \ldots, \neq x_k$ we can set $kx = S(x_a)$ and $x_a = x(1 + \beta_a)$, $x_b = x(1 + \beta_b)$, etc. Then our estimate of k is

$$K_2 = \frac{k}{1 + V_\beta}.$$

As an estimate of k, K_2 is superior to K_1 in not being subject to reduction by incomplete concentration. Since, however, $x_a = \frac{1}{2}d_a^2 + \frac{1}{4}h_a^2$, the variation in β, measured by V_β, will be greater than the variation in α measured by V_α. This is partly because V_β measures the variation of d_a^2 as opposed to d_a, and partly because V_β must be inflated by any variation in the ratio $h/2d$.

It may be noted in this latter connection that an estimate of k, similar to K_2, can be obtained from the variances of the second backcrosses. Where $_H \bar{V}_{2B}$ stands for the sum of the mean heritable variances of the two second backcrosses (to the two parental lines) of single first backcross plants, we find that

$$_H \bar{V}_{2B} = {_H \bar{V}_{B11}} + {_H \bar{V}_{B12}} = {_H \bar{V}_{B21}} + {_H \bar{V}_{B22}} = \frac{1}{4}S(d_a^2 + h_a^2),$$

the mean being taken over the variances of the progenies of all the plants in the first backcross. Similarly $_H V_{V2B} = \frac{1}{16}S(d_a^2 + h_a^2)^2$. Then we can estimate k as $_H \bar{V}_{2B}^2 / _H V_{V2B}$ and the reduction in value due to inequality of $(d_a^2 + h_a^2)$, $(d_b^2 + h_b^2)$, etc. will depend on the variation in value of d^2, in the same way as with K_2, but also on the variation in h/d, which should exceed in general the variation in h/d, not $h/2d$ as with K_2. If this new estimate is lower than K_2 the difference must then be ascribed to the greater effect of variation in h/d, which should exceed in general the variation in $h/2d$ through which K_2 is reduced. This comparison, therefore, provides a test, though not a sensitive test, of the variation in the proportionate dominance, i.e. the ratio borne by the h increment to the d increment, from gene to gene. The means of estimating k from backcross data is, of course, especially valuable with animals and dioecious plants where true F_3 families cannot be raised. An estimate similar in principle to K_2 can also be obtained from biparental progenies of the third generation, though by a somewhat different approach, as we shall see in the next section.

If F_4 data are available, the lowering effect of variation in the $h/2d$ ratio can be avoided. Each F_2 individual can be regarded as an F_1 and the D and H increments separated within its descendance by the means already described. In this way we should have the value of D available separately for each F_3 family, and the process of estimation could proceed using D and V_D in place of

$_H V_{2F3}$ and $_H V_{VF3}$, so eliminating the distracting influence of H. The process could also, of course, be carried out with H and V_H, and the discrepancy between the two estimates would reflect the difference between the magnitudes of variations of d and h increments.

Even, however, when the consequences of variation in h/d have been avoided, K_2 will be less than k by an amount depending on V_β. Now if V_α is small, $V_\beta \simeq 4V_\alpha$; but if V_α is not small (or, of course, if variation in h/d is having its effect), $V_\beta > 4V_\alpha$. In either case $K_2 < K_1$, unless $V_\beta = V_\alpha = 0$ when $K_2 = K_1 = k$, or unless the K_1 is reduced by incomplete concentration of alleles in the parents.

Where the distribution of alleles is isodirectional in the parental lines so that K_1 is not reduced in value by lack of full concentration, we have

$$k = K_1(1 + V_\alpha) = K_2(1 + V_\beta).$$

Then putting

$$r = \frac{V_\beta}{V_\alpha}$$

$$V_\alpha = \frac{K_1 - K_2}{rK_2 - K_1}$$

and

$$k = \frac{K_1 K_2 (r - 1)}{rK_2 - K_1}.$$

When V_α is small and $r = 4$

$$V_\alpha = \frac{K_1 - K_2}{4K_2 - K_1} \text{ and } k = \frac{3K_1 K_2}{4K_2 - K_1}.$$

To assume that $r = 4$ when V_α is not small would lead to an underestimate of k, a negative value being obtained where $K_1 > 4K_2$.

Thus when both K_1 and K_2 are available, something may be learned of the variation in magnitude of d from gene to gene; or if the estimate based on the departure of F_1 from mid-parent is used in place of K_1, we learn something of the variation of h.

When, as judged by comparison with K_2, a useful value of K_1 or of its counterpart is not available, K_2 may still be put to good use. It is then possible to calculate from K_2 and D the difference between two true breeding strains which would give a K_1 equal to the observed K_2, i.e. to calculate the difference between two strains which would respectively contain all the $+$ and all the $-$ alleles. Since the mid-parent value is independent of the concentrations of alleles, the mean measurements of these two hypothetical strains could be estimated by adding and subtracting half their difference from the mid-parent, and they would represent the limits to which selection could push the measurement in either direction. Since in general $K_2 < K_1$, these estimates must be minimal. They would be exceeded by even more if some linkage in repulsion, so tight as to have escaped detection, were broken by a fortunate recombination in one of the later generations. Nevertheless, such minimal selective limits would have their use in showing the immediate selective progress that could be expected; for if these limits lay outside the actual parental range, selection would clearly be capable of leading to real progress. If they fell inside the parental range, then equally clearly the concentration of alleles in the two parents would be such that selection would be unlikely easily to produce anything markedly transgressing what is already available. If selection were judged to be worth while, and the experiment was continued in a form permitting in each generation the calculation of D and K_2 (both of which must be diminishing), the prospects of further advance could be estimated at each stage. The selection could then be discontinued either when D and K_2 fell to zero, or when the prospective progress ceased to be worth while (see also Section 60).

Though K_1 and K_2 are estimates of k, they cannot be consistent estimates unless d and h are the same for all genes. Any variation in these increments will lead to underestimation of k. A second limitation of the methods of estimating k is imposed by the assumption that the genes contribute to the heritable variance independently of one another. Now their contributions will not be independent unless the genes act additively and recombine freely, and the validity of the estimates must consequently be conditional on the scaling tests revealing no evidence of interaction and the linkage test no evidence of dependent segregation.

If two genes are, in fact, linked they will appear as less than two in the contribution they make to the variation. With recombination in the region of 0·10 to 0·30 we have a good chance of detecting the linkage and adjusting our estimates of k accordingly. With a higher recombination frequency the linkage may escape detection. In that case, however, the departure from independence of the contributions made by the genes to the variance will be smaller, and the disturbance in estimating k will not be serious. But tight linkage may also escape detection, and the two genes will then appear more nearly as one in the segregation. Under such circumstances the true number of genes may be virtually impossible to find. Nor need we concern ourselves unduly about the actual number of genes. Statistically, less error is involved in treating a closely linked pair as one unit than as two.

The same essential difficulty is, of course, encountered in Mendelian genetics when no recombination occurs between two genes distinguishable by their different effects on the phenotype. They must then be treated as one gene of pleiotropic action. The present case differs only in that rare cases of recombination cannot be recognized individually. To be detected, recombination must be sufficiently frequent for its statistical consequences to emerge clearly, and it must be quite free if the genes are to be recognized as fully distinct in their contributions to the variation. This difference in the level of the recombination frequencies, necessary for the recognition of the genes as distinct units in Mendelian and biometrical genetics, respectively, must however mean that genes which would be separated in Mendelian genetics may fall within the same effective unit of inheritance in biometrical genetics. The biometrical units need not, therefore, be ultimate genes. In consequence they will be referred to by the more empirical term of effective factors. The consequences of linkage and the nature of these factors will become clearer as we consider some examples.

56. THE PROCESS OF ESTIMATION

Relatively few sets of data are available from which estimates of the K_2 type can be obtained, but data from which values of K_1 can be found are not uncommon. Among the most instructive of these are the observations of Croft and Simchen (1965) and Simchen (1966) on the fungi *Collybia velutipes* and *Schizophyllum commune*, respectively. These fungi are basidiomycetes and the observations were made on monokaryons. Genically, therefore, they are haploids and as such offer even more direct estimates of K_1 than do diploid organisms. The difference between two parental haploids will be $2S(d)$, given that the genes by which they differ are distributed isodirectionally, and the heritable portion $(_H V)$ of the variance of the haploids derived from the diploid produced by their mating will be $S(d^2)$ as has been seen in Section 53. Then

$$K_1 = \tfrac{1}{4} \frac{(\bar{P}_1 - \bar{P}_2)^2}{_H V}$$

without any complication from dominance. The character observed in both fungi was the growth in mm of the mycelium per 10 days, as measured in a special growth tube. Each monokaryon, whether parental or progeny, was grown in duplicate, the whole experiment taking the form of two randomized blocks, and the variation between duplicates, as measured by the block × monokaryon interaction item in the analysis of variance, supplied an estimate of the non-heritable variation. The variance among the progeny monokaryons could thus be reduced by the estimate of the non-heritable variation to yield an estimate of the heritable variation $(_H V)$ among them. It should be observed that in *Collybia* at any rate there was evidence of heritable cytoplasmic variation within individual monokaryons. This is prospectively a complicating factor but it was not large and so will be ignored.

In the case of *Collybia*, Croft and Simchen observed behaviour in seven different crosses, but only four will be considered here because two of the seven were complicated by the segregation of major genes which grossly distort the estimates of k, and the data recorded in their paper are incomplete for a further cross. The number of progeny monokaryons observed was in no case less than 68 and in one was as high as 95. Estimates of $_H V$ and the mean difference in growth rate among the parental monokaryons are given in Table 124, together

TABLE 124
The number of effective factors (K_1) in the fungus *Collybia velutipes*

Cross	$_H V$	Parental difference	K_1	Progeny range	K_1	d
9	69·0	9·6	0·3	55·0	11·0	2·5
11	44·3	21·6	2·6	35·0	6·9	2·5
12	95·5	40·3	4·2	48·5	6·2	3·9
13	164·7	38·9	2·3	66·5	6·7	4·8

with estimates of K_1 derived from them. Thus with cross 9, the parental difference was found as 9·6 mm and $_H V$ was estimated as 69·0. Then

$$K_1 = \frac{(\bar{P}_1 - \bar{P}_2)^2}{4_H V} = \frac{9 \cdot 6^2}{4 \times 69 \cdot 0} = 0 \cdot 33,$$

and similarly for the other crosses.

Croft and Simchen, however, carried the genetical analysis a step further. They argued that, with such large numbers of progeny, the fastest growing and slowest growing of them would have a good chance of carrying all or nearly all the + and − alleles, respectively, even if these had not been isodirectionally distributed in the parental monokaryons. The squared difference between the growth rates of the fastest and slowest monokaryons, divided by $4_H V$, should thus give an estimate of K_1 undisturbed by departures or at any rate serious departures, from the isodirectional distribution of alleles. The choice of the extreme monokaryons may tend to exaggerate the difference and hence to overestimate K_1; but such a tendency would be offset by any failure to pick up the extreme genetical types within the range of progeny monokaryons available.

The values of K_1 obtained from the progeny in this way are also shown in Table 124, from which it now becomes clear that all the values of K_1 found from the parental monokaryons must have been underestimates. Evidently in every case there has been some balancing of − and + genes in the parents and in one case (cross 9) this is very striking. One further point emerges from the progeny estimates of K_1. Three of the crosses yield almost identical estimates despite the fact that the range shown by $d = \frac{1}{2}$(Progeny range)/K_1 is nearly 2:1 over these three cases. Even when we take all four crosses into account, the range of values shown by K_1 is proportionately lower than that shown by d, albeit not greatly lower.

This same point emerges even more clearly from Simchen's study of *Schizophyllum*. For technical reasons it was impossible to recover the parental monokaryons in most of the crosses in this species. All the values of K_1 are estimated, therefore, from the progeny ranges and so will be disturbed little, if at all, by departures from isodirectional distribution, the number of progeny recorded being never less than 59 and generally 90 to 100 and in one case even more. Again the values of K_1 from the different crosses are surprisingly alike (Table 125). So, however, are the values of $_H V$ and the progeny ranges in most of the crosses. However, four lines were selected from the progeny of cross 2, two (L_1 and L_2) for slow growth rates and two (H_1 and H_2) for high growth rates. These four selected lines were crossed in all four possible combinations of L × H. The results from these are shown on the right side of Table 125, together with two sets of results yielded on different occasions by cross 2, from which the selections were made. Despite the responses to selection in both directions which Simchen observed, and which are reflected as would be expected in both $_H V$ and progeny range from the L × H crosses, the values of K_1 from the parental cross and from the L × H crosses are very similar. The number of factors segregating among the selected lines is thus no greater than the number segregating in their original cross. Thus the differences fixed by selection cannot be due to the introduction of new factors into the system by mutation or any other means: it must be due to changes in the magnitude of effect of the factors themselves, as indeed is further indicated by the values of d shown in the furthest right column of the table. The implications of this finding will be discussed further in Section 58.

TABLE 125
The number of effective factors (K_1) in the fungus *Schizophyllum commune*

Cross	$_H V$	Progeny range	K_1	Cross	$_H V$	Progeny range	K_1	d
1	37·8	36·5	8·8	2(a)	25·3	27·0	7·2	1·9
3	21·5	24·0	6·7	2(b)	32·4	31·5	7·7	2·0
4	14·3	26·0	11·8	$L_1 \times H_1$	39·9	33·5	7·0	2·4
5	28·4	31·5	9·3	$L_1 \times H_2$	61·7	49·5	9·9	2·5
6	22·8	30·0	9·9	$L_2 \times H_1$	119·9	54·0	6·1	4·4
				$L_2 \times H_2$	164·2	60·5	5·6	5·4

The observations on these fungi unfortunately cannot be made to yield K_2 estimates of k, the number of effective factors. A few sets of observations in diploid organisms are, however, available from which K_2 estimates can be obtained and compared with K_1. Mather (1949a) has analysed the data of Quisenberry (1926) on the inheritance of grain length in a cross between two varieties of oats. The mean grain lengths of the two varieties, Victor and Sparrowbill, were 16·361 and 11·484 mm, respectively. The deviation of each from the mid-parent value was therefore 4·877/2 or 2·4385. D was found by Mather to be 1·3211. Hence $K_1 = 2·4385^2/1·3211 = 4·501$.

The mean variance of F_3 was observed to be 0·7878, but this includes a non-heritable component, which must obviously be eliminated before K_2 can be estimated. The heritable portion of V_{2F3} is $\frac{1}{4}D + \frac{1}{8}H$, which from the values of D and H found in the analysis is $\frac{1}{4}(1·3211) + \frac{1}{8}(1·0694)$ or 0·4640. The variance of the F_3 variances was found by direct calculation to be 0·1022 and 0·0609 in the two halves of the experiment, or 0·08154 over all the data. V_{VF3} must, however, be inflated by a sampling item, since each F_3 variance is but an estimate whose accuracy depends on the size of the family from which it was calculated. This sampling component will be $2V_{2F3}^2/(n-1)$ (Fisher, 1946), and here V_{2F3} must be taken as including its non-heritable portion since this will serve as well as the heritable portion to inflate the sampling error. In this formula n represents the number of plants per F_3 family. The number of plants was not, in fact, constant in Quisenberry's F_3s, and so the harmonic mean of the family sizes, namely 32·387, must be used. The pooled value observed for V_{2F3} was 0·7878, so giving a correction of $(2 \times 0·7878^2)/(32·387-1)$ or 0·03955, which on subtracting from 0·08154, the value observed for V_{VF3}, leaves 0·04199.

This value is, however, still too high as V_{VF3} will reflect variation of the non-heritable component of V_{2F3} as well as that of the heritable component with which we are concerned. We can calculate a correction for the effect of the non-heritable component from the variance of the variances of the 36 samples of each parental variety as grown by Quisenberry. The average variance of the variances observed for the two varieties was $\frac{1}{2}(0·005 + 0·017) = 0·011$, but this must be reduced by an item representing the sampling variance of the mean variance E_w, just as was done for V_{VF3}. In Mather's analysis E_w was estimated as 0·3427 and with the samples of the two parents averaging 34·85 individuals each, the sampling variance becomes

$$\frac{2}{34·85}0·3427^2 = 0·006\ 74$$

leaving as our estimate of the correction for non-heritable effect 0·0110 $-0·0067 = 0·0043$. Subtracting this, in its turn, from the corrected value of V_{VF3} we obtain as the variance of the heritable portion of V_{2F3}, $0·0420 - 0·0043 = 0·0377$. Thus with

$$\tfrac{1}{4}D + \tfrac{1}{8}H = 0·4640$$

$$K_2 = \frac{0·464^2}{0·0377} = 5·71.$$

The denominator of this fraction, which yields the estimate of K_2, was arrived at after the application of two corrections and one at any rate of these corrections was of roughly the same size as the estimate eventually obtained for $_HV_{VF3}$. Even a moderate change in the estimate of $_HV_{VF3}$ would have a noticeable effect on the estimate of K_2. It is clear, therefore, that this estimate must be used with some caution.

K_2 from the pooled data is 5·71, as compared with a K_1 of 4·50. There are apparently ·some + alleles in the smaller parent Sparrowbill, and some − alleles in Victor. On the face of it, however, the concentration of like alleles cannot be far from complete and so selection is not likely to produce lines exceeding Victor or falling short of Sparrowbill by much in grain length. In fact $K_2 D = 5\cdot71 \times 1\cdot32 = 7\cdot54$, and hence the immediate selective limit would be a departure in either direction of $\sqrt{(K_2 D)}$ or $2\cdot75$ mm from the mid-parent. Since the parents already depart by $2\cdot44$ from their mid-value, selection seems likely to be virtually ineffective unless some undetected close linkage in repulsion be broken.

The lowest F_3 mean observed was $11\cdot70$ and the highest $16\cdot18$, i.e. departures from the mid-parent of $2\cdot22$ mm and $2\cdot26$ mm, respectively. These bear out the expectations quite well, because a family homozygous for all five $+$ alleles, or all five $-$ alleles, is not likely to occur in a group of 150 F_3s. These two departures of $2\cdot22$ and $2\cdot26$ are therefore likely to represent families homozygous for four like alleles out of the five factors.

The reasonably close agreement of the extreme F_3 means with the expectations based on K_2 and D suggests that the value of K_2 is not over low, or in other words that V_β is not very large. One is, therefore, led to conclude that the five effective factors do not differ much amongst themselves in magnitude of action.

The data from the *Drosophila* experiment, reported by Cooke *et al.* (1962) and Cooke and Mather (1962) and described in Section 27, also yield estimates of the number of effective factors concerned in the inheritance of the number of sternopleural chaetae. In *Drosophila*, however, we cannot raise a true F_3 generation but instead have an S_3 consisting of families derived from biparental matings among flies of the F_2. Thus, although the calculation of K_1 will follow just the same lines as with Quisenberry's oats, the method of finding K_2 must be modified, and in fact in contrast to the F_3 generation the square of the heritable portion of V_{2S3} divided by the variance of the heritable portion of V_{2S3} does not yield an estimate of k. Following a suggestion by R. M. Jones, however, Cooke and Mather derived a means of finding K_2 from the S_3 generation which, although different in detailed approach, is similar in principle to that used for the F_3.

Table 126 sets out the contribution of each gene pair to the heritable

TABLE 126
The contribution of a gene pair to the variance of the variances in S_3

Family	Frequency	V	$V - \bar{V}$
AA × AA	$\frac{1}{16}$	0	$-\frac{1}{4}d^2 - \frac{3}{16}h^2$
AA × Aa	$\frac{4}{16}$	$\frac{1}{4}(d-h)^2$	$-\frac{1}{2}dh + \frac{1}{16}h^2$
AA × aa	$\frac{2}{16}$	0	$-\frac{1}{4}d^2 - \frac{3}{16}h^2$
Aa × Aa	$\frac{4}{16}$	$\frac{1}{2}d^2 + \frac{1}{4}h^2$	$\frac{1}{4}d^2 + \frac{1}{16}h^2$
Aa × aa	$\frac{4}{16}$	$\frac{1}{4}(d+h)^2$	$\frac{1}{2}dh + \frac{1}{16}h^2$
aa × aa	$\frac{1}{16}$	0	$-\frac{1}{4}d^2 - \frac{3}{16}h^2$
	1	$\bar{V} = \frac{1}{4}d^2 + \frac{3}{16}h^2$	0

$(V - \bar{V})^2 = \frac{1}{256}(8d^4 + 40d^2h^2 + 3h^4)$

variances of the six different types of S_3 progeny together with their frequencies. The average of all these contributions is of course $\frac{1}{4}d^2 + \frac{3}{16}h^2$, denoted as $\bar{V}$ in the table. Deducting this from the contribution (V) to each type of family gives $V - \bar{V}$ in the last column of the table, and if these differences are squared, multiplied by their frequencies of occurrence, and then summed, we find the contribution of this gene pair to the variance of the variances of S_3 as $\frac{1}{256}(8d^4 + 40d^2 h^2 + 3h^4)$. Summing over k genes assumed to be of like d and h.

$$_\text{H}V_{VS3} = \frac{k}{256}(8d^4 + 40d^2 h^2 + 3h^4).$$

Now, again assuming all gene pairs to have like d and h, $D = S(d^2) = kd^2$, and $H = S(h^2) = kh^2$, and

$$\tfrac{1}{256}(8D^2 + 40DH + 3H^2) = \frac{k^2}{256}(8d^4 + 40d^2 h^2 + 3h^4).$$

Thus
$$K_2 = \frac{8D^2 + 40DH + 3H^2}{256_\text{H}V_{VS3}},$$

and is a consistent estimate of k if all genes are of like effect. It will, of course, be an underestimate if d and h vary and this underestimate will depend on a quantity like V_β.

K_1 will of course be estimated just as in the oats. It will be seen from Table 55 that the parent lines differed by 4·27 chaetae in females and 4·41 in males, giving an average difference of 4·34. Now D was estimated from the overall data as 3·048 by the unweighted method and 2·897 by the weighted (Table 65). Then

$$K_1 = \frac{4 \cdot 34^2}{4 \times 3 \cdot 048} = 1 \cdot 54 \text{ using the unweighted estimate of } D$$

$$\text{and } K_1 = \frac{4 \cdot 34^2}{4 \times 2 \cdot 897} = 1 \cdot 63 \text{ using the weighted estimate of } D.$$

Cooke and Mather rejected the data deriving from the BS females in their calculations because of the internal inconsistencies which these data show and which have been noted in Section 27. The values of D which they used are thus slightly lower than those used above and they consequently found $K_1 = 1 \cdot 68$ using the unweighted D and 1·85 using the weighted. No significance can be attached to the difference between these values and those derived above, in view of the variation in D derived from the remaining three parts of the experiment (see Table 65).

The data from the BS females were also omitted from the calculation of K_2 (which they call K_S) by Cooke and Mather, and since they do not record all the data from which overall values can be found, their values are used in Table 127. The actual calculation of this estimate of k can, however, be illustrated by reference to the SB females which Cooke and Mather set out in detail as an example of how the calculation is carried out.

The variance of the variances was found to be $V_{VS} = 3{\cdot}7804$. This must however be corrected, just as in finding K_2 for Quisenberry's oat results, first for the sampling variance introduced by V_{2S3} and secondly for the effects of the non-heritable component in the variances of the S_3 families. Taking first the sampling variance, V_{2S3} from these females was $3{\cdot}1315$ (see Table 56) and since the variance of each family was calculated from observations on 10 flies, the sampling variance is found as $(2/n - 1)V_{2S3}^2 = \frac{2}{9}3{\cdot}1315^2 = 2{\cdot}1792$. Then the value of V_{VS} corrected for sampling variation is $3{\cdot}7804 - 2{\cdot}1792 = 1{\cdot}6012$.

The contribution of the non-heritable component to V_{VS} is found, of course, from the variance of the variances of the cultures of the parental lines and F_1s raised during the experiment. Since the variances in the two parents and their reciprocal F_1s were not the same, they have been averaged, and the variance of the variances obtained in this way by Cooke and Mather was $1{\cdot}8110$. This, like V_{VS}, is of course inflated by the sampling variance arising from the variation within cultures, i.e. from E_w. The correction for this sampling variance is $(2/n - 1)E_w^2$, just as in the case of the correction made to V_{VS} but, of course, with E_w in place of V_{2S3}. All the variances were found from 10 flies, and using the average E_w from the whole experiment, the sampling variance is $\frac{2}{9}2{\cdot}3354^2 = 1{\cdot}2120$, leaving as the non-heritable element in V_{VS}, $1{\cdot}8110 - 1{\cdot}2120 = 0{\cdot}5990$. Then

$$_H V_{VS} = 1{\cdot}6012 - 0{\cdot}5990 = 1{\cdot}0022.$$

Now the unweighted analysis of the data from SB females gave $D = 2{\cdot}955$ and $H = 3{\cdot}012$, while the weighted analysis gave $D = 2{\cdot}750$ and $H = 2{\cdot}548$ (Table 65). Then $8D^2 + 40DH + 3H^2 = 453{\cdot}1$ using the unweighted estimate and $360{\cdot}2$ using the weighted. We can thus calculate

$$K_2 = \frac{8D^2 + 40DH + 3H^2}{256 \times {}_H V_{VS}} = \frac{453{\cdot}1}{256 \times 1{\cdot}0022} = 1{\cdot}77$$

from the unweighted analysis

and $\qquad K_2 = \dfrac{360{\cdot}2}{256 \times 1{\cdot}0022} = 1{\cdot}40$ from the weighted analysis.

TABLE 127
The number of effective factors in the *Drosophila* experiments

Estimate	Observed		Expected with $k = 3$
	Unweighted	Weighted	
K_1	1·54	1·63	1·76
K_2	1·88	1·15	1·52

328 · GENES, EFFECTIVE FACTORS AND PROGRESS

Averaging the estimates obtained from these SB females and those from the two kinds of males, Cooke and Mather obtained $K_2 = 1.88$ unweighted and $K_2 = 1.15$ weighted, as shown in Table 127. It will be seen that using the lower weighted value for D raised the estimate of K_1 relative to that obtained using the higher unweighted D, since D is the divisor in finding K_1. But D and H appear in the numerator of K_2, and so the weighted values of D and H give lower estimates of K_2 than do the unweighted.

Now there are three major chromosomes in *Drosophila*. Table 66 shows the ds associated with these chromosomes as revealed by Cooke and Mather's whole chromosome analysis. The effect of the X chromosome was small but those of chromosomes II and III significant. We should therefore expect k the number of effective factors, to be at least 2 and almost certainly at least 3 because Cooke and Mather adduced other evidence for action by the X chromosome. It is, however, clear that the d effects of the three chromosomes differ from one another and we must therefore expect K_1 and K_2 to be underestimates. By how much?

Table 66 shows d_1 when averaged over the sexes, to be -0.030, while $d_2 = 0.603$ and $d_3 = 1.138$ are both positive. There will thus be some underestimation of K_1 because of failure of isodirectional distribution of the ds, but it will be a very small effect. The underestimation springing from inequality of the ds will be greater. Assuming that each chromosome is acting as a unit (which is of course not wholly accurate) we can set

$$d_1 = d(1 + \alpha_1) = -0.030; \quad d_2 = d(1 + \alpha_2) = 0.603;$$
$$d_3 = d(1 + \alpha_3) = 1.138$$

giving $d = 0.570$; $\alpha_1 = -1.053$; $\alpha_2 = 0.058$; $\alpha_3 = 0.996$.

Then $S(\alpha^2) = 2.104$ and $V_\alpha = \dfrac{S(\alpha^2)}{k} = \dfrac{2.104}{3} = 0.701$

giving $K_1 = \dfrac{k}{1 + V_\alpha} = \dfrac{3}{1.701} = 1.76$.

This is entered in the column headed Expected in Table 127.

Turning to K_2, there is of course no underestimation from failure of isodirectional distribution: underestimation can arise only from inequality among the effects of the factors. In seeking to assess this underestimation we should bring the hs into account as well as the ds, but the low values of h in Table 66 provide no good basis for assessing the variation in the contributions of the three chromosomes to dominance variation. So the values of β necessary for calculating the extent to which K_2 is made an underestimate will be derived from the ds alone. In view of the low values of h in Table 66 and the inconsistency of the evidence for dominance in the analysis whose results are recorded in Table 65, we need not fear that we shall grossly distort the picture by proceeding in this way.

Then $\quad d_1^2 = d^2(1+\beta_1) = 0.001;\ d_2^2 = d^2(1+\beta_2) = 0.364;$
$$d_3^2 = d^2(1+\beta_3) = 1.296;$$

giving $\quad d^2 = 0.554;\ \beta_1 = -1.000;\ \beta_2 = -0.343;\ \beta_3 = 1.342$

and $\qquad S(\beta^2) = 2.916;\ V_\beta = \dfrac{S(\beta^2)}{k} = \dfrac{2.916}{3} = 0.972$

so that $K_2 = \dfrac{k}{1+V_\beta} = \dfrac{3}{1.972} = 1.52$ as entered in Table 127.

The figures observed for K_1 and K_2 are compared in Table 127 with the values expected if $k = 3$. The table requires little comment: the agreement between the values observed and expected is surprisingly good when we remember the error variances to which the observed values must be subject, and the simplifying assumptions which have been used in calculating the expectation. Agreement is especially good with K_1 and this requires no further comment. The values observed for K_2 straddle the value expected and the values observed on the basis of the weighted and the unweighted analyses thus differ more from one another than either does from the expected. This is very encouraging. Clearly the values found for K_1 and K_2 using the D and H as estimated in the biometrical analysis of the data from this experiment are reasonable. We may thus have some confidence that with proper safeguards and corrections estimates of the number of effective factors yield reasonable results. In particular in the present case K_1 and K_2 have similar values, as indeed they should where there is no great departure from isodirectional distribution. This similarity in value between K_1 and K_2 warns us, as it did with Quisenberry's oats, that selection would produce lines transcending the parental values only in so far as recombination took place of genes within the same linkage group—in other words that the parental lines would not be transcended to any great extent during at least the next few generations. It is worth noting in this connection that it is much more important for K_2 to be a reasonable predictor of the maximum K_1 (i.e. the K_1 yielded when gene distribution between the parents is isodirectional) than for it to be a close estimate of k: the important result from the *Drosophila* experiment is that K_1 and K_2 came reasonably close together even though both are less than 2, despite the fact that k most likely equals or even exceeds 3.

57. ESTIMATION BY GENOTYPE ASSAY

An alternative approach to the estimation of k, by genotype assay, has been described by Jinks and Towey (1976). The central feature is the determination of the proportion of individuals in the F_n generation of a cross between two true-breeding lines that are heterozygous at one locus, at least, by an assay of their F_{n+2} grandprogeny families. The observed proportion can then be compared with theoretical expectations which for any generation n are functions of k. Hence by matching observed and expected proportions an estimate of k can be obtained which we shall call K_3.

For a combination of statistical and genetical reasons, not all heterozygotes present in the F_n generation will be detectable in practice. To detect a heterozygote we must prove that its F_{n+1} progeny has segregated, by demonstrating a significant difference among the family means or variances of its F_{n+2} grandprogenies. The probability of misclassifying a heterozygote is clearly a function of the number of siblings (p) in the F_{n+1} family that are progeny tested in the F_{n+2} and the size (l) of these F_{n+2} grandprogeny families. The larger p the less likely that all the siblings in the sample drawn from the F_{n+1} family will have the same genotype, while the larger l the more sensitive will be the tests for differences in the means and variances of the p F_{n+2} grandprogeny families. In practice an increase in p will, of course, compete with an increase in l for resources. The effect of p can, however, be completely allowed for in the theoretical expectations (Towey and Jinks, 1977). Thus for any value of p the expected proportion of detectable heterozygotes in the F_n generations can still be expressed as a function of k.

In a polygenic system not all genotypic differences lead to phenotypic differences and hence they cannot contribute to differences in family means and variances. The principal causes are dominance and the balancing of increasing and decreasing effects where alleles at different loci are dispersed. As a consequence there are maximum and minimum proportions of detectable heterozygotes in the F_n corresponding with any one value of k. The minimum proportion (P_{Min}) is obtained if the dominance relations and dispersion are such as to minimize the phenotypic expression of the genotypic differences while the maximum proportion (P_{Max}) is obtained if all of the latter are fully expressed in the phenotype.

Table 128 contains a sample of the estimates of K_3 obtained by Jinks and Towey for flowering-time and plant height in the cross of varieties 1 and 5 of *Nicotiana rustica* that we have already referred to a number of times. Estimates

TABLE 128
The observed proportions of heterozygotes in the F_2, F_4 and F_6 generations of the cross of varieties 1 and 5 of *N. rustica* for flowering-time and plant height and the estimates of K_3 derived from them.

Character F_n		Flowering time			Plant height		
		F_2	F_4	F_6	F_2	F_4	F_6
Observed P_{Het}	5%	0·200	0·314	0·306	0·400	0·257	0·306
	1%	0·200	0·086	0·166	0·200	0·174	0·250
K_3 from P_{Max}	5%	1	5	19	2	4	19
	1%	1	2	10	1	3	15
K_3 from P_{Min}	5%	2	9	> 20	3	7	> 20
	1%	2	2	16	2	5	> 20

are given for three generations ($F_n = F_2$, F_4 and F_6) based upon comparisons of the means and variances of pairs of $F_{n+2} = F_4$, F_6 and F_8 grandprogeny families ($p = 2$). The number of F_n individuals assayed (m) was 20, 35 and 36 in the F_2, F_4 and F_6, respectively. The proportion of heterozygotes (P_{Het}) is obtained as the proportion of pairs of F_{n+2} families which differed significantly for their means or variances. Two proportions are given, one where these differences are significant at the 5 % level and the other at the 1 % level. There are, therefore, four sets of estimates of K_3, namely, a P_{Max} and a P_{Min} estimate corresponding with each of the two sets of observed proportions of heterozygotes.

The general formulae relating k to the maximum (P_{Max}) and minimum (P_{Min}) proportions of detectable heterozygotes in the F_n are as follows:

$$P_{Max} = 1 - \left(1 + \frac{2^p + 2 - 4^p}{2^{n + 2p - 1}}\right)^k$$

$$P_{Min} = 1 - \left(1 - \frac{1}{2^{n-1}}\right)^k \overset{k}{\underset{r=0}{S}} \frac{\overset{r}{\underset{s=0}{C \; S}} \; 3^{ps} \left(\overset{r}{\underset{s}{C}}\right)^p}{(2^{n+2p-1} - 2^{2p})}$$

For flowering-time in the F_2 the proportion of heterozygotes detected in a sample of 20 was 0.2 for both the 5 % and 1 % levels of significance. Substituting $n = 2$ and $p = 2$ into P_{Max} it becomes

$$P_{Max} = 1 - \left(1 + \frac{4 + 2 - 16}{2^5}\right)^k.$$

To determine the value of k which will lead to the observed proportion of 0.2 we must equate it to the theoretical expectation P_{Max}, i.e.

$$1 - \left(\frac{11}{16}\right)^k = 0.2.$$

Since for $k = 1$ the left hand side equals 0.3125 the estimate of k, K_3 cannot be more or less than one.

Following the same procedure for P_{Min} we find that for $k = 1$ the expected proportion is 0.1875 and for $k = 2$ it is 0.3223. The observed proportion 0.2 falls between these two values, hence K_3 must be two. These and the corresponding estimates for plant height and for the F_4 and F_6 generations are given in Table 128.

The most striking feature of the estimates of K_3 is that they steadily increase over successive generations of selfing following the initial cross. Jinks and Towey were able to confirm the relatively low estimates in the F_2 by three different applications of genotype assay which used all of the data from the later generations in one form or another. By so doing they were able to show that the low estimate in the F_2 is independent of the generation from F_4 to F_8 used to assess it.

Since *N. rustica* naturally sets about 80 % of its seed by self-pollination we neither expect nor indeed do we observe, either for major genic or for polygenic variation, any evidence of selection for heterozygotes during inbreeding. Since this is the only possible source of an upward bias in estimates of K_3 obtained from the later generations we must accept that the estimates in these generations are genuinely higher. The explanation of this increase rests upon the nature of effective factors and the fact that strong linkage disequilibria can be initially generated in a cross between a pair of lines and subsequently resolved during successive rounds of recombination (see next section).

58. THE INTERPRETATION OF EFFECTIVE FACTORS

As we have seen, K_1 and K_2 must generally be underestimates of k, though as we have also seen the approximation of K_2 to the maximum K_1 is more important in the prediction of response to selection than is their joint departure from k. We must now observe that k itself is not a measure of the number of ultimate genes contributing to the variation, and indeed must generally be less than this number.

In deriving the methods of estimating k, we have tacitly assumed that the k units are independent of one another in the contributions they make to the variation. This implies that they not only show no interaction in producing their effects but also no linkage in their transmission from parent to offspring.

Now where a chromosome pair differs in a number of genes scattered along its length and all affecting the character in question, any chiasma which may form between the two homologues will, as it were, serve to distinguish two super-genes. The gross aggregate of all the genes within each of the two segments which have recombined at the chiasma will be transmitted as a unit. If two chiasmata form, three such super-genes will appear. The chromosome will thus appear to consist on the average of super-genes to a number greater by one than its mean chiasma frequency. The greatest number of effective factors, i.e. the greatest k which can be found, will thus be given by what Darlington (1958) has called the recombination index, i.e. the haploid number of chromosomes plus the mean chiasma frequency of the nucleus. The effective factor, as we have been using it, is in the general case therefore not an ultimate gene, but merely a segment of chromosome acting as a unit of inheritance and separated from other such units by an average recombination frequency of 50 %.

Since the positions in which chiasmata form may vary from nucleus to nucleus even within the same individual, the super-genes which the chiasmata distinguish will not be of constant content. They will be variable even within a generation, and V_β will in consequence be high. The value of K_2 should, therefore, be distinctly lower than that of either k or even of K_1 where there are polygenes scattered all along the chromosomes and chiasmata can form.

Finally, since chiasmata vary in position, a further breakdown of the effective factors must occur in later generations. The total number of factors found in these later generations will generally be greater than the first estimate. The effective unit of inheritance is thus a unit only for one generation, and even within this period it may well be a statistical rather than a physical unit. This cannot, of course, prevent us using the unit for our statistical purposes. It merely serves to emphasize that the 'gene' of biometrical genetics, unlike the gene of Mendelian genetics, is not an ultimate unit. It is more like the 'gene' of structural change (Darlington and Mather, 1949).

Our factors may be treated as units for the purpose of calculation. If the chiasmata were invariable in the way they separated the factors within the chromosome, the factors would, in fact, be constant units and could be treated as final genes. With variation in the points at which the factors are delimited by chiasma formation, the factor will itself be variable, but we may still regard it as unitary within a generation provided that we recognize the existence of variation in the effect of the unit. In doing so we are, as it were, replacing the factors which potentially merge into one another along the chromosome by a series of separate genetical centres of gravity, one for each factor. Each centre of gravity will have an average effect which reflects the mean genic content of the factor it represents, and a variance of effect which reflects the variation in genic content of that factor.

The basis of the effective factor is thus a segment of chromosome whose limits are as variable as the positions of the chiasmata which mark it off. Variation in the positions of the chiasmata restricts the extent to which we can regard the effective factor as a unit over time: the number of effective factors will thus be expected to rise over the generations, as we see in fact is the case in *Nicotiana rustica* (Table 128). Variation also restricts the extent to which we can regard it as a unit of action. The action that we observe to be associated with the factor reflects the whole genic content of the segment of chromosome, and this has certain consequences for our interpretation of any apparent dominance and pleiotropy that we may detect in our effective factors, as we shall see (Section 63). It also means, however, that the apparent magnitude of effect of the factors we observe segregating can differ in two crosses even though we may be confident that the actual genes segregating are identical, or nearly so, in the two. This is well illustrated by Simchen's observations on *Schizophyllum*. His original isolate 2 gave evidence that some seven to eight effective factors were segregating, the average effect of the factors being about 2 (Table 125). In the crosses between the high and low lines derived from this isolate by nine generations of selection (during which time mutation can have added but little to the variation) he still found an average of about seven effective factors, but now each with an average effect nearly twice as great. Clearly the factors are being delimited in the same sort of way, but their gene-content has changed: recombination has resulted in genes which showed repulsion linkages and therefore balancing actions, in the original cross appearing with coupling linkages, and therefore reinforcing actions, in the crosses between the selected lines. The effective number of chromosome segments is the same but the genic differences associated with these segments have been rearranged by recombination during the nine generations over which selection was practised. Furthermore, since the behaviour of the low lines L_1 and L_2 do not behave in the same way, we may infer that this redistribution of the genic differences has differed between the two of them – as indeed it well might do because of differences in the patterns of recombination arising from differences in the positions of the chiasmata which actually formed in the immediate ancestors of lines L_1 and L_2. It will be observed from Table 124 that the *Collybia* results from crosses 11, 12 and 13 also suggest similar numbers of effective factors with different genic contents, though obviously the evidence is less convincing than that from *Schizophyllum*.

Loosely linked genes might be expected to appear in different effective factors and closely linked genes in the same one. Such loose linkages and close linkages are also, as we have seen in Section 31, difficult to detect by the changes they produce in the genetical components of the statistics of different rank. Linkage of intermediate strength produces the greatest change with rank in these components and so is the easiest to detect. The changes in the genetical components also offer us an opportunity of learning something further about the number of genes or factors involved.

In V_{1F2}, $D = S(d_a^2) \pm 2S[d_a d_b(1 - 2p)]$; and in V_{2F3}, $D = S(d_a^2) \pm 2S[d_a d_b(1 - 2p)^2]$, the sign depending on the phase of linkage between the individual pairs of genes. In general, with preponderant coupling D_2, the second rank D, will then be less than D_1, the first rank D, and the magnitude of the fall from F_2 to F_3 will depend on the relative magnitudes of d_a, d_b, etc., on the recombination frequencies, and on the number of genes involved. The fall will be maximal when all genes are both coupled and of equal action, and when adjacent genes are equally spaced along the genetical chromosome. We cannot know, of course, that our factors are either localized in distribution or invariable in genetic content like single genes. As we have seen, however, we may regard them as represented by genetical centres of gravity spaced out along the chromosomes like the genes in terms of which the discussion has been carried on. The genetic weight at each centre will be variable, but the variation should do no more than make our estimates of the fall in D, calculated on the assumption of invariable genes, too high, and the number of factors correspondingly too low. This underestimation will not be seriously misleading because even if single genes were involved we could aim only at a minimal estimate.

The precise recombination frequency giving maximal fall from D_1 to D_2 will vary with the number of factors, but it can clearly be calculated for any given number of them. One point must, however, be made clear before proceeding to the calculation, namely that maximal must be taken as meaning maximal in relation to the value observed for D_1 or D_2. Otherwise there is no standard of reference. We shall take D_1 as the basis for our determinations of maximal change.

The fall in D is $S(d_a^2) + 2S[d_a d_b(1 - 2p)] - S(d_a^2) - 2S[d_a d_b(1 - 2p)^2]$ and its ratio to D_1 is thus

$$\frac{2S[d_a d_b(1 - 2p - (1 - 2p)^2)]}{S(d_a^2) + 2S[d_a d_b(1 - 2p)]},$$

which becomes

$$\frac{4d^2 S[p(1 - 2p)]}{d^2[k + 2S(1 - 2p)]}$$

when k factors all having effect d are concerned. With $k = 2$ this is $4p(1 - 2p)/2 + 2(1 - 2p)$. The maximal fall ratio is then 0.17 when p is 0.29. Assuming the absence of interference, and equal spacing of adjacent factors, the maximum fall ratio with $k = 3$ is 0.25 when adjacent factors show $p = 0.27$ recombination. Similarly the ratio is 0.30 at $p = 0.26$ when $k = 4$. Thus a fall of, say, 0.22 requires a minimum of three factors. A fall ratio of 0.28 would similarly require at least four factors. With interference between chiasmata, both the maximum fall ratio and the value of p which gives it are lower than when it is absent (Table 129).

TABLE 129
Maximum fall ratios in D and H with $k = 2, 3$ and 4 in coupling and reinforcement respectively

	k	p	Fall in D $\left(\dfrac{D_1 - D_2}{D_1}\right)$	$\dfrac{S(d^2)}{D_1}$	Fall in H $\left(\dfrac{H_1 - H_2}{H_1}\right)$	$\dfrac{S(h^2)}{H_1}$
No interference	2	0·29	0·17	0·71	0·06	0·85
	3	0·27	0·25	0·58	0·09	0·77
	4	0·26	0·30	0·50	0·12	0·71
Complete interference	3	0·20	0·22	0·52	0·11	0·67
	4	0·16	0·24	0·40	0·13	0·54

The experiment concerned with ear-conformation in barley (Mather, 1949a) discussed in Section 32 can be analysed further in this way. Coupling linkage was demonstrated and D_1 and H_1 were estimated at $10\,389$ and $13\,169$ respectively. In the absence of data from F_4, D_2 cannot, on the other hand, be separated from H_2 and indeed the exclusive analysis was undertaken by letting the value observed for V_{2F3} become its own expectation. There will, however, be a fall in H, similar to, though less than that in D, if the h increments are preponderantly reinforcing (Table 129). If we therefore use $D + \frac{1}{2}H$ in place of D for calculating the fall, the result will be slightly to underestimate its magnitude. This only means a slightly lower value for k, a minimal estimate of which is being aimed at in any case. In F_2, $D_1 + \frac{1}{2}H_1 = 10\,389 + \frac{1}{2}(13\,169)$ $= 16\,974$. In F_3, $D_2 + \frac{1}{2}H_2$ can be found by subtracting E_w (as estimated when allowance is made for linkage) from V_{2F3} and multiplying by 4. It is, therefore, $4(4314 - 1224) = 12\,360$. The fall ratio is then

$$\frac{16\,974 - 12\,360}{16\,974} = \frac{4614}{16\,974} = 0.27.$$

Three coupled factors will not quite suffice, since their maximum fall ratio is 0·25. Four will be enough, since their maximum fall ratio is 0·30.

We have already seen, however, that there must be at least two linked groups of genes, each with preponderant coupling but balancing one another's actions almost exactly in the parents. Now two coupled groups each of k factors will give the same maximum fall ratio when segregating simultaneously that each does by itself. With unequal numbers of factors, the joint fall ratio will be intermediate in value between the individual ratios. Thus in the present case we must suppose, with two groups acting, that more than six factors are operating, with at least three of them lying in each group. (Two in one group and four in the other gives no better a fit than three in each, the fall ratios being much alike in each case.) Eight factors, four in each group, would be ample to explain the results. Seven factors, four in one group and three in the other, would also explain the fall, but might not account for the close balance of effects in the parents.

Now with coupling, the value of D_1 exceeds $S(d^2)$ by an amount depending on the number of factors and their linkage relations. When $k = 2$ and $p = 0.29$, the value giving the maximum fall ratio of 0·17, $D_1 - S(d^2)$ is 0·29 of D_1. A factor of 0·71 must therefore be applied to D_1 in order to give $S(d^2)$. The corresponding factors for $k = 3$ and $k = 4$ under the conditions of maximum linkage fall are given in Table 129. Thus the linkage fall ratio enables us to estimate the minimum number of factors which must be involved, and also the correction necessary for reducing D_1 to $S(d^2)$.

In the barley case, if we take the underestimate of $k = 6$, three in each group, then $S(d^2)$ is 0·58 of D_1, i.e. 0·58 × 10 389 or 6025. If we regard each group as comprising four factors, giving eight in all, $S(d^2)$ is 0·50 × 10 389 or 5195. Then in the former case $kS(d^2) = 6 \times 6025 = 36\,150$ and in the latter $kS(d^2)$ $= 41\,560$. Since the minimal limit to selective change in either direction is given by $\sqrt{[kS(d^2)]}$, it will be $\sqrt{36\,150}$ or 190 with six factors, and 203 with eight. The mid-parent was 314, and so the minimal limits should be 124 and 504 if six factors are assumed, or 111 and 517 if eight are assumed. Actually the means of the extreme F_3 families were 181 and 568 (Mather, 1949a), the former inside and the latter outside the limits calculated on both assumptions. We have, however, already seen that the scale was not fully adequate (Section 32) and its distortion is such that the expected upper limit would be exceeded before the expected lower limit was attained. The difference between the extremes observed in F_3 is $568 - 181$ or 387, which lies between 2×190 and 2×203, the differences between the limits expected on the two assumptions. Thus there is no ground for expecting that much selective advance could be achieved on the extremes observed in F_3, though some progress might be made if the units of inheritance were either not so constituted or not so related as to give the maximum possible fall ratio. Since, however, the linkages are preponderantly in coupling, 100 F_3s would be very likely to contain families reaching the limits. Everything points, in this cross, to the F_3 families having already reached the limits of selective progress which might be envisaged as easily obtainable.

A further use may be made of the estimate of the number of factors in the present case. The F_1 with a mean of 400 showed heterosis. H_1 was found to be 13 169. This value will, of course, exceed $S(h^2)$ by the item $2S[h_a h_b(1 - 2p)^2]$. The value of $S(h^2)$ will, therefore, be less than H_1, though not so small a part of it as $S(d^2)$ is of D_1. With the values of p found in calculating the maximum fall ratios of D (assuming, of course, complete reinforcement of the h increments, in the same way as complete coupling of the factors was assumed when considering the d increments), the values of $[S(h^2)]/H_1$ for $k = 2, 3$ and 4 are found to be as shown in Table 129.

If we take $k = 6$, three in each group, $S(h^2)$ is 0·77 of H_1, i.e. 0·77 × 13 169 or 10 140. With $k = 8$, $S(h^2)$ is 0·71 of H_1 or 9350. Then $kS(h^2)$ is 60 840 in the former and 74 800 in the latter case. The degree of heterosis to be expected if all the factors were reinforcing will thus be $\sqrt{60\,840}$ or 247, with $k = 6$; and 274 with $k = 8$. The heterosis observed was $400 - 314$ or 86. Thus, even when the large error of estimation of H_1 is taken into account, the factors cannot be regarded as all reinforcing one another. New inbred lines could be extracted from the cross, which would give greater heterosis through the fixation of the + alleles of these factors whose h increments were in opposition to the heterosis.

Such calculation has little value in barley where homozygous varieties are used in practice; but it might well be of use in certain crops in deciding the value of inbred lines for hybrid seed production and assessing the prospects of their improvement.

59. THE SPEED OF ADVANCE UNDER SELECTION

Not only will the minimal limit of selective advance be determined by the number of units of inheritance which contribute to D and H, but the speed of selective advance must clearly also depend on this number. Where only one gene is segregating, $\frac{1}{4}$ of the F_2 individuals will be homozygous for each allele. In the absence of obscuring agents, the selection of any number of individuals up to 25 % at either end of the scale will lead to immediate fixation of the homozygous type in the next generation. All the progress possible will be achieved immediately.

With two genes, segregating independently, 6.25 % of the F_2 individuals will be homozygous for each of the extreme genotypes. Thus the chance of recovering an extreme genotype from an F_2 of any given size is reduced. Furthermore, the progress made by selecting, say, 10% of the individuals at one end of the scale in F_2 cannot be complete even in the absence of obscuring agents. Some genetically less extreme types must also be taken for breeding.

Where four genes are segregating only one individual in 256, i.e. 0.39% will be homozygous for each of the extreme genotypes. Not merely is the chance of immediate fixation under selection very small, it is even unlikely that any extreme homozygote will be picked up in an F_2 of small or medium size. The group selected for further breeding must, therefore, include individuals of various genotypes showing various degrees of heterozygosity in proportions depending, among other things, on the magnitudes of the h increments. Progress will be incomplete, and only as the extreme homozygotes come to form an increasing proportion of the segregates in later selected generations will the full advance under selection become realized.

Furthermore, where the relevant genes are linked, the speed of advance under selection in the progeny of a cross between inbred lines will depend on the way the linkage relations affect the number of effective factors that we can recognize, since we must expect the consequent linkage disequilibrium to be a feature of the early segregating generations from such a cross. We may note, however, that since linkage of itself does not affect the relative frequencies of the various genotypes in a randomly breeding population (insofar as it is at equilibrium) the linkage relations of the segregating genes can generally be neglected; or, to put it another way, the number of effective factors in such a population approximates to the number of genes.

The efficacy of selection must also be reduced by the effects of non-heritable agents, which make the phenotypic expression of the various genotypes overlap and so reduce the chance of effective separation of the genotypes in selection. The greater the non-heritable variation, therefore, the less effective the selection. On the other hand, only the additive portion of the heritable variation is fixable (Chapter 3), hence, in general, the greater the proportion of the heritable variation that is of the additive kind, the greater the effectiveness of the selection. Indeed, it can be readily shown (see Falconer, 1960, for details) that the expected response to selection is proportional to the narrow heritability (ht_n) which in the general case of a randomly mating population equals $\frac{1}{2}D_R/(\frac{1}{2}D_R + \frac{1}{4}H_R + E_w + E_b)$ (Sections 38 and 39) and for the special case of an F_2 equals $\frac{1}{2}D/(\frac{1}{2}D + \frac{1}{4}H + E_w + E_b)$. Thus where a simple additive-dominance model is adequate, the response to selection (R) is equal to the product of the narrow heritability and the selection differential (S), that is

$$R = ht_n \times S$$

where S is the difference in the mean expressions of the original population and the sample selected as parents and R is the difference in the mean expressions of the original population and of the progeny produced by the selected parents. The same relationship also holds for the cumulative response and the cumulative selection differential over a number of successive cycles of selection. However, because the relative values of the components of heritability (D, H or D_R, H_R and E_w and E_b) depend on the gene and genotype frequencies, types of gene action, the level of environmental variation and the experimental design, if any of these change the heritability will also change. For selection to be effective, gene and genotype frequencies must obviously change, hence the more effective the selection the fewer the cycles of selection over which reliable predictions of the response can be expected. Nevertheless, the relative magnitudes of the heritabilities of the same character in different populations or of different characters in the same population may provide a guide to the relative responses that are likely to be achieved for the same selective effort.

The variation in proportion of the different genotypes in F_2 consequent on sampling error may also reduce the efficacy of selection by reducing the frequencies of the more extreme and desired types. The efficacy of selection might, however, also be increased by corresponding chance excess of these types in F_2.

Thus taking all these factors into account the speed of progress under selection following a cross of two lines will depend on:

(a) the rigour of selection (i.e. proportion of F_2 chosen for breeding);
(b) the number of genes, as organized into effective factors in the way we have discussed;
(c) the variation in magnitude of action of the genes or factors (variation in magnitude of d increments);
(d) their dominance relations (magnitudes of h increments);
(e) their linkage relations;
(f) the size of the contribution of non-heritable agents to the F_2 variation, or to put it another way, the heritability of the character in the F_2;
(g) the sampling variances of the genotype frequencies.

Some of these agents can be assessed in the ways already discussed. On the others information will be less certain or even absent. The practical problem is to decide how much can be foreseen about the speed of progress from the statistics we have learned to calculate. This problem was tackled by Panse (1940a, b), who developed a method of analysis and showed how it works out in one set of cases.

The character Panse considered was staple-length in three cotton (*Gossypium arboreum*) crosses. The F_2 families were grown in one year and F_3 families in the next. It is impossible from Panse's data to undertake analyses of the kind necessary to separate D, H and E in his crosses. Panse himself (1940a), however, estimated what he terms the genetic variance of two of the crosses. This was found by the regression of F_3 mean on F_2 parent value, i.e. as $(\frac{1}{2}D + \frac{1}{8}H)/(\frac{1}{2}D + \frac{1}{4}H + E)$. Values of 1·543 and 1·516 were obtained in his two cases, leaving residual variances (which he regards as non-heritable, though these, like the so-called genetic variances, will contain items depending on H) of $1{\cdot}472$ and $1{\cdot}697$, respectively. For simplicity Panse approximated by setting the total F_2 variance at 3, of which $1{\cdot}5$ was supposedly genetic and $1{\cdot}5$ not genetic, in each case.

The number of effective factors was estimated as K_2 for each cross, the one giving $K_2 = 1{\cdot}64$ and the other $K_2 = 2{\cdot}77$. Adequate corrections were not used in making this estimate which must thus be over-low. Panse, however, chose to take $K_2 = 3$ as the basis of his further consideration. In order to consider the effects of both dominance and unequal action of the genes involved he set up five genetical models. In three of them only three genes were involved. In the first the genes were equal in action and without dominance. In the second, two of the three genes were equal in action but showed complete dominance $(h = d)$ in opposite directions, the third gene showing no dominance but contributing equally to the variance with the other two, i.e. $\frac{1}{2}d_c^2 = \frac{1}{2}d_a^2 + \frac{1}{4}h_a^2 = \frac{1}{2}d_b^2 + \frac{1}{4}h_b^2$. The third model also had two genes with complete dominance in opposite directions and one without dominance but it put $d_a = d_b = d_c$, so that the genes contributed unequally to the variance. The other two models involved an infinite number of genes, in the one case showing no dominance and with the values of d descending in geometric progression, and in the other case falling into two series of equal action, each with complete dominance but in opposite directions, the values of d descending geometrically within each series. Linkage was supposed to be absent from all the models, and the average contributions of the genes were adjusted to give $K_2 = 3$.

Taking these five models the effect was considered of selecting the top 10% of the F_2 plants for further breeding by self-pollination. The mathematical method used for this purpose is over-elaborate for presentation here; it will be found in detail in Panse (1940b). Briefly, however, it consists of setting down a moment generating function for the simultaneous distribution of the F_2 parent values, the F_3 means and the heritable variances of the F_3s. On expanding, quartic regression equations are found relating the F_3 statistics to the F_2 parent values. These equations are then integrated between limits appropriate to the method of selection. The results of the mathematical operations are shown in Table 130, but with the scale changed from that used by Panse, so as to put the heritable portion of F_2 variance ($_HV_{1F2}$) equal to 1. This scalar change should enable the table to be used more easily with other genetical data, since if $_HV_{1F2} = a$, the various columns of the table can be multiplied by the factors shown in its bottom row to give the values appropriate to the case.

As would be expected, the selective advances and the F_3 variances change with the particular genetical model used, but the changes are surprisingly small. The advance achieved is at least 36% and at most 51% of the maximum possible. The different variances show an even smaller relative range. When the errors of estimation of the F_3 quantities are borne in mind, it is clear that the variation between models is not likely to lead to serious misjudgement. The

TABLE 130
Effects of raising F_3s from the upper 10% of F_2 whose mean is zero, with heritable variance = non-heritable variance = 1 in F_2, and $K_2 = 3$. (Panse's models)

Model	Limit of selective advance	Advance achieved in F_2	F_2 advance as proportion of that possible
I No dominance geometric series	3·415	1·235	36%
II No dominance 3 equal units	2·449	1·235	50%
III Balanced dominance 2 geometric series	2·642	1·034	39%
IV Balanced dominance 3 genes of equal variance	2·150	1·100	51%
V Balanced dominance 3 genes of equal action	2·121	1·084	51%
Where $_HV_{1F2} = a$ multiply by	$\sqrt{a}$	$\sqrt{a}$	1

results of selection can, therefore, be forecast with useful accuracy a generation or so ahead by this means.

In regard to the advance achieved, expressed as a ratio of the maximum advance possible (and this ratio is the most important feature from the point of view of practical forecasting), the chief difference in the table is that between the three models assuming three units on the one hand, and the two assuming an infinite number on the other. The first group shows about $\frac{1}{2}$ the possible advance as achieved, the other just over $\frac{1}{3}$ of it. Since the units are assumed to be unlinked an infinite number is in practice an impossibility, so that the lower limit of advance achievable in F_3 must actually lie somewhat above the percentage shown for the two models with infinite series.

Panse developed his models in terms of genes. We have seen, however, that there can be no certainty that the units of segregation with which we are concerned will, in fact, consist of ultimate genes. Rather we must consider the possibility of groups of genes appearing as our factors. The surprisingly small effects of dominance and number of genes on the rate of advance under selection encourage us to believe that Panse's results will still apply broadly to systems of such factors, recombining freely with one another through lying in different chromosomes or through being separated by chiasmata if within the same chromosome, even though the separation will not be into uniform factors in the latter case. In so applying Panse's results we are, of course, once again making the assumption that the full effect of each factor is concentrated at its genetical centre of gravity, which is recombining freely with other such centres. This assumption is artificial but provided its limitations are recognized, it enables us to see Panse's calculations in their application to a wider field of experimental situations.

Panse's approach offers a valuable means of predicting the advances expected in F_3 relative to those ultimately attained by continued selection. The case in question, where $_HV_{1F2} = {}_EV_{1F2}$, $K_2 = 3$ and 10% selection, is of course only one of the many which might be encountered; but it is possible, by the use of Panse's method, to tabulate the advances expected where other selective rigours are used, where K_2 takes other values, and where $_HV_{1F2}$ is differently related to $_EV_{1F2}$. Such tables would be a powerful aid to the geneticist and breeder in the use of his biometrical data.

Panse did not take into account the effects of linkage in his calculations but its consequences are not difficult to see in general terms. Linkage in either phase reduces the value of K_2. The speed of advance under selection, relative to the limiting advance as immediately foreseeable from the data, must therefore be greater than it would be in the absence of linkage. There is, however, an all-important qualification in the case of repulsion linkage. The above statement is true in so far as the limits of advance are calculated on the basis of the existing linked combinations; but crossing-over will give new combinations, which must widen the ultimate limits of selective advance. Thus while the speed of advance relative to the limits set by the existing combinations may be greater than in the absence of linkage, relative to the limits set by the new combinations which follow recombination, the speed of advance will generally be slower than in the case of no linkage. This is the reason for the favouring by natural selection of balanced polygenic combinations with repulsion linkages (Mather, 1941, 1943). With coupling linkages, on the other hand, recombination can act only to narrow the ultimate limits of advance under selection. The sole effect of linkage in coupling is consequently to lead to a greater speed of advance towards these limits, as we have seen in the case of ear-conformation in barley, discussed in the previous section.

60. THE RANGE OF ADVANCE UNDER SELECTION

The limits to the range of the true-breeding lines that can be derived from a cross between true-breeding parents will be given by the two homozygotes in which all the increasing and decreasing alleles at the k loci, at which the parents differed, are completely associated. The range will therefore be $m \pm Sd$.

The method of predicting this range devised by Mather (1949a) and illustrated in Section 56, depends on using the number of effective factors, found as K_2 or K_3, as an estimate of k, in conjunction with D. Thus if there are k loci with mean additive genetic effect d and $D = Sd^2$, the range $\pm Sd = \pm kd$
$= \pm \sqrt{(kD)}$, where $d_a = d_b = d_c \ldots = d$. Hence we can estimate the range from estimates of k and D.

This estimate is, of course, no better than the estimate of k, and as we have seen (Sections 55 and 57) the number of effective factors as obtained from the early generations of a cross must always be an underestimate of k. An example of the use of this approach to prediction has already been given in Section 56 for grain length in oats, and a further example using K_3 instead of K_2 is afforded by data from the cross of varieties 1 and 5 of *Nicotiana rustica*. From the F_2 and backcross families raised in 1950, 51 and 52 the mean of the estimates of D for plant height is 164.10. The K_1 estimate of k (Section 55) is less than one and tells us only that the range of the derived inbreds must transcend the parental values. The K_3 estimate, however, ranges from 2 to 3 in the F_2 to 4 to 7 in the F_4 (Towey and Jinks, 1977). These give estimates of the range which vary between ± 18.12 when $K_3 = 2$ to ± 33.89 when $K_3 = 7$. The latter compares favourably with the mean range of ± 33.49 observed in a random sample of 20 true breeding lines derived from the same F_2 when assessed in nine environments between 1953 and 1961, and also with the mean range of ± 33.58 observed in a subsequent sample of 82 such lines derived from another sample of the F_2 and assessed in 1970, 71 and 73.

An alternative method based upon the inverse dominance ratio avoids the problems associated with the estimation of k (Jinks and Perkins, 1972). Thus

$$\pm \sqrt{(H/D)} = \pm \sqrt{\left(\frac{Sh^2}{Sd^2}\right)}$$

which has the same numerical value as $\pm Sh/Sd$ where the dominance ratio is constant at all loci.

This becomes clearer if we substitute $h/d = f$ where f is the constant dominance ratio. Then

$$\pm \sqrt{H/D} = \pm \sqrt{\frac{f^2 Sd^2}{Sd^2}} = \pm \frac{f Sd}{Sd}.$$

If we assume a constant dominance ratio this equation is equivalent to $\pm Sd = \pm [h]\sqrt{(D/H)} = \pm [\bar{F}_1 - \frac{1}{2}(\bar{P}_1 - \bar{P}_2)]\sqrt{(D/H)}$, which estimates the range from components that can themselves be readily estimated from the early generations of a cross.

Again from the cross of varieties 1 and 5 in *N. rustica* we have an estimate of the dominance ratio for plant height, from the F_2 triple test cross, of 0.37 (Section 27). This gives $\pm \sqrt{(D/H)} = \pm 2.70$ as an estimate of the inverse dominance ratio. The estimate of $[h]$ can be obtained from the comparison of F_1 and mid parental value. On average over 16 seasons this was 13.94 (Section 19). These lead to an estimate of ± 37.64 for the range, which is somewhat higher than the observed mean range of ± 33.58 quoted earlier, but is within the sampling distribution of the latter over seasons.

A further kind of prediction can also be made, using a different approach (Jinks and Pooni, 1976). The value of knowing the range is, of course, that it tells us whether true breeding lines can be extracted which lie outside some pre-existing limits, e.g. the parental range or their heterotic F_1. The probability of doing so can be estimated directly from the expected mean, m, and genetic

344 · GENES, EFFECTIVE FACTORS AND PROGRESS

variance, D, of the phenotypic distribution of the derived true-breeding lines.

Now, after a cross between a pair of true breeding lines, and in the absence of non-allelic interaction, the mean, variance and skewness of the heritable components of the phenotypic distribution in any generation, n, derived from it by inbreeding can be written in the general form (Jinks, 1981).

$$(\text{mean})_n = m + \beta_n[h]$$
$$(\text{variance})_n = (1 - \beta_n)D + \beta_n(1 - \beta_n)H$$
$$(\text{skewness})_n = -3\beta_n(1 - \beta_n)Sd_i^2 h_i + \beta_n(1 - \beta_n)(1 - 2\beta_n)Sh_i^3$$

where β_n is the frequency of heterozygotes in that generation. For regular systems of inbreeding we can give a general specification for the value of β_n (see Section 13), and for other situations it may be estimated from the pedigree. All that we normally require for predicting the phenotypic distribution are, therefore, estimates of the genetical components of these statistics.

Following inbreeding, by whatever means, β rapidly approaches 0, with the skewness vanishing and the phenotypic distribution having a mean of m and a variance of D. Hence the probability of obtaining a true breeding line that falls outside the parental range $m \pm [d]$ is the sum of the normal probability integrals

$$\int_{-\infty}^{\bar{P}_2} = \int_{\bar{P}_1}^{\infty}$$

corresponding with the value of

$$\frac{[d]}{\sqrt{D}} = \frac{\bar{P}_1 - m}{\sqrt{D}} \text{ or } \frac{\bar{P}_2 - m}{\sqrt{D}}.$$

If we again used the estimate of $D = 164 \cdot 10$ for plant height from the F_2 and backcross families of the cross of varieties 1 and 5 raised in 1950, 51 and 52 and the estimate of $[d]$ from the parental means in the same experiments of $3 \cdot 28$, the normal probability integral corresponding with the value of $[d]/\sqrt{D} = 0 \cdot 256$ is $0 \cdot 39$. There is therefore a probability of $0 \cdot 39$ of obtaining true breeding lines from this cross with means equal to or greater than P_1 or equal to a smaller than P_2. The parents were compared with the true breeding lines extracted from the F_2 of this cross in 1953, 54, 57 and 61 and on average a proportion of $0 \cdot 45 \gg \bar{P}_1$ and $0 \cdot 30 \leqslant \bar{P}_2$ giving an average of these homogeneous estimates of $0 \cdot 375$.

Similarly, the probability of obtaining a true breeding line that deviates from m more than the $\bar{F}_1 = m + [h]$ are the normal probability integrals

$$\int_{\bar{F}_1}^{\infty} \quad \text{if } [h] \text{ is positive}$$

and
$$\int_{-\infty}^{\bar{F}_1} \quad \text{if } [h] \text{ is negative}$$

corresponding with the value of

$$\frac{[h]}{\sqrt{D}} = \frac{\bar{F}_1 - m}{\sqrt{D}}.$$

In the presence of non-allelic interactions the predictions are more complex, nevertheless many of the predictions can still be made. For example,

$$(\text{mean})_n = m + \beta_n[h] + \beta_n^2[l]$$

which on prolonged inbreeding reduces, as in the simpler case, to $\bar{F}_\infty = m$. The corresponding variances and skewnesses are, as we have already seen for a number of special cases (Section 30), too complex to be used for estimation and the estimates of D and H obtainable from them are biased to differing extents depending on their sources. Nevertheless, they all become simple when $\beta_n = 0$ as a result of prolonged inbreeding or dihaploidy. For example, the mean, variance, and skewness of the resulting F_∞ generation of true breeding lines are

$$F_\infty \text{ mean} = m$$
$$F_\infty \text{ variance} = D + I$$
$$F_\infty \text{ skewness} = 6Sd_i d_j i_{ij} + 6Si_{ij} i_{js} i_{is}.$$

Although the means of P_1, P_2 and F_1 are now $m + [d] + [i]$, $m - [d] + [i]$ and $m + [h] + [l]$, respectively, the probability that any of these true breeding lines will fall outside the parental range or deviate more from the mean m than the F_1 will be given by the same probability integrals as in the simpler case, if we ignore the skewness, but their expectations become

$$\int_{-\infty}^{\bar{P}_2} \text{ corresponding with the value of } \frac{[d] + [i]}{\sqrt{(D+I)}} = \frac{\bar{P}_1 - m}{\sqrt{(D+I)}}$$

$$\int_{\bar{P}_1}^{\infty} \text{ corresponding with the value of } \frac{-[d] + [i]}{\sqrt{(D+I)}} = \frac{\bar{P}_2 - m}{\sqrt{(D+I)}}$$

and
$$\int_{-\infty}^{\bar{F}_1} \text{ if } [h] + [l] = \bar{F}_1 - m \text{ is negative}$$

or
$$\int_{\infty}^{\bar{F}_1} \text{ if } [h] + [l] = \bar{F}_1 - m \text{ is positive,}$$

both corresponding with the value of $[h] + [l]/\sqrt{(D+I)}$.

The limits of the range of true breeding lines can be obtained from an estimate of k, D and I. These limits are

$$\pm \left(\underset{k}{Sd} + \underset{\frac{1}{2}k(k-1)}{Si} \right)$$

Sd is always positive but *Si* takes sign according to the type of non-allelic interaction and the net direction of dominance (Sections 16 and 17).

They may be estimated as

$$\pm (\sqrt{kD} \pm \sqrt{\tfrac{1}{2}k(k-1)I})$$

if we assume that, as before, $d_a = d_b = d_c \ldots = d$ and also that $i_{ab} = i_{ac} = i_{bc} \ldots = i$. On substitution of these equalities the estimator becomes

$$\pm (\sqrt{k^2 d^2} \pm \sqrt{[\tfrac{1}{2}k(k-1)]^2 i^2})$$
$$= \pm (kd \pm \tfrac{1}{2}k(k-1)i).$$

Again less stringent assumptions are required if the alternative prediction of the limits is used. Thus

$$\pm \left(\frac{Sd}{k} + \frac{Si}{\tfrac{1}{2}k(k-1)} \right) = \pm [h] \sqrt{\frac{D}{H}} \pm [l] \sqrt{\frac{I}{L}} \quad \text{providing that}$$

$h/d = f$ and $l/i = f'$, where f and f' are constant over all loci and pairs of loci, respectively. Since we will rarely have reliable estimates of I and L it may be necessary also to assume that $f = f'$, in which case the estimate becomes

$$\pm ([h] + [l]) \sqrt{\frac{D}{H}} = \pm (\overline{F}_1 - m) \sqrt{\frac{D}{H}}$$

The cross of varieties 2 and 12 of *Nicotiana rustica* is the most heterotic for plant height of the many examined and non-allelic interactions are major contributors (Section 21). In the presence of these interactions we can neither estimate D unbiased by them nor $D + I$. Similarly we cannot estimate H or $H + L$ (Section 34). Nevertheless, we expect and in practice observe that the estimates of D and H obtained from an F_2 triple test cross approximate more closely to $D + I$ and $H + L$ than those obtained from other sources (Pooni and Jinks, 1979). For example, for plant height in the 2×12 cross the heritable variation among 60 F_7, F_8 families, which is expected to equal $D + I$, was 421·07. Biased estimates of D from the early generations raised in 1970 were:

290·02 from F_2, B_1 and B_2 families

654·85 from F_3 families

and   492·84 from an F_2 triple test cross.

The estimate of D from the triple test cross is clearly the best predictor of the heritable variation among the true breeding lines and not surprisingly the predictions themselves are good. The expected frequency of true breeding lines that have a height equal to or greater than the latter parent (P_1 = variety 12) or the heterotic F_1 are approximately

$$\geqslant \bar{P}_1 = \int_{\bar{P}_1}^{\infty} \text{ for } \frac{[d]+[i]}{\sqrt{D^*}} \text{ and } \geqslant \bar{F}_1 = \int_{\bar{F}_1}^{\infty} \text{ for } \frac{[h]+[l]}{\sqrt{D^*}}$$

where D^* is the estimate of D inflated by the non-allelic interaction bias. Estimates of the components of the means from the 1970 experiment were

$$[d]+[i] = \bar{P}_1 - m = 14.49$$
$$[h]+[l] = \bar{F}_1 - m = 47.49.$$

Using the estimate of D^* from the triple test cross we would therefore expect a proportion 0.258 of true breeding lines to be as tall as, or taller than P_1, this being the probability corresponding with the value of $14.49/\sqrt{492.84} = 0.65$. In the random sample of 60 F_7, F_8 families we would, therefore, expect 15.48 to fall into this class and 17 were observed to do so. These and similar observed and expected frequencies are summarized in Table 131. With the exception of the predictions based upon an estimate of D^* from F_2 and backcrosses the agreement between observed and expected is exceptionally good.

TABLE 131
The number of F_7, F_8 families that are predicted and observed to fall within the specified phenotypic classes for final height in a cross of Nicotiana rustica (Jinks and Pooni, 1981).

Phenotypic class	Source of D^*	Predictions			Observed
		$F_2, B_1 + B_2$	F_3	TTC	
$\geqslant F_1$		0·16	1·93	0·97	3
$\geqslant P_1$		11·88	17·04	15·48	17
$< P_1$ or F_1		48·12	42·96	44·52	43

The same experiment yielded an estimate of the dominance ratio from the F_2 triple test cross of ± 0.89 giving an inverse of ± 1.12. We, therefore, expect the tallest inbred line to deviate from the mean m by $+1.12 (\bar{F}_1 - m) = +1.12 \times 47.49 = +53.19$. In an assessment of the 60 inbred lines in ten environments the tallest deviated from m on average by 49.32. Both observed and expected are, of course, taller than the F_1 which is the most heterotic of all of those examined.

Prediction is less precise in the presence of linkage because D and H are no longer expected to remain constant over ranks, generations and mating systems (Sections 31 and 32). We can, however, detect its presence and determine its predominant phase in the early generations and therefore have early warning of the direction of the bias on our predictions. For example, if the predominant phase is coupling an estimate of D from the early generations will overestimate the frequency of true breeding lines in later generations with extreme phenotypes. Predictions will, therefore, be over optimistic about obtaining true breeding lines superior to the parental lines or their heterotic F_1. Similarly if the predominant phase is repulsion such predictions will be over pessimistic.

In general, the D estimated from the early generations e.g. F_2 and backcross families, F_2 triple test cross, will have the rank 1 expectation of

$$Sd^2 \pm S_R^C 2(1 - 2p)d_a d_b.$$

At the other extreme, the D at equilibrium after inbreeding by selfing has the expectation

$$Sd^2 \pm S_R^C \frac{2(1 - 2p)}{1 + 2p} d_a d_b$$

which has a linkage bias that is smaller but in the same direction. For loose and tight linkages the difference between these two Ds is very small. The maximum difference occurs for pairs of genes showing intermediate linkages ($p = 0.25$) but for a pair of genes of equal effect ($d_a = d_b$) this maximum is only 33 per cent (or 15 % for $\sqrt{D}$). It is perhaps not surprising therefore, that the small repulsion linkage biases detected in the early generations of *Nicotiana rustica* crosses (Section 31) do not appear to invalidate prediction of the properties of the lines derived from them by selfing.

Prediction in the presence of linkage of the range of true breeding lines based upon estimates of k and D was illustrated in Section 58 for ear conformation in barley where there was substantial coupling linkage (Mather, 1949a). The prediction used the theoretical fall ratios to estimate the number of effective factors linked in coupling and to correct the estimate of D for its inflation due to the linkage bias.

For predictions based on the dominance ratio linkage may lead to serious discrepancies. The dominance component H estimated from an F_2 triple test cross has the expectation $Sh^2 + S2(1 - 2p)h_a h_b$ while from F_2 and backcross variances it is $Sh^2 + S2(1 - 4p^2)h_a h_b$. Both are inflated by linkage between loci whose dominance deviations are in the same direction. Since repulsion linkages deflate D, heterosis arising from dispersed, unidirectionally dominant, linked alleles will lead to an inflated H and deflated D (Section 31). The dominance ratio will therefore be inflated and if dominance is complete or nearly complete the ratio will be greater than one, indicating overdominance. It might wrongly be inferred, therefore, that no true breeding line as good as or better than the heterotic F_1 is obtainable. Furthermore, because the dominance ratio is inflated, prediction of the range of true breeding lines based on its inverse will be an underestimate. These predictions will be considerably more pessimistic than those based on the expected mean m and variance D of the phenotypic distribution of the true breeding lines. The latter should, therefore, be given preference when there is evidence of repulsion linkages and directional dominance.

Proof of a spuriously inflated dominance ratio can only be obtained by re-estimating it after opportunities have been given for the linkage biases in the F_2 to be reduced by random mating or inbreeding. Examples where this has been used to detect spurious overdominance have been given by Gardner (1963) and Jinks (1981).

In the absence of linkage, the expected phenotypic distribution of true breeding lines derivable from any cross or population is independent of the inbreeding system used. In the presence of linkage this is clearly no longer the case as the breeding system will determine the magnitude of the residual linkage bias. The faster the rate of inbreeding the greater will be this residual. The fastest rate achievable is the production of dihaploids either from the haploids derived from anther culture of F_1 hybrids or from the elimination of the chromosomal complement of one parent of a species hybrid. A random sample of such dihaploids will have a heritable variation equal to a D of rank 1 (Table 70). To reduce the linkage bias to a level approaching that achieved under inbreeding by selfing it is necessary to delay the production of the dihaploids until the F_3 generation of a cross or to interpose a generation of random mating (Jinks and Pooni, 1981).

In making predictions there are usually two environments or two sets of environments to be considered; those from which the predictors are estimated and those in which the generations whose properties have been predicted are assessed. Genotype × environment interactions across these two environments or sets of environments will obviously reduce the agreement between prediction and observation. This occurrence can be monitored by growing a common set of genotypes, for example, the parents and their F_1 in all environments used. The presence of interactions between the two environments or sets of environments will not, however, invalidate the predictions if they result mainly from a linear expansion or contraction of the scale with no changes in ranking of the genotypes. In practice this is the more likely form that the interactions will take unless the environments are deliberately varied and, therefore, fall outside of the normal range.

Where, of course, the reaction of a genotype to deliberately varied environments is itself a character of interest it can be analysed, predicted and selected like any other character (Jinks et al., 1977; Jinks and Pooni, 1980).

The predictions can be extended to randomly mating populations with unequal gene frequencies provided that estimates of the diallel form of D and H, that is $4Suvd^2$ and $4Suvh^2$, can be obtained. This restricts it to populations for which there are adequate true breeding testers (Section 40). In general, therefore, predictions in such populations will be confined to those that are possible with the aid of an estimate of the narrow heritability (Section 57 and Falconer, 1960).

Finally, all of the predictions can be extended to two or more characters simultaneously irrespective of the genetical correlations between them providing that the magnitudes of the latter are known. The genetical correlations between a pair of characters attributable to the additive and dominance effects of the genes can be estimated directly from an F_2 triple test cross as the covariance of sums $(L_{1i} + L_{2i} + L_{3i})$ for the two characters and the covariance of differences $(L_{1i} - L_{2i})$ for the two characters, respectively. For example, the bivariate normal distribution for a pair of characters (1 and 2) in the F_∞ generation can be predicted from estimates of m_1, m_2, D_1, D_2 and the additive genetic covariance, D_{12} (Jinks and Pooni, 1980; Pooni and Jinks, 1978).

12   *Experiments and concepts*

61. EXPERIMENTAL DESIGN

When the first edition of this book appeared thirty years ago (Mather, 1949a), few experiments were available whose design and scope were such as to allow full analysis and confident interpretation. Since that time many adequate experiments have been carried out, ranging widely in both the material they drew on and statistical designs they utilized. As their results have come available we have learned more about both the genetical situations with which we are faced in understanding continuous variation and the experimental layouts that are needed for their exploration.

The first requirement of any experiment in continuous variation is that it should supply a measure of the non-heritable variation to which the individuals it contains are subject and a means of separating this from the heritable component of variation, in so far as this is possible. It may be necessary, as we have seen, to measure the non-heritable variation between families (E_b) separately from that within families, and to do so makes its own demands for replication and randomization.

With *Nicotiana rustica*, from which so much of our own experimental material has been drawn, randomization adequate for the separation of heritable and non-heritable components of variation within families can be secured by pricking off seedlings at random from the seed pans and taking the young plants at random from the seedling boxes for planting out in the field. We have seldom had cause to doubt that this simple procedure was satisfactory. To separate these components of variation in comparisons between families, however, introduces the further need for replication, particularly of the genetically uniform families within the experiment, in order to supply a direct estimate of E_b. In the early days of these experiments the replication was achieved simply by growing enough plants of the family from a sowing of seed in a single seed pan and treating the family as a unit in all the cultural operations up to the time when the young plants were set out in the necessary groups in randomized positions in the field. It quickly became apparent, however, that, even as adults, plants raised when young in the same seedling box were more alike than plants of the same genotype from different seedling boxes. Thus, with genetically uniform families, the comparison of plot or group means did not supply a measure of E_b suited for use in the analysis of, for example, V_{1F3} or V_{1S3} when the families had been raised in this way; on the contrary, the variance of such plot means underestimated E_b because the individuals, though going into different plots in the field, had been raised in the same seedling box. Equally, the differences between the means of F_3 families, which had necessarily to be kept separate in all the early cultural operations, contained elements which sprang from the differences between the environments which the various families had experienced in their different seedling boxes, yet which became inextricably confounded with the genetical differences between the families whose effects were to be measured. In other words, this method of handling the plants inevitably overestimated genetical variation between families in segregating generations.

What was less readily established was that even the sowing of a family in a single seed pan, from which all the plants of that family needed in the experiment were raised, had the same effect, admittedly on a smaller scale but one nevertheless sufficiently great to introduce prospective bias into the genetical analysis. Eventually, however, this also became clear, and the only safe way to handle the replication of the plots of genetically uniform families needed for the direct measurement of E_b was seen to be the derivation of each replicate plot from a separate sowing treated as a separate cultural unit, just as the different families in F_3, S_3 etc. were treated. The necessary replication and randomization had, in fact, to be introduced right at the start of the experiment. When this need was eventually established it was decided to take the process a stage further by sowing and raising single plants each in its own disposable pot, randomly distributed as individuals in the glass-house. These plants were then further set out in the field in individually randomized positions, with the result that plots, as such, were eliminated from the experimental design except in so far as each individual could be regarded as a separate plot. With the elimination of plots E_b, as distinct from E_w, vanished from the analysis. A common E was estimated from the differences among the individuals of the genetically uniform families. This, of course, appears as the E component in such statistics as V_{1F2}, V_{2F3}, V_{2S3}, and the non-heritable component of differences between the means of families each of n plants becomes simply E/n and appears as part of the sampling variance in, for example, V_{1F3} and V_{1S3}. Thus a further simplification of the analysis and conservation of a degree of freedom is achieved. This simplification and conservation is not of course achievable in all cases. It is impossible, for example, in *Drosophila* where each family must be raised as a separate entity in its own culture bottle or vial: the design of an experiment must always reflect the biology of the species it utilizes.

Not all the problems of arriving at an estimate of the non-heritable variation to be expected in the segregating generation can be overcome by adequate experimental design. It was noted by Mather (1949a) that, when measured on an otherwise satisfactory scale, the variance of corolla length of the F_1 between the species *Petunia axillaris* and *P. violacea* was lower than that of either parent. Evidently the hybrid is subject to less non-heritable variation than are its parents. This property of heterozygotes is very widespread (Mather, 1946c; Lerner, 1954) and while it is not observed in naturally inbreeding species (Jinks and Mather, 1955), it would appear to be the rule that, in general, more homozygous lines are subject to higher non-heritable variation than are more heterozygous lines in naturally out-breeding species. The non-heritable variation as measured from parents and F_1 will thus commonly differ in such species and none of the estimates can be expected to be directly utilizable as E in the segregating generations. Some combination of the estimates can be used (and indeed was so used by Mather in the analysis of the *Petunia* cross) and, if desired, the combination can be adjusted to take account of the average degree of heterozygosity of individuals in the generation in question.

Even with such adjustment, however, the value so obtained for E is only an approximation, for the amount of non-heritable variation in a given set of environmental circumstances depends not just on the degree of heterozygosity of the individuals, but on their precise genetical constitution of which heterozygosity is but one aspect (Breese and Mather, 1960 and see Section 33). The difficulty of arriving at a direct estimate of E in such a case is thus genetical; it cannot, in principle, be overcome by statistical means (nor, we may observe, by any obvious or acceptable transformation of the scale) though in practice an appropriate combination of the estimates from parents and F_1 may be expected generally to prove adequate. Where individuals are capable of clonal multiplication the difficulty need not, of course, arise, because clonal replication of the segregating generations can be used to supply a direct estimate of E without recourse to parents and F_1 (Section 53).

Such differences in capacity for non-heritable variation provide an example of genotype × environment interaction which cannot readily be removed by transformation of the scale on which the character is measured unless it leads to a high correlation between family mean and variance. We have already seen (Chapter 5) that differences between family means, as well as variances, may reflect the effects of genotype × environment interactions also. Nor can these effects of the interactions be removed readily by scalar transformation. It would appear, however, from the cases described and cited by Bucio Alanis (1966), Bucio Alanis and Hill (1966), Perkins and Jinks (1968a, b), Bucio Alanis *et al.* (1969) and Breese (1969) that the effect of the interaction may well be widely if not generally, related linearly to the direct effect of the environment itself. Bucio Alanis *et al.* (1969) have gone further and shown that, at any rate in one cross of *Nicotiana rustica*, the linear relations observed in parents and F_1 can be used to predict successfully the relation to be found in the segregating generations. The appropriate theoretical elaboration has thus provided the basis for fruitful analysis and confident prediction in a class of cases with which simple scalar transformation cannot cope.

Experiment has also shown that the interaction of non-allelic genes may produce disturbances which cannot be removed entirely by scalar transformation. For example, a set of diallel crosses in *Nicotiana rustica*, described by Jinks (1954), revealed evidence of non-allelic interactions which it was possible, by later examination of segregating generations, to pin down to the behaviour of certain specific crosses among the inbred lines used in the experiment (Jinks, 1956). The effects of the interactions could have been removed, most probably completely, by scalar transformation in these crosses; but any such transformation, though removing the effects of interaction in the crosses for which it was devised, merely served to introduce interactive disturbances into crosses hitherto free from them. In such a case transformation is of limited use and recourse must be made to appropriate elaboration of genetical theory to cope in any final way with the disturbances caused by the interaction. This theoretical elaboration is now available, as we have seen in earlier chapters.

Experience has thus taught us how to design better experiments from the statistical point of view. It has also stimulated development of the genetical theory on which analysis and understanding must be based. However, difficulties, some of which are essentially statistical in nature, still remain to be overcome. The second degree statistics, variances and covariances, which are the raw material of much of the analysis, are themselves subject to high error variances and these will, of course, vary both with the kind of statistic in question and the number of individual observations from which it is derived (see Chapter 6). Given a sufficiently large number of observations a variance can be estimated with any desired precision; but the availability of time and facilities set limits to the precision that can be achieved in this way. Furthermore, within the limited scope of an experiment higher precision of one statistic is commonly achievable only by loss of precision of another. Thus where a given number of F_3 individuals can be raised the decision must be taken as to how this number shall be distributed between and within families. The more F_3 families that are raised, the higher in principle the precision of the estimate of V_{1F3} that is obtained; equally, the larger each family is the higher the precision of V_{2F3}. Obviously some compromise must be effected. If the aim of the experiment is to estimate all variances with roughly equal precision, a total n of F_3 plants is best divided into $n/2$ F_3 families each of two plants, for V_{1F3} will then be estimated from $(n/2) - 1$ degrees of freedom and V_{2F3} from $n/2$ degrees of freedom. Genetical and biological considerations may, however, enter in. If we are concerned to estimate the number of effective factors, K_2, we may well be prepared to sacrifice some precision in V_{1F3} in order to obtain more reliable values for the variances of the F_3 families, and we should reduce the number of families accordingly. Or, on the other hand, if we are not concerned to test for linkage and interaction, but only to separate D, H and E, we might in conjunction with appropriate numbers of parents, F_1 and F_2, prefer to raise n F_3 families each of one individual and combine, as it were, V_{1F3}

and V_{2F3} into a single datum. Similarly, as we have already observed, an experimental plan applicable to plants would be impossible, or at least very wasteful, if applied to *Drosophila*, where the raising of a large number of small S_3 families could pose problems of technique which do not arise when larger S_3 families are raised in conventional containers by conventional methods. This is, of course, but one more example of the general principle that experimental design must reflect the biology of the species under investigation. By the same token backcrosses will, in general, be avoided in cereals where crossing is difficult but where the selfed generations are easy to secure, whereas in dioecious species F_3s are unobtainable and backcrosses easy, while in a species like maize neither selfs nor crosses normally present any difficulty.

Such problems of statistical design in biometrical genetics are worthy of more investigation than they have yet received. Indeed much of what we know has emerged incidentally or even accidentally in the course of carrying out or analysing experiments. Nevertheless, much insight has been gained in this way and we have been able to point out the advantages over others of certain designs, for the particular aims specified, of experiments described in the earlier chapters. Recently, however, a more direct approach to the problem has been initiated drawing on both the accumulated information on the control of particular characters in particular crosses and computer simulations of more general situations.

An example of the use of accumulated information from past experiments may be illustrated by the design of the experiment described by Jinks and Perkins (1969). Information on the generations that can be derived by selfing, backcrossing and sib-mating following a cross between inbred varieties 1 and 5 of *Nicotiana rustica* for the two characters, time of flowering and final height, has been accumulated over 35 years. Previous experiments had shown that the selfing series alone was relatively insensitive to non-additive sources of variation, both dominance and epistasis (Mather and Vines, 1952), whereas recurrent backcrosses and backcrosses of the F_2 to both parents and to the F_1 were particularly sensitive to these sources of variation (Hill, 1966; Mather and Vines, 1952; Opsahl, 1956). In order to investigate the relative contributions of all genetic sources of variation it was decided, therefore, to include representatives of all these mating systems in a single experiment. In all, 21 kinds of families were produced simultaneously by making all possible selfings, backcrossing and sib-matings among the parental, F_1, F_2 and first backcross generations of the cross. Although, as we have previously noted, families differ in the amount of information they provide on any particular kind of gene action there is no reason to believe that they differ in the information they provide in total on all the kinds of gene action present. In these circumstances, therefore, it is appropriate to use a design in which the total amount of information provided by each kind of family is constant as this is likely to maximize the total amount of information in the experiment as a whole.

To achieve this for the analysis of the family means it is necessary to design the experiment so that all such means have the same variance. Since these variances will differ because of their different expectations in terms of the genetic and environmental components of variation the variances of the family means can be equalized only by varying the family size: all family means will have the same variance if the family size is proportional to the observed variance within the family. The appropriate family sizes can, therefore, be calculated from the relative magnitudes of the family variances observed in previous experiments. Where, as in the experiment under discussion, a few types of family had never previously been grown, their expected variances were used, the latter being estimated by substituting the values of the genetic and environmental components of variation (D, H, E_w and E_b) into the expectations in terms of these components. Because of the extent of the accumulated information about this cross, the adoption of this procedure led to estimates of the variances of the 21 family means, and hence to amounts of information of these means, which were virtually identical apart from four kinds of families based on the F_2, whose sizes had been overestimated by a factor of about 2.

As a different example of designing an experiment for a specific aim we can take the case of human populations. We have seen (Section 37) that twins are particularly valuable for determining the relative contribution of heritable and environmental agencies to human characteristics and that in the usual kinds of data available monozygotic twins raised apart (MZA) are absolutely essential for separating heritable and environmental components within and between families. Jinks and Fulker (1970), however, have shown that there are three minimal sets of human families which permit the complete separation of heritable and environmental components. These are:

(i) Monozygotic twins raised together (MZT), monozygotic twins raised apart (MZA) and full-sibs or dizygotic twins raised together (FST or DZT);

(ii) MZT, FST or DZT and full-sibs or dizygotic twins raised apart (FSA or DZA);

(iii) MZT, FSA (or DZA) and half-sibs raised together (HST).

In set (i), which is the standard kind of data collected for studying human populations, MZA, which are difficult and expensive to find and are available only in small numbers, are an essential requirement. In sets (ii) and (iii) these are replaced by FSA (or DZA) which are more readily available. Since the solution obtained by equating observed variances with their expectations in terms of the biometrical genetical model differs among the three sets, it is important to know whether they differ in the amount of information they provide. Thus it would be of no advantage to substitute FSA for MZA families if, for example, considerably larger numbers of the former were required to provide the same amount of information. The three sets of families have been compared by Eaves (1970) for the relative amounts of information they provide on three questions of interest. These are:

EXPERIMENTAL DESIGN · 357

(i) The relative magnitudes of the heritable and environmental components of variation, i.e. the broad heritability (Chapter 8);

(ii) The relative magnitudes of the additive, D_R, and dominance, H_R, components of variation (Chapter 8);

(iii) The relative magnitudes of the within, E_w, and the between, E_b, family components of the environmental variation.

Unlike the previous case we are no longer concerned with the amount of information contained in the observed variances or means but in the amounts of information they contain about specific sets of parameters which can be estimated from them. Our criteria, therefore, are no longer the variances of the observed statistics but the variances of the estimated parameters and of various comparisons among them. The method used was to conduct weighted least squares analyses generated from a hypothetical population of unit total variance for all values of the heritability from 0·1 to 0·9, by intervals of 0·1, and three relative values of $\sqrt{(H_R/D_R)}$, namely 0·1, 0·5 and 1·0 for all relative proportions of the three kinds of families within the sets. The weights used in the analysis were the reciprocals of the theoretical variances of the 'observed' variances (Section 28).

With such a large number of possible combinations it is necessary to use computer simulation to sort out the most informative set of families for any particular combination of the heritability and dominance ratio. The variances of the estimates of the parameters in each combination were obtained from the inverted matrix which in this case is a variance–covariance matrix (Chapters 4 and 6). We can then compare the efficiencies of the different designs as the reciprocals of the variances of the estimates. To take the simplest case as an example: suppose we wish to compare the designs for their ability to separate D_R and H_R, then an index of efficiency can be written as

$$I = \frac{1}{V_{(D_R - H_R)}}$$

where $V_{(D_R - H_R)} = V_{D_R} + V_{H_R} - 2W_{D_R H_R}$, the two variances and the covariance being obtainable from the variance–covariance matrix. Using this index of efficiency, sets (ii) and (iii) prove to be the better designs. So for this purpose it is better to use FSAs and DZA than the more difficult to obtain MZAs. This conclusion holds for all values of heritability and all relative values of D_R and H_R. For any value of these components we can say precisely which proportions of families within each set gives the maximum efficiency. For example, for a heritability of 0·5 the best combination for set (i) is 20 % MZT, 30 % MZA and 50 % FST while for set (ii) the best combination·is 20 % MZT, 50 % FSA and 30 % FST.

In contrast to this finding set (i) provides the most efficient collection of data for estimating the heritability and the relative magnitudes of E_w and E_b. Thus the best design is, as expected, dependent on the aim of the experiment and clearly, if the aim is to obtain the maximum amount of information about all aspects of the control of the phenotype, a compromise design combining families from sets (i), (ii) and (iii) will prove to be desirable.

In an extension of this approach Martin *et al.* (1978) have developed a method based on the non-central χ^2 distribution to calculate the sample sizes required to achieve 95% rejection at the 5% probability level of models of variation which are wrong. When applied to simple alternative models of variation in the classical twin study of monozygotic twins reared together and dizygotic twins reared together (Section 37) they found that frequently encountered situations, for example, deciding between the purely environmental (E_w, E_b) and one involving a genetic component (E_w, D_R) required more than 600 pairs. On the other hand, they were able to demonstrate that tests for skewness and the regression of family variance on family mean in DZ twins were more sensitive to non-additivity, arising from genotype × environment interactions, directional dominance and unequal gene frequencies, than the standard model fitting procedures.

Kearsey (1970) has also compared the relative efficiencies of those designs such as North Carolina 2 and 3 and diallels (Chapters 8 and 9) in which an analysis of variance leads to tests of significance of the genetic components of variation as well as providing direct estimates of their magnitudes. We have already seen (Sections 27, 38, 39 and 42) that with these designs the relative magnitudes of the mean squares in the analysis of variance and their degrees of freedom can be written in terms of the number of parents sampled from the population, the progeny family size and the relative magnitudes of D_R, H_R and E. It is, therefore, possible to determine the optimal combinations of sample sizes and family sizes and the minimum experimental size that will give significance for the mean squares testing for the presence of D_R and H_R. Kearsey, using this approach, compared the different designs on the basis of the minimum experimental sizes required to detect H_R at a probability level of 5% with 95% certainty. He found that for a wide range of relative magnitudes of D_R, H_R and E and of the mating systems in the population sampled, design 2 and diallels are similar in efficiency but design 3 is invariably the most efficient. Furthermore, over most of the situations considered, the minimum experimental size of all the designs was large, usually too large to be used in practice to make comparisons between the dominance variation in two or more populations and larger than the experiments frequently used to investigate it in single populations. As a consequence dominance variation would not have been detected in these experiments unless the heritability and dominance ratio were both high.

Two of the commonest designs currently used for investigating a cross between a pair of true breeding lines, namely (i) F_2 and backcross families, and (ii) and F_2 North Carolina design 3 (Section 27), have also been compared by Kearsey (1980) with respect to the estimates of D and H they provide. He found the ideal design for (i) was replication of the P_1, P_2, F_1, F_2, B_1 and B_2 families in proportion to their expected variances. This is the design used by Jinks and Perkins (1969) and Pooni and Jinks (1980) for the experiments they conducted with varieties 1 and 5 and 2 and 12 of *Nicotiana rustica* discussed in Sections 18 and 21. For design (ii) the aim should be to include at least 20 F_2s. In all circumstances design (ii) has by far the greater efficiency. For example, with low heritability design (ii) requires only one-fifth the experimental size of design (i) for estimating D and one-seventh for estimating H.

A type of disturbance about which more information would be of great value is illustrated by the results which Cooke *et al.* (1962) obtained with *Drosophila* and which have been described and analysed in Sections 29 and 32. It will be recalled that the experiment was derived from a pair of reciprocal crosses B × S and S × B, and the sexes were recorded and analysed separately in the two halves of the experiment, thus allowing four replicate sets of observations to be recognized and used. Of these four replicates, three gave results in good accord with one another; but the fourth (BS females) was not in good accord with the rest. Furthermore, there were internal inconsistencies in this replicate series, for example W_{1S23} came out nearly as high in value as V_{1S3}. Why these B × S females should give results differing from their brothers as well as from the other half of the experiment, when all the other replicates were in mutual accord, is far from obvious. Was it due to some genotype × environment interaction? (Any interaction would have to be of a most curious and complex type to account for such results.) Was it just an extreme example of the effects of the high sampling variation to which second degree statistics are subject? One can only speculate, and because one can only speculate one feels disturbed. This experiment with *Drosophila* is a striking example of the kind of inconsistency under discussion; but it is by no means unique. Many sets of data in the literature show changes and even inconsistencies from one replicate to another, or one statistic to another, not always as difficult to understand as those in Cooke *et al.*'s data, but all suggesting the value of further investigation.

Second degree statistics are perhaps more prone to this type of disturbance than are first degree statistics because of their higher error variation. It is because of this higher error variation that Gilbert (1961) has denied the value of any analyses in biometrical genetics other than an analysis of means. If the analyses of first and second degree statistics were mutually exclusive alternatives, one could understand Gilbert's argument; but they are not. We have seen in earlier chapters not only how both types of statistics can be made to yield information meaningful in terms of the same genetic parameters, but also how the types of information derived from the two kinds of statistics are genetically complementary to one another. The analysis of neither type of statistic can replace or obviate that of the other. To seek, as Gilbert would do, to deny the value of one because it is statistically more troublesome is merely to overlook this genetical complementarity.

We have concentrated so far on the statistical problems of experimental design in biometrical genetics. There are genetical problems too. These are broadly of two kinds: the problems of disentangling the various genetical phenomena, and the parameters by which they are represented, in the tests of significance and estimates to which our experiments naturally lead; and the decisions as to whether we should aim at a more general knowledge of these parameters as they appear over a wider range of material or at a more thorough and detailed knowledge over a narrower range.

Because of the low efficiency of all of the designs available for detecting and estimating the dominance component in a random sample drawn from a population, Kearsey (1970) suggested that it is too ambitious to attempt to investigate this component in terms of population parameters that have applicability to the whole population. He therefore, concluded that it would be more useful to take advantage of the more sensitive approaches available with restricted non-random samples. For example, F_1, F_2 and backcrosses derived from crosses between extreme phenotypes in the population or lines derived by selection.

A start to the investigation of the genetical requirements in the design of experiments has been made by Jinks and Perkins (1969) and Perkins and Jinks (1970) in the experiment mentioned earlier where 21 types of family consisting of the selfing series as far as the F_3, the backcrossing series to the second backcrosses, sib-mating of the F_2 and first backcrosses and the triple test cross carried out on both the F_2 and combined first backcrosses, were raised simultaneously. This experiment which provided 21 different means and 36 variances permitted the detection and estimation of the contribution to first and second degree statistics of linked genes interacting in pairs, linked genes which are interacting with the environment and the simultaneous effects of linkage, genotype × environment interactions and non-allelic interactions. Two general conclusions emerged in addition from this experiment. First, special tests based on statistically simple comparisons between means or variances were more sensitive indicators of the presence of these complex effects than model fitting. Second, these tests could all be carried out on the nine types of family provided by P_1, P_2, F_1, F_2, B_1, B_2 and the L_1, L_2 and L_3 sets of families of an F_2 triple test cross. As a result these nine types of families have become the basic design for subsequent investigations. Current emphasis is now on the circumstances and assumptions under which this design can be further reduced to increase the number of crosses that can be investigated simultaneously.

The efficiency of the F_2 and F_∞ triple test cross design for detecting variation due to non-allelic interactions has been investigated by Pooni and Jinks (1976) as a development of the approach of Kearsey for dominance. For modest levels of non-allelic interactions (one-tenth of the additive and dominance components of variation) the optimal experimental sizes are impractically large except when heritability and the dominance ratio are high and the degree of association (Section 13) in the P_1 and P_2 testers is 0·50 or more. In general, smaller experimental sizes are required to detect duplicate than complementary interactions of the same magnitude and this difference is more pronounced for lower heritabilities and dominance ratios.

There is scope, too, for further investigation into the problem of separating the effects of linkage from those of non-allelic interaction. In principle, linkage can be recognized because the changes it brings about in the values of D and H appear between variances, etc. of different ranks, while those stemming from interaction appear between statistics from different generations. Since, however, an nth rank variance cannot be found before the $n + 1$th generation, there is commonly a measure of confounding of the two types of effect, with the result that linkage and interaction are generally not easy to separate clearly in an analysis (see Chapter 7). Van der Veen (1959) has shown that they can be tested independently of each other, given the right breeding programme and the right genetical composition of an experiment. The advantages of this approach over the standard methods of detecting the components of a variation depending on linkage and non-allelic interaction, has been demonstrated by Perkins and Jinks (1970). In the presence of both sources of variation Van der Veen's approach proved to be better for detecting and clarifying their contribution. There is, however, the need for still more investigation before we can see how best to set about fully disentangling these phenomena in experiment. The equivalent problems arise in diallel analysis in separating the effects of gene association from that of interaction.

One of the great problems of designing experiments in biometrical genetics is that of securing sufficient, and sufficiently different, comparisons to provide estimates of all the genetical parameters, especially where non-allelic interaction is involved. Such interactions are, however, generally detectable by the scaling tests, and their chief effect on the second degree statistics is to bring about changes in the values of D and H over generations. Commonly, therefore, for most purposes it is sufficient to obtain an estimate of the uncertainty that is thus introduced into the estimates of D and H, and the design will be adequate if it provides for such an estimate, even where it is impossible to take the analysis to completion by estimating all of the interactive parameters. There should, in any case, be a consistent trend in the changes of D and H over the generations, as there should be over ranks in the case of linkage, thus opening up the possibility of arriving in a more empirical way at the ultimate values that D and H would attain under the system of mating in use. Replication over generations, and by derivation over ranks, should thus be sufficient to yield not only a measure of the uncertainty introduced by interaction and linkage into the estimates of D and H, but also some notion of the trends that the values of these parameters are showing.

The same considerations apply to genotype × environment interactions, but with one important difference. These interactions also affect the sizes of D and H as they appear under any given set of conditions. The uncertainty so engendered in the estimates of D and H can thus be detected and measured by replication over varying conditions. Furthermore, as we have seen in Chapter 5, trends can be detailed and specified in the changes of D and H with altering environments as measured by mean performances of a genetically constant range of material in the various environments. In so far, however, as the environment changes are not specifiable in advance, as for example is the case with seasonal changes in the weather, the effects of these trends cannot be predicted, but only identified after the observations have been made, and the status of the environment thus established empirically. The measurement of effect of the genotype × environment interactions is then chiefly of use in introducing an appropriate component into the estimates of error of D and H and so appropriately inflating the measure of uncertainty of the values to be expected for these parameters the next time that the experiment is repeated or equivalent material is raised. Where, on the other hand, the changes in D and H can be related to environment factors that can be detected and measured in ways other than by the use of mean measurements of performance of the experimental material in question, the observed trends can be related to these factors. Prediction then becomes possible and measurement of the interaction takes on a value beyond that of merely introducing an appropriate component into the error variation of the parameters. Thus with plants in the field, any interaction relatable to, for example, the nitrogen status of the soil can be used directly in predicting the D and H components of variation in any soil whose nitrogen status has been ascertained. So can interaction relatable to meteorological differences that are relatively regular, as where two areas of different average rainfall are involved. But where the chief changes in environment are unpredictable, as with seasonal changes of meteorological conditions in a given place, the chief value of the knowledge of the genotype × environment interactions must lie in enhancing our knowledge of the error variation expected to be shown by the chief genetical parameters.

It is this kind of consideration which presumably underlies the doubts Hayman (1960a) and others have expressed about the use of covariances between parents and offspring grown in different years in the analysis of experiments designed to yield estimates of the genetical components of variation. Clearly genotype × environment interaction could affect such covariances and if, as we would expect with, for example, variation of our highly bred domestic plants like the cereals, the interactions are large, the covariances could be materially affected. Equally, however, with much experimental material the interactions, and with them the effect on the covariances, could be relatively small. To reject them from the analysis would then be unnecessarily to abandon statistics in an analysis where there are never too many, and seldom enough, statistics for all that we are seeking to do. Covariances over years can be abandoned too lightly as easily as they can be trusted too readily. Again, further investigation is needed into their use. The start that has been made in recent years on the study of genotype × environment interactions promises a better understanding of experimental design in biometrical genetics, as well as a fuller appreciation of the application of its results to practical problems of plant and animal improvement.

62. DOMINANCE AND INTERACTION

The study of continuous variation demands a combination of genetical theory with biometrical analysis. The hereditary determinants of continuous variation are carried on the chromosomes. Like other genes they segregate and they show linkage. If therefore we are to understand the variation which they control we must seek to measure their segregation and their linkage; and in doing so we must take account of their physiological relations in the form of dominance and interaction.

The partition of variation into its components would be impossible without the foundation of Mendelian genetics upon which to build. We could no more devise the means of partition than we could interpret its results, if we were not familiar with the phenomena of segregation and linkage, of dominance and interaction. But our biometrical methods do not, and indeed cannot, lead to measures of their effects exactly like those measures which we use in Mendelian genetics. In place of the segregation ratios of genes with individually identifiable effects, we observe their pooled effects in the form of D. In the same way the pooled dominance is measured by H, and all the linkages act together, as do interactions, in changing the values of D and H over the generations. The genes are always dealt with in the aggregate and we therefore learn nothing of their individual properties. The first results of analysis show us in fact only the combined effects of all the genes, though if we can find the number of effective factors it is of course possible to determine average effects and sometimes even the variances of the distribution of effects.

Thus biometrical analyses yield results which are in a sense complementary to those of Mendelian analysis. Where the genes can be followed individually we can discover their individual properties, but we cannot be sure without further tests that all the genes affecting variation have in fact been traced. Biometrically we cover the whole of the variation, but we do not measure individual properties. We are examining the same basic genetical phenomena by the two methods, but we are examining them in different ways. Not only will our measures of their effects be different, but properties of the genes which are finally distinguishable in Mendelian analysis may not be so when we are analysing variation biometrically (see Mather, 1971).

There exists an ambiguity of this kind between the effects of dominance and genic interaction in biometrical analysis. Dominance itself is a form of interaction, or non-additiveness in effect, between alleles of the same gene. It was, of course, observed, and its consequences for genetical analysis were understood, by Mendel. In his peas the offspring of a cross so closely resembled one of the parents in each character that the effect of the other parent 'either escaped observation completely or could not be detected with certainty'. A single dose of the dominant allele had the same effect as the double dose. Complete dominance of this kind was generally accepted in the early days of genetics as the standard for judging the relations between the phenotypes of heterozygous and homozygous individuals. All other cases were pooled under the heading of imperfect dominance.

As used in this way the concept of dominance is precise but its application is limited. The need for separating degrees of imperfect dominance gradually became felt and the standard of reference was moved to the point where dominance was absent. The degree of dominance could then be measured, and complete dominance would be merely the extreme of a continuous range of possible manifestations.

The new standard of reference, that of no dominance, introduces a new problem. At what relation between heterozygotes and homozygotes can dominance be regarded as absent? It is easy to see that the heterozygotes should have a phenotype midway between those of the homozygotes, but the problem is not thereby solved, for we must ask upon what scale it should be midway. So soon as we cease to be satisfied with seeing merely that the phenotype of the heterozygote falls between those of the homozygotes, and wish to see how far it lies from each, we introduce the need for a metric and raise the question of how an appropriate scale can be devised and justified. The choice of scale can in fact determine whether we regard dominance as existing at all and, if it does, the direction in which it is shown, as we have already seen in Section 9.

The history of interaction between genes at different loci is much the same. The idea of genic interaction was originally used by Bateson in an all-or-none fashion like that of dominance. He found cases where two alleles of one gene led to different phenotypes only in the presence of particular constitutions in respect of a second gene. Complementary action, epistasis, and so on, are all variants of the same principle, differing only in the individual dominance relations of the interacting genes and in the symmetry, or lack of it, in the relations between the genes. Whatever the special relations might be, interaction, as Bateson used the term, implied an absence of difference in effect between the alleles of at least one of the loci under particular circumstances in respect of the other. As with dominance there has been a shift in the standard of reference from that of complete interaction to that of no interaction. This has, of course, raised the same question: that of the metric to be used in measuring the interaction; for it is easy to see that, as with dominance, the presence, direction and degree of interaction must depend upon the scale used in representing the phenotypes which are to be compared. Furthermore, though a change in scale may affect the estimates of dominance and interaction in the same way, it need not do so: indeed, under some circumstances the change of scale may cause the measure of dominance to increase as that of interaction decreases, and *vice versa* (see Section 9).

Our findings both with regard to the presence of any non-additiveness in effect, and with regard to its distribution between dominance and interaction must thus be conditioned by the scale used in measuring the expression of the character. This need cause us little trouble where the genes can be followed individually. The various genotypes can be built up at will, and their phenotypes measured on any scale or scales which seem useful. The effects of the genes in their various combinations can then be completely and unambiguously specified on any or all of the scales and a firm basis obtained for any calculation or prediction. No such rigorous specification is possible where the genes cannot be followed and manipulated as individuals. It is then necessary to resort to biometrical analysis of the types we have been discussing, and the question immediately arises of the scale to be used. An analysis can be carried out of measurements made on any given scale; but that is not to say that all scales are of equal value for this purpose.

We have already seen in Chapter 3, that in our present state of knowledge no scale can be regarded as having a special theoretical merit in terms of the physiology of gene action. This may be expressed in another way. The relative magnitudes of D, H, I etc. and of G measuring genotype × environment interactions, will reflect the scale used for the measurements. They will change as the scale changes, and we have no reason to regard the values obtained on any one scale as presenting a physiologically truer picture of the genetical situation than those yielded by any other scale. We are at liberty, therefore, to choose a scale which gives values of those components most suited to our needs.

Now the partition of variation must remain little more than an exercise in statistics unless we can put its results to some use in understanding the genetical situation in such a way as to enable us to see the consequences of inbreeding, selection or whatever other adjustment of the system we may have in mind. Sometimes this understanding is little affected by choice of scale. To take an example, Mather and Harrison (1949) showed that selection for the number of abdominal chaetae in *Drosophila* resulted in the appearance of interaction between genes in different chromosomes, which interaction had not been detectable in the lines from whose cross the selection had been started. Furthermore, the greater the progress under selection, the stronger the interactions that appeared. Similar interactions were reported and analysed in some detail by Spickett and Thoday (1966) after selection for the number of sternopleural chaetae. Clearly selection can result in the rise of non-allelic interaction whose strength depends on the efficacy of the selection; and, as Mather and Harrison (page 10) noted, this finding which is of importance for our understanding of the genetical architecture of characters (Mather, 1960b, 1966) does not spring from an unfortunate choice of scale, since no justifiable transformation of the scale would eliminate it. A change of scale could, of course, affect our estimate of the strength of the interaction; but this is secondary to the demonstration of the rise of interactions as a consequence of selection.

At the other extreme the data themselves sometimes point clearly to the value of a particular type of transformation for their interpretation. Kearsey and Barnes (1970) have studied natural selection in a population of *Drosophila* and have shown that stabilizing selection must be acting in respect of flies with different numbers of sternopleural chaetae. The curve representing the pressure of selection is, however, asymmetrical when plotted against the expression of the character as represented by sternopleural chaeta number itself: the fitness of the flies is highest near the modal number of chaetae but falls away much more slowly as chaeta number rises than as it falls. At the same time, when measured on the direct scale of chaeta number, there is evidence of unidirectional dominance of the decreasing alleles, whereas previous evidence had indicated that, with a character under stabilizing selection, dominance is expected to be ambidirectional if it is present at all. Kearsey and Barnes have also shown, however, that a transformation of the scale used in representing the phenotype in respect of chaeta number, which foreshortens the upper end of the distribution sufficient to make the fitness curve symmetrical, also removes all the indications of dominance. The transformation thus clears up two anomalies simultaneously, and we can hardly doubt that the new scale to which it gives rise is preferable for the interpretation of these experimental results.

Thus the data themselves may tell us that the scale used is not of the essence in interpretation of the results, or alternatively point to the kind of scale on which interpretation is most fruitful. This will, however, not always, or perhaps even generally, be the case. We may then adopt the more empirical procedure. We seek to understand a genetical situation in order that we can foresee its consequences. The reason that it is profitable to distinguish between D, H, I etc., is that these components depend on phenomena whose consequences differ. Generally, the D component will be of most importance to us, and indeed our interest in the other components is very often confined to their elimination for the purpose of rendering predictions based on D the more precise. At other times, however, as when considering heterosis, we may have a more direct interest in non-additive effects. But whatever our interest may be, the analysis will be the simpler the smaller the number of components there are to deal with.

The very fact that heterosis is of widespread occurrence warns us that a component of variation depending on non-additive effects of the genes must be expected, no matter what scale is used. But even so, we are still at liberty to explore the possibilities of adjusting the proportions which will appear as dominance and as interaction. Now dominance is more limited in its scope and simpler in its genetical consequences than interaction. Prediction is correspondingly difficult in the presence of interaction: neither the d nor the h increments of the various genes will be additive and the correction necessitated by their non-additiveness must vary with the kind or kinds of interaction which are in operation. It is therefore preferable to have the non-additive effects represented as dominance rather than interaction, and this is what the method of scaling developed in Chapter 3 seeks to accomplish. If applied, for example, to the data of Table 7 the scaling criterion would reveal the log scale as adequate but the direct scale as inadequate for our purpose, which is, of course, to eliminate interaction and leave only D and H as the components of heritable variation, separable both in measurement and in prediction. We must, however, remember that, as noted earlier, it is not always possible to remove non-allelic and genotype × environment interactions by rescaling, though of course we may always seek to minimize their effects.

63. THE EFFECTIVE FACTOR

In examining the phenomena of dominance and interaction as they appear in biometrical genetics, we have seen that while we must continue to regard them as distinct in their genetical consequences, in the way made familiar by Mendelian analysis, they lose some of their distinction, even becoming partly interchangeable, in their biometrical measurement. We can no longer regard any given degree of dominance and interaction as final properties of the genes in question, because we can adjust them by an artificial manipulation of the scale. Thus although the concepts derived from Mendelian genetics must remain, our notions as to the ways in which they are used must be extended and adjusted to the circumstances of biometrical analysis. In just the same way, we must still seek to explain our biometrical results in terms of the genes which Mendelian studies have led us to infer, but the new circumstances of observation make us apply the idea of genes in a modified fashion.

The fine gradation in manifestation of a character showing continuous variation suggests that the number of genes, whose differences control the variation, is generally large; certainly of the order of 10^1 and possibly of the order of 10^2. The various studies of abdominal chaeta number in *Drosophila* reported in the literature suggest that at least 15 to 20 genes must be involved in its determination, and other characters probably agree, though following selection the chief contributions to the variation may appear to come from a smaller number of 'genes' which have presumably been built up by the selection itself. The values found for the number of effective factors are in striking contrast with this expectation, for they are generally smaller and often surprisingly small. The reason for this discrepancy is, of course, clear. As we have already seen, the effective factor is not of necessity the ultimate gene, and the estimate of the number of factors is further reduced by any variation in the size of their effects.

The gene is not the only unit to which genetical investigations lead. The whole nuclear complement, the haploid set of chromosomes, the single chromosome and the chromosome segment can all behave as units in one way or another. The gene occupies its unique position solely because it is the smallest unit into which we are able to see the heritable material as divisible by classical genetic means. We know too that even the gene as a unit of action can be broken down into smaller units, but these units show special spatial relations which still lead us to recognize the gene, or 'cistron' as it is called in this connection, as the entity within which these spatial relations apply.

Operationally we can recognize a gene only by the difference to which it gives rise. If the gene does not vary and, in varying, produce a detectable phenotypic difference, we cannot know that it is there. The inference of a gene is therefore limited by the means available for detecting the effect of its variation. If the difference produced in the phenotype is too small to be picked up by the means at our disposal we cannot identify the determinant, though we may be able to detect a group of such determinants of small effect when they are acting together.

Having detected the difference, the second condition that must be satisfied before it can be ascribed to a single gene is that it cannot be broken down into two or more smaller differences; or, amounting to the same thing, that there can be found no evidence of two or more smaller genetic differences which can combine to give the one in question. Again, since genetic differences can be followed only by their effects on the phenotype, the means of detecting these effects will limit the operational test of divisibility.

Where two gene differences affect characters that are separately recordable, it is relatively easy to show that the genes are distinct from one another. Re-association of the genes in the chromosomes will then be regularly detectable by the re-association of the characters in the phenotype. The re-association of the genes may arise by recombination or by mutation (which is not formally distinguishable from rare recombination), but no matter how rare an occurrence it may be, the re-association of the genes can always be recognized in these circumstances as an individual event. A single recombination or mutation is therefore sufficient to establish the two genes as distinct units.

The test is less precise where the two genes are less readily distinguishable by the differences they produce in the phenotype. Suppose that the two have complementary actions such that the genotypes aabb, aaB and Abb give indistinguishable phenotypes, only the combination AB having a different effect. On crossing two true breeding lines, AABB and aabb, the F_2 will contain only two distinguishable classes, but we may recognize that two genes are involved by the characteristic departure from a 3:1 ratio. If the two genes recombine freely the F_2 ratio will be 9:7, and there can be found types, aaB and Abb, which while phenotypically like aabb can be separated from it by their breeding behaviour in crosses with one another. If the two genes are linked, however, the test is less precise. The two phenotypic classes are expected in the ratio $3 - 2p + p^2 : 1 + 2p - p^2$, where p is the frequency of recombination, assumed to be the same on both male and female sides. The smaller p becomes the more difficult it will be to detect departure from the 3:1 ratio. Thus with $p = 0.01$, the ratio will be $2.981 : 1.019$, which will be very troublesome to distinguish from 3:1. Furthermore, the classes Abb and aaB, which afford material for a confirmatory breeding test, will be rare. They will occur with frequencies of only $2p - p^2$ each compared with $1 - 2p + p^2$ for aabb, or $0.019 : 0.019 : 0.981$ when $p = 0.01$. There is, therefore, no easy means of recognizing single cases of recombination between the two genes and in consequence the test is basically statistical. A unique case of recombination is insufficient: recombination must be sufficiently common for us to detect the deviations it causes from certain statistical expectations, and we cannot therefore be sure of always isolating the ultimate unit of inheritance.

Mutation might appear to help in such a case. We might obtain aaBB and AAbb by independent mutation from AABB. They would appear alike but on crossing would give an F_1 like the parental stock, AABB. A situation like this is known at the yellow locus in *Drosophila melanogaster*, but even so there is still doubt as to whether this should be regarded as two genes or one, for the mutation which would be, in our present two-gene notation, from AB → ab is more common than AB → aB or AB → Ab. The final conclusion is thus still dependent on statistical considerations: the unique observations of mutation to aB and Ab is not enough (Mather, 1946d). The successful analysis of the fine structure of genes, initially by Benzer (1955), in micro-organisms has depended on the large numbers of individuals that can be raised in these organisms, making possible the accumulation of many rare mutants and the measurement of very low recombination values.

The test of the unitary nature of a genetic difference is therefore conditioned by the type of effect on the phenotype, even with genes of major effect. This limitation of observation shows us that we can never be sure of tracing the ultimate polygene. Polygenes, or polygenic complexes, may be built up by selection to have effects large enough for them to be located in the chromosomes by special techniques (Thoday, 1961); but generally polygenes have effects sufficiently small by comparison with the residual variation to prevent them being followed as individuals in genetic analysis. Their effects are also similar to one another. It is therefore, in general, impossible to detect the recombination of two polygenes as a unique event by its effect in the phenotype. The test of recombination must be statistical.

There are two reasons why, in the absence of special techniques and marker genes, the test of recombination between polygenes will be even less sensitive than the test which we have considered in the case of linked complementary genes. The first reason is that recombination must be sufficiently frequent for us to detect a change in variance depending at best (Section 31) on the difference between $(1 - 2p)$ and $(1 - 2p)^2$. This change is from 0.9800 to 0.9604 when $p = 0.01$ and is only from 0.80 to 0.64 when p is as high as 0.1. Secondly, we have no means of separating the effects on variation of all the pairs of genes which are segregating, so that change in some of them may be partly swamped by lack of it in others. The extent to which we can push the analysis of a polygenic system into the ultimate units which genetics has taught us to recognize, must thus be limited by the conditions under which our observations are made. The less the extraneous variation relative to the effect of the gene, and the fewer the genes which are segregating, the further the analysis can be taken. Where the genes or gene complexes are of relatively large effect, such as may be built up by selection, the individual genes or complexes may be distinguished, isolated and manipulated by the use of special tests (Spickett and Thoday, 1966) but under ordinary circumstances where the individual units are of small effect, relative to the residual variation, no means are available of recognizing recombination as an individual event. Thus in the general case, while we must interpret the properties of a polygenic system in terms of ultimate genes, we must be prepared to use biometrical units specified by analytical criteria different from those of Mendelian genetics, and depending on statistical relations rather than unique events. The classical test for distinguishing one gene from another is no longer available and we must distinguish our units by other criteria. And since the criteria have changed we must expect the units to do so also. Though we cannot detect recombination as a unique event, we may proceed by assuming some convenient frequency of recombination by which our units in polygenic variation are distinguished from one another, while recognizing of course that no matter how low we set this frequency we can still have no certainty that the factors so distinguished will be ultimate genes. The number of factors into which a chromosome can be divided in this way cannot, of course, exceed the ratio its total genetic length bears to the recombination frequency assumed between adjacent factors. Since we may assume that factors in different chromosomes recombine freely, their numbers are additive over chromosomes. The maximum number of factors we can find is thus dependent on the assumption made as to the frequency of recombination between them. If we assume that the factors are independent in inheritance, statistically the easiest assumption to make (see Chapter 7), the maximum number is given by the recombination index, i.e. the haploid number of chromosomes plus the mean number of chiasmata per nucleus.

This assumption of independent inheritance is equivalent to supposing that there is no variation in either the number of chiasmata formed or in the position in which each forms within the chromosome, at least in respect of chiasmata falling in parts of the chromosomes carrying genetic differences. On such a basis pairs of genes will either recombine freely or will not recombine at all, according to whether they do or do not lie astride a point of chiasma formation. The factor is then clearly the same as the 'gene' which could be detected by classical Mendelian methods, since chiasma formation, upon which recombination within a chromosome depends, is either all or none. This situation, or something approximating to it, is encountered in such species as *Stethophyma grossum* and *Fritillaria meleagris* where the chiasmata can be observed to be localized in position. In heterokaryotic fungi, too, where, apart from para-sexual behaviour, no recombination occurs within a nucleus and the variation arises from the sorting out of whole nuclei, recombination is all or none. The nucleus is then the final unit of recombination and transmission, as is the segment delimited by the positions of chiasma formation in *Stethophyma* and *Fritillaria*. It is at once the 'gene' and the effective factor. The latter will therefore be as constant a unit as the gene in these cases.

Generally, however, chiasmata vary in their position of formation. The linkage test of Chapter 7 is in fact a test of such variation. If all pairs of genes either recombined freely or not at all, each would contribute either as two independent units or only as one to D and H. There could then be no change in D and H over the generations. Such changes have been clearly observed in barley and they are strongly suggested in other experiments also. Even, however, where we fail to observe the type of change in D and H, which recombination values between 50 % and 0 % bring about, it may mean no more than that the genes were so distributed as not to offer the means of detecting the linkage. However this may be, the linkage test offers us a means of deciding whether we may reasonably assume the factors to be inherited independently. When linkage is detected we cannot fairly make this assumption, and any count of the number of factors based on the assumption of independence must be incorrect.

Variation in the position of chiasma formation has a further consequence. The segment of chromosome, on the average 50 recombinational units long, which on the basis of independent inheritance is recognized as an effective factor in transmission from parent to offspring, will not be constant: because of variation in the position of chiasma formation, the segment delimited by chiasma formation in one individual and, in consequence, passed on as a block to one of its offspring will be subject to breakdown in transmission from that offspring to its own progeny. Thus the factors arrived at must in a sense be statistical abstractions. Each is a piece of chromosome which will not only vary in its genetic content within a generation but will also be broken down by further recombination as the generations succeed one another and indeed has been observed doing so (Towey and Jinks, 1977). The demonstration of polygenic linkage, which is essentially a demonstration that the positions of chiasmata vary, carries with it the demonstration that no permanent system of effective factors can be derived.

This breakdown of the effective factors by recombination has three consequences. First of all, the number of factors must appear to increase as the experiment progresses. Thus with four factors segregating in F_2 and no change by recombination within them, F_3 families should average two segregating factors, F_4s one segregating factor, and so on. But with breakdown the fall in number per family in the succeeding generations will be less, so that the number which must be assumed in the experiment as a whole will increase. At the same time, of course, the average effect of each factor must diminish correspondingly. In other words, the gap between gene and effective factor will be narrowing as the experiment proceeds, and if this process is continued for sufficiently long we might hope that the biometrical factor would ultimately approximate to the individual polygene. The experiment would, however, need to be very lengthy for individual polygenes to be counted in this way.

A second consequence of the breakdown of the factors is that they can be used as units only in a temporary sense. They enable us to forsee behaviour in the near future, to predict minimal limits and rates of advance under selection for one or two generations. Where the linkage test reveals little evidence of change in the units, the period over which prediction can be made may be increased; but even in this case, the longer the term of the prediction the more hazardous it must be. The number of effective factors must therefore be used with reserve: unlike genes, they are not final units.

Nevertheless, it is worth noting that the factors are the only units we have. We can arrive at some estimate of their number; whereas we can arrive at no estimate of the number of genes, except possibly with great labour. And even could we know how many genes were involved the information would benefit us little, for their organization within the chromosomes is all-important in determining the contributions they make to the components of variability and to the selective responses. It is of more use to us to consider the system in terms of contributions to the variability than in terms of individual genes. The factors to which the analysis of variability leads us are admittedly neither permanent nor constant, so that the predictions which they permit have only a short range validity; but a knowledge of the number of genes without any knowledge of their organization into factors would permit no prediction at all.

The third consequence of recombination within the effective units is that just as the unit is broken down in this way, it can also be built up. Genes themselves change only by the rare process which we term mutation, and we know that polygenes are no exception (Mather, 1941; Mather and Wigan, 1942; Clayton and Robertson, 1955; Paxman, 1957). Recombination, on the other hand, is much more frequent in its occurrence, and it therefore offers much greater possibility of useful change in the factor. The effect, however, of intra-genic mutation and inter-genic recombination will not be directly distinguishable where the genes have physiologically small, similar and supplementary actions. We can recognize change through recombination within the factor only by comparison of the speed of the change observed with that known to be due directly to mutation as measured in an inbred line (Mather, 1941). From the point of view of the effective factor, viewed as a whole, the two processes are therefore inseparable.

The changing genic content of effective factors, as derived from biometrical analysis, and the way in which the content of the factors can be built up, is well illustrated by Simchen's (1966) observations on *Schizophyllum* described in Section 56. He found the same number of effective factors in both the cross from which he started his experiment and the cross between the extreme individuals selected from the progeny of that first cross. But the action of the factors, and hence we infer their genic content, had changed. Following selection the allelic factors produced markedly larger differences in their effects on the individuals carrying them. The genes must have been redistributed between the factors, repulsion relationships of the kind $+ -/- +$ being broken down and replaced by coupling linkages $+ +$ and $- -$, as a result of recombination. Such changes in linkage phase under selection have, in fact, been demonstrated experimentally by Thoday (1960) in *Drosophila*.

The way in which linked groups of polygenes built up by recombination under selection tend to stick together in segregation and so appear more or less as one large factor is well shown by the selection experiment with *Drosophila* described in Section 4 (Mather and Harrison, 1949). The first selection line advanced slowly and fairly steadily for 20 generations, the mean abdominal chaeta number increasing by about 20 over that period (Fig. 5). Such a change is too slow to be wholly, or even largely, due to recombination of whole chromosomes: there are only three large chromosomes in *Drosophila melanogaster*, and all their combinations as whole chromosomes would appear and be exposed to selection in the first few generations. The main advance must have been due rather to recombination within chromosomes replacing repulsion by coupling linkage.

This recombination also upset the balance of other systems of polygenes one of which affected fertility. As we saw in Section 4, fertility fell as chaeta number rose, until finally it became so low and the flies so few that selection could no longer be practised and mass cultures had to be used. With selection relaxed for chaeta number, the fertility system took charge and as fertility rose under natural section, chaeta number fell. But it fell at such a speed that 80 % of the gain in chaeta number, which took 20 generations to achieve, was lost in only 4 more. When selection was again practised for increased chaeta number, the lost ground was regained once more in 4 generations. The coupling linkages so laboriously built up, had persisted with corresponding tenacity, so that the original segregation for factors which, by reason of their balanced genic content, were of small effect was replaced by segregation for factors of larger effect. The original advance had been conditioned by the slow building up of the very polygenic combinations whose behaviour as effective factors in transmission and segregation permitted later advance at four times the original speed.

The persistence of these units built up by selection was inferred from the speed of decline when the selection was relaxed, and of advance when it was reimposed in Mather and Harrison's experiment. When analysing a line of *Drosophila melanogaster* whose number of sternopleural chaeta had been very materially increased by selection, Spickett and Thoday (1966) were able to go further, by locating within the chromosomes a number of genes or gene complexes whose effects on the character were measured and found to be large. These behaved as units in the genetical manipulations necessary for their action and interaction to be measured: they could be handled like genes and in this sense could be described as genes. Yet they could hardly have been present as such in the original cross from which the selection line was derived or selection would not have taken so long as it did to bring about the advance that was finally achieved. How they in fact came into being is not known; but it is, to say the least of it, likely that these 'genes' too were built up by the reshuffling through recombination of closely linked polygenic complexes. For if the constituents of the complex are but a few recombination units apart, they will be reshuffled only slowly with the result that, on the one hand, the complexes of large effect may take several generations to emerge while, on the other, once built up recombination will break them down so seldom that they will appear permanent units whose breakdown will escape notice in all but large experiments designed specifically to detect it. Thoday (1973) also subscribes to this interpretation.

In these experiments, as indeed must be the case in all such experiments, the speed and immediate limit of advance under selection was not determined by the number of individual polygenes but by their organization into linked combinations. An experiment covering only a few generations will therefore discover the properties of these combinations rather than those of their constituents. In consequence the factors, in terms of which we interpret biometrical experiments, must be related to the combinations rather than to individual genes. The individual polygene must remain, however, as the final unit of action, change and recombination upon which the properties of the combinations or factors are based. We are, in fact, working at two levels. In the immediate sense the combination is all-important, but it is not a final unit. Genetically it is composite and it can therefore change as its constituent genes change or become re-associated.

This distinction between the gene and the effective factor, which we recognize in biometrical segregation, is important to our understanding and use of the latter concept. The factor must derive its properties from the genes of which it consists, but these properties need not be so limited as those of a single gene. We have seen that change in a factor depends more on the re-association of its constituent genes than on change within the genes themselves. We must now observe that we cannot even be sure that a change which is not to be traced to recombination is due to mutation of one gene. Structural change, leading, for example, to duplication of several genes within a factor, cannot be distinguishable from single gene mutation, and indeed the apparent mutations of Jones (1945) in inbred maize may well be of this kind.

In the same way, super-dominance ($h > d$) of an effective factor, should it be observed, could always be interpreted in terms of normal dominance ($h \leqslant d$) of individual genes of like action which were linked, perhaps very closely, in the repulsion phase (see Section 21). This is indeed a classical interpretation of heterosis (East and Jones, 1919) whose validity has now been amply demonstrated by experiment (Jinks, 1981). Equally, too, apparent pleiotropy of the effective factor could always be interpreted as due to its containing within the chromosome segment which is its physical basis, polygenes belonging to unlike systems. Genetical experiments have in fact shown that genes of unlike effect are characteristically mingled along the chromosomes.

This is well illustrated by the observations of Breese and Mather (1957, 1960), who analysed the differences shown by chromosome III in *Drosophila melanogaster* as they affected the number of abdominal chaetae and the viability of female flies. Two chromosomes III, from lines selected for high and low number of abdominal chaetae respectively, were recombined in such a way as to give combinations of four segments of the chromosome, and the composite chromosomes so produced were tested in a set of diallel crosses. The W_r/V_r graphs are shown in Fig. 23 for both chaeta number (a) and female viability (b). The distribution of the points are quite different on the two graphs: all segments are active for both characters, but segment B has a much greater effect than any other on chaetae, whereas it is in no way outstanding, being in fact less in effect than D, in respect of viability. Furthermore, while there is evidence of dominance for both characters, there is no indication of non-allelic interaction for chaeta number, but good evidence for duplicate type interaction for viability. Clearly, we have two systems of genes, one affecting chaeta number and the other viability, each with its own balance and

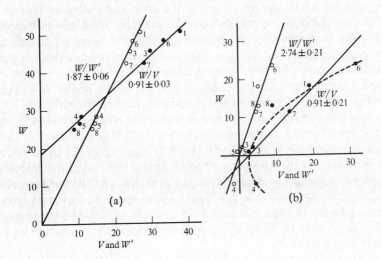

Figure 23 Results of a diallel experiment involving seven composite chromosomes III in *Drosophila melanogaster*. The W_r/V_r (solid circles) and W_r/W_r' (open circles) graphs relate to (a) number of abdominal chaetae and (b) viability of the flies. The straight lines are the best fitting regression lines, for which the calculated slopes are shown. The dotted curve in (b) has no special significance being included only to bring out the curvilinear nature of the W_r/V_r graph. The numbers by the points indicate the chromosomes to which the points relate. Note that both the shape of the W_r/V_r graph and the slope of the W_r/W_r' graph differ between (a) and (b) and that the distribution of the corresponding points along the lines differs between (a) and (b), thus showing that corresponding pairs of chromosomes have different relative effects on the two characters. (Reproduced by permission from Breese and Mather, 1960)

properties, mingled along the four segments into which the chromosome has been broken by recombination. Each segment, or effective factor as it was in this experiment, is having a pleiotropic effect, not because the genes it carries are pleiotropic in action, but because it simultaneously contains member genes of different polygenic systems each affecting its own particular character. The genes are not pleiotropic but, because of its composite nature, the effective factor is.

Such pleiotropic action, springing from linkage of genes not themselves pleiotropic, must be expected to be a regular property of effective factors. The observations on correlated responses in such experiments as that of Mather and Harrison vindicate this expectation. These correlated responses may in fact be very complex. In the selection experiment in question, fertility, spermatheca number and mating behaviour all changed as a result of selection for chaeta number. Doubtless other characters not recorded in the experiment behaved similarly. Pleiotropy in the classical sense is therefore almost useless as a concept for application in biometrical genetics. We know that apparent pleiotropy can be due to linkage, and we know that units of inheritance we can recognize are linked groups of genes. We have, and indeed in biometrical genetics we can have, no proven case of pleiotropy of a single polygene.

In order to account for continuous variation we have had to postulate systems of genes having small, similar and supplementary effects. We have been able to show that these genes must behave in transmission in the same way that the major genes of Mendelian genetics behave, and on this basis we have been able to develop a means of analysing continuous variation. But we have also come to see that, in biometrical genetics, observations can be made only on linked groups of the genes, on in fact the total genic content of factors whose physical basis is to be found in whole segments of the chromosomes. The properties of the groups or effective factors will reflect not merely the behaviour of each individual constituent gene, but also the mechanical and physiological relations of the genes with one another. And since we have no means of predicting these genic relations in detail, we can learn the properties of the factors only by direct observation and experiment, as indeed we have already begun to do.

References

ALLARD, R. W. (1956), The analysis of genetic environmental interactions by means of diallel crosses, *Genetics*, **41**, 305–318.

BARIGOZZI, C. (1951), The influence of the Y-chromosome on quantitative characters of *Drosophila melanogaster, Heredity*, **5**, 415–432.

BARNES, B. W. (1968), Maternal control of heterosis for yield in *Drosophila melanogaster, Heredity*, **23**, 563–572.

BATESON, W. (1909), *Mendel's Principles of Heredity*, University Press, Cambridge.

BEADLE, G. W. (1945), Biochemical genetics, *Chem. Rev.*, **37**, 15–96.

BEALE, G. H. (1954), *The Genetics of Paramecium aurelia*, University Press, Cambridge.

BEDDOWS, A. R., BREESE, E. L. and LEWIS, B. (1962), The genetic assessment of heterozygous breeding material by means of a diallel cross: I Description of parents, self- and cross-fertility and early seedling vigour, *Heredity*, **17**, 501–512.

BENZER, S. (1955), Fine structure of a genetic region in bacteriophage, *Proc. Nat. Acad. Sci. Wash.*, **41**, 344–354.

BIRLEY, A. J., COUCH, P. A. and MARSON, A. (1981), Genetical variation for enzyme activity in a population of *Drosophila melanogaster* VI. Molecular variation in the control of alcohol dehydrogenase (ADH) activity, *Heredity*, **47**, 185–196.

BREESE, E. L. (1969), The measurement and significance of genotype–environment interactions in grasses, *Heredity*, **24**, 27–44.

BREESE, E. L., HAYWARD, M. D. and THOMAS, A. C. (1965), Somatic selection in perennial ryegrass, *Heredity*, **20**, 367–379.

BREESE, E. L. and MATHER, K. (1957), The organisation of polygenic activity within a chromosome in *Drosophila*: I Hair characters, *Heredity*, **11**, 373–395.

BREESE, E. L. and MATHER, K. (1960), The organisation of polygenic activity within a chromosome in *Drosophila*: II Viability, *Heredity*, **14**, 375–400.

BRINK, R. A. (1960), Paramutation and chromosome organisation, *Quart. Rev. Biol.*, **35**, 120–137.

BRITTEN, R. J. and DAVIDSON, E. H. (1969), Gene regulation for higher cells: a theory, *Science*, **165**, 349–357.

BROADHURST, P. L. and JINKS, J. L. (1965), Parity as a determinant of birth weight in the rhesus monkey, *Folio primat*, **3**, 201–210.

BRUMBY, P. J. (1960), The influence of the maternal environment in mice, *Heredity*, **14**, 1–18.

BUCIO ALANIS, L. (1966), Environmental and genotype–environmental components of variability: I Inbred lines, *Heredity*, **21**, 387–397.

BUCIO ALANIS, L. and HILL, J. (1966), Environmental and genotype–environmental components of variability: II Heterozygotes, *Heredity*, **21**, 399–405.

BUCIO ALANIS, L., PERKINS, J. M. and JINKS, J. L. (1969), Environmental and genotype–environmental components of variability: V Segregating generations, *Heredity*, **24**, 115–127.

BUTCHER, A. C. (1969), Non-allelic interactions and genetic isolation in wild populations of *Aspergillus nidulans*, *Heredity*, **24**, 621–631.

CALIGARI, P. D. S. (1981), The selectively optimal phenotypes of coxal chaetae in *Drosophila melanogaster*. *Heredity*, **47**, 79–85.

CALIGARI, P. D. S. and MATHER, K. (1975), Genotype–environment interaction. III. Interactions in *Drosophila melanogaster*, *Proc. R. Soc. Lond. B.*, **191**, 387–411.

CALIGARI, P. D. S. and MATHER, K. (1980), Dominance, allele frequency and selection in a population of *Drosophila melanogaster*, *Proc. R. Soc. Lond. B.*, **208**, 163–187.

CATEN, C. E. (1979), Quantitative genetic variation in fungi. In *"Quantitative Genetic Variation"* (eds. J. N. Thomson, Jr. and J. M. Thoday) pp. 35–60. Academic Press, New York.

CATEN, C. E. and JINKS, J. L. (1976), Quantitative Genetics, in *Second International Symposium on the Genetics of Industrial Microorganisms* (ed. K. D. Macdonald), pp. 93–111, Academic Press, London.

CAVALLI, L. L. (1952), An analysis of linkage in quantitative inheritance, *Quantitative Inheritance* (eds. E. C. R. Reeve and C. H. Waddington), pp. 135–144, HMSO, London.

CHARLES, D. R. and GOODWIN, R. H. (1943), An estimate of the minimum number of genes differentiating two species of golden-rod with respect to their morphological characters, *Amer. Nat.*, **77**, 53–69.

CHOVNIK, A., GELBART, W., McCARRON, M., OSMOND, B., CANDIDO, E. P. M. and BALLIE, D. L. (1976), Organisation of the rosy locus in *Drosophila melanogaster*: Evidence for a control element adjacent to the xanthine dehydrogenase structural element, *Genetics*, **84**, 233–255.

CLARKE, C. A. and SHEPPARD, P. M. (1960), The evolution of dominance under disruptive selection, *Heredity*, **14**, 73–87.

CLAYTON, G. and ROBERTSON, A. (1955), Mutation and quantitative variation, *Amer. Nat.*, **89**, 151–158.

COCKS, B. (1954), Polygenic systems controlling the expression of major mutant genes which affect chaeta number in *Drosophila melanogaster*, *Heredity*, **8**, 13–34.

COMSTOCK, R. E. and ROBINSON, H. F. (1952), Estimation of average dominance of genes, *Heterosis* (ed. J. W. Gowen), pp. 494–516, Iowa State College Press, Ames, Iowa.

CONNOLLY, V. and JINKS, J. L. (1975), The genetical architecture of general and specific environmental sensitivity, *Heredity*, **35**, 249–259.

CONNOLLY, V. and SIMCHEN, G. (1968), Linkage to the incompatibility factors and maintenance of genetic variation in selection lines of *Schizophyllum commune*, *Heredity*, **23**, 387–402.

COOKE, P., JONES, R. M., MATHER, K., BONSALL, G. W. and NELDER, J. A. (1962), Estimating the components of continuous variation: I Statistical, *Heredity*, **17**, 115–133.

COOKE, P. and MATHER, K. (1962), Estimating the components of continuous variation: II Genetical, *Heredity*, **17**, 211–236.

COOPER, J. P. (1961), Selection and population structure in *Lolium*: V Continued response and associated changes in fertility and vigour, *Heredity*, **16**, 435–453.

COUGHTREY, A. and MATHER, K. (1970), Interaction and gene association and dispersion in diallel crosses where gene frequencies are unequal, *Heredity*, **25**, 79–88.

CRICK, F. H. C. (1967), The genetic code, *Proc. Roy. Soc. B*, **167**, 331–347.

CROFT, J. H. and SIMCHEN, G. (1965), Natural variation among monokaryons of *Collybia velutipes*, *Amer. Nat.*, **94**, 451–462.

DARLINGTON, C. D. (1937), *Recent Advances in Cytology* (2nd edn), Churchill, London.

DARLINGTON, C. D. (1950), *Chromosome Botany*, Allen & Unwin, London.

DARLINGTON, C. D. (1958), *Evolution of Genetic Systems* (2nd edn), Oliver & Boyd, Edinburgh.

DARLINGTON, C. D. and MATHER, K. (1949), *The Elements of Genetics*, Allen & Unwin, London.

DARLINGTON, C. D. and THOMAS, P. T. (1941), Morbid mitosis and the activity of inert chromosomes in *Sorghum*, *Proc. Roy. Soc. B*, **130**, 127–150.

DAVIDSON, E. H. and BRITTEN, R. J. (1973), Organisation, transcription and regulation in the animal genome, *Quart. Rev. Biol.*, **48**, 565–613.

DAVIES, D. R. (1962), The genetical control of radiosensitivity: I Seedling characters in tomato, *Heredity*, **17**, 63–74.

DAVIES, R. W. (1971), The genetic relationship of two quantitative characters in *Drosophila melanogaster*. II. Location of the effects, *Genetics*, **69**, 363–375.

DAVIES, R. W. and WORKMAN, P. L. (1971), The genetic relationship of two quantitative characters in *Drosophila melanogaster*. I. Responses to selection and whole chromosome analysis, *Genetics*, **69**, 353–361.

DAWSON, P. S. (1965), Estimation of components of phenotypic variance for development rate in *Tribolium*, *Heredity*, **20**, 403–417.

DICKINSON, A. G. and JINKS, J. L. (1956), A generalised analysis of diallel crosses, *Genetics*, **41**, 65–77.

DOBZHANSKY, TH. (1927), Studies in the manifold effect of certain genes in *Drosophila melanogaster*, *Z.I.A.V.*, **43**, 330–388.

DOBZHANSKY, TH. (1930), The manifold effects of the genes stubble and stubbloid in *Drosophila melanogaster*, *Z.I.A.V.*, **54**, 427–457.

DOBZHANSKY, TH. (1951), *Genetics and the Origin of Species* (3rd edn), Columbia University Press, New York.

DURRANT, A. (1958), Environmental conditioning of flax, *Nature*, **181**, 928–929.

DURRANT, A. (1962), The environmental induction of heritable change in *Linum*, *Heredity*, **17**, 27–61.

DURRANT, A. (1965), Analysis of reciprocal differences in diallel crosses, *Heredity*, **20**, 573–607.

DURRANT, A. and JONES, T. W. A. (1971), Reversion of the induced changes in amount of nuclear DNA in *Linum*, *Heredity*, **27**, 431–439.

DURRANT, A. and MATHER, K. (1954), Heritable variation in a long inbred line of *Drosophila*, *Genetica*, **27**, 97–119.

DURRANT, A. and TYSON, H. (1964), A diallel cross of genotypes and genotrophs of *Linum*, *Heredity*, **19**, 207–227.

EAST, E. M. (1915), Studies on size inheritance in *Nicotiana*, *Genetics*, **1**, 164–176.

EAST, E. M. and JONES, D. F. (1919), *Inbreeding and Outbreeding*, Lippincott, Philadelphia.

EAVES, L. J. (1970), The genetical analysis of continuous variation: A comparison of experimental designs applicable to human data, *Brit. J. Math. Statist. Psychol.*, **22**, 131–147.

EAVES, L. J. (1976a), A model for sibling effects in Man, *Heredity*, **36**, 205–214.

EAVES, L. J. (1976b), The effect of cultural transmission on continuous variation, *Heredity*, **37**, 41–57.

EAVES, L. J., LAST, K. A., MARTIN, N. G. and JINKS, J. L. (1977), A progressive approach to non-additivity and genotype–environmental covariance in the analysis of human differences, *Brit. J. Math. Statist. Psychol.*, **30**, 1–42.

EAVES, L. J., LAST, K. A., YOUNG, P. A. and MARTIN, N. G. (1978), Model fitting approaches to the analysis of human behaviour, *Heredity*, **41**, 249–320.

EGLINGTON, E. G. and MOORE, C. A. (1973), The nature of the inheritance of permanently induced changes in *Nicotiana rustica*. II. F_4 and F_5 generations of selected crosses, *Heredity*, **30**, 387–395.

EMERSON, R. A. and EAST, E. M. (1913), The inheritance of quantitative characters in maize, *Bull. Agr. Exp. Sta. Nebraska*, Res. Bull. 2.

EVANS, G. M. (1968), Nuclear changes in flax, *Heredity*, **23**, 25–28.

FALCONER, D. S. (1960), *Introduction to Quantitative Genetics*, Oliver & Boyd, Edinburgh.

FALCONER, D. S. (1971), Improvement of litter size in a strain of mice at a selection limit, *Genet. Res.*, **17**, 215–235.

FISHER, R. A. (1918), The correlations between relatives on the supposition of Mendelian inheritance, *Trans. Roy. Soc. Edin.*, **52**, 399–433.

FISHER, R. A. (1936), The use of multiple measurements in taxonomic problems, *Ann. Eugenics*, **7**, 178–188.

FISHER, R. A. (1938), The statistical use of multiple measurements, *Ann. Eugenics*, **8**, 376–386.

FISHER, R. A. (1946), *Statistical Methods for Research Workers* (10th edn), Oliver & Boyd, Edinburgh.

FISHER, R. A., IMMER, F. R. and TEDIN, O. (1932), The genetical interpretation of statistics of the third degree in the study of quantitative inheritance, *Genetics*, **17**, 107–124.

FISHER, R. A. and MATHER, K. (1943), The inheritance of style-length in *Lythrum salicaria*, *Ann. Eugenics*, **12**, 1–23.

GALTON, F. (1889), *Natural Inheritance*, Macmillan, London.

GARDNER, C. O. (1963), Estimates of genetic parameters in cross-fertilizing plants and their implications in plant breeding. In *Statistical Genetics and Plant Breeding* (eds. W. D. Hanson and H. F. Robinson), pp. 225–252, *Nat. Acad. Sci.*, Washington.

GILBERT, N. (1961), Polygene analysis, *Genetic Res.*, **2**, 96–105.

GOODWIN, R. H. (1944), The inheritance of flowering time in a short-day species, *Solidago sempervirens*, *Genetics*, **29**, 503–519.

GRAFIUS, J. E. (1961), The complex trait as a geometric construct, *Heredity*, **16**, 225–228.

GRIFFING, B. (1956), A generalised treatment of the use of diallel crosses in quantitative inheritance, *Heredity*, **10**, 31–50.

GRÜNEBERG, H. (1938), An analysis of the 'pleiotropic' effects of a new lethal mutation in the rat (*Mus norvegicus*), *Proc. Roy. Soc. B*, **125**, 123–143.

GRÜNEBERG, H. (1952), Genetical studies on the skeleton of the mouse: IV Quasi-continuous variations, *J. Genet.*, **51**, 95–114.

GRÜNEBERG, H. (1963), *The Pathology of Development. A Study of Inherited Skeletal Disorders in Animals*, Blackwell Sci. Publ., Oxford.

HALEY, C. S., JINKS, J. L. and LAST, K. A. (1981), The monozygotic twin half-sib method for analysing maternal effects and sex-linkage in humans, *Heredity*, **46**, 227–238.

HANKS, M. J. and MATHER, K. (1978), Genetics of coxal chaetae in *Drosophila melanogaster*. II. Response to selection, *Proc. R. Soc. Lond. B*, **202**, 211–230.

HARBORNE, J. B. (1967), *Comparative Biochemistry of the Flavenoids*, Academic Press, London.

HARLAND, S. C. (1936), The genetical conception of the species, *Biol. Revs.*, **11**, 83–112.

HARRIS, H. (1969), Genes and isozymes, *Proc. Roy. Soc. B*, **174**, 1–31.

HAYMAN, B. I. (1953), Components of variation under sib-mating, *Heredity*, **7**, 121–125.

HAYMAN, B. I. (1954a), The analysis of variance of diallel tables, *Biometrics*, **10**, 235–244.

HAYMAN, B. I. (1954b), The theory and analysis of diallel crosses, *Genetics*, **39**, 789–809.

HAYMAN, B. I. (1957), Interaction, heterosis and diallel crosses, *Genetics*, **42**, 336–355.

HAYMAN, B. I. (1958), The theory and analysis of diallel crosses: II, *Genetics*, **43**, 63–85.

HAYMAN, B. I. (1960a), Maximum likelihood estimation of genetic components of variation, *Biometrics*, **16**, 369–381.

HAYMAN, B. I. (1960b), The separation of epistatic from additive and dominance variation in generation means: II, *Genetica*, **31**, 133–146.

HAYMAN, B. I. and MATHER, K. (1955), The description of gene interaction in continuous variation, *Biometrics*, **11**, 69–82.

HAYWARD, M. D. and BREESE, E. L. (1966), Genetic organisation of natural populations of *Lolium perenne*: I Seed and seedling characters, *Heredity*, **21**, 287–304.

HENDERSON, C. R. (1952), Specific and general combining ability, *Heterosis* (ed. J. W. Gowen), pp. 352–370, Iowa State College Press, Ames, Iowa.

HIGHKIN, H. R. (1958), Transmission of phenotypic variability in a pure line (Abstr.), *Proc. Xth Int. Cong. Genets., Montreal*, **2**, 120.

HILL, J. (1964), Effects of correlated gene distributions in the analysis of diallel crosses, *Heredity*, **19**, 27–46.

HILL, J. (1966), Recurrent backcrossing in the study of quantitative inheritance, *Heredity*, **21**, 85–120.

HILL, J. (1967), The environmental induction of heritable changes in *Nicotiana rustica*: Parental and selection lines, *Genetics*, **55**, 735–754.

HILL, J. and PERKINS, J. M. (1969), The environmental induction of heritable changes in *Nicotiana rustica*: Effects of genotype–environment interactions, *Genetics*. **61**, 661–675.

HOGBEN, L. (1933), *Nature and Nurture*, Allen & Unwin, London.

JACOB, F. and MONOD, J. (1961), Genetic regulatory mechanisms in the synthesis of proteins, *J. Mol. Biol.*, **3**, 318–356.

JAYASEKARA, N. E. M. and JINKS, J. L. (1976), Effect of gene dispersion on estimates of components of generation means and variances, *Heredity*, **36**, 31–40.

JINKS, J. L. (1954), The analysis of continuous variation in a diallel cross of *Nicotiana rustica* varieties, *Genetics*, **39**, 767–788.

JINKS, J. L. (1956), The F_2 and backcross generations from a set of diallel crosses, *Heredity*, **10**, 1–30.

JINKS, J. L. (1957), Selection for cytoplasmic differences, *Proc. Roy. Soc. B*, **146**, 527–540.

JINKS, J. L. (1978), Unambiguous test for linkage of genes displaying non-allelic interactions for a metrical trait, *Heredity*, **40**, 171–173.

JINKS, J. L. (1981), The genetic framework of plant breeding, *Phil. Trans. R. Soc. Lond. B.*, **292**, 407–419.

JINKS, J. L. and BROADHURST, P. L. (1965), The detection and estimation of heritable differences in behaviour among individuals, *Heredity*, **20**, 97–115.

JINKS, J. L., CATEN, C. E., SIMCHEN, G. and CROFT, J. H. (1966), Heterokaryon incompatibility and variation in wild populations of *Aspergillus nidulans*, *Heredity*, **21**, 227–239.

JINKS, J. L. and FULKER, D. W. (1970), A comparison of the biometrical genetical, MAVA and classical approaches to the analysis of human behaviour, *Psychol. Bull.*, **73**, 311–349.

JINKS, J. L., JAYASEKARA, N. E. M. and BOUGHEY, HILARIE (1977), Joint selection for both extremes of mean performance and of sensitivity to a macroenvironmental variable. II. Single seed descent, *Heredity*, **39**, 345–355.

JINKS, J. L. and JONES, R. M. (1958), Estimation of the components of heterosis, *Genetics*, **43**, 223–234.

JINKS, J. L. and MATHER, K. (1955), Stability in development of heterozygotes and homozygotes, *Proc. Roy. Soc. B*, **143**, 561–578.

JINKS, J. L. and PERKINS, J. M. (1969), The detection of linked epistatic genes for a metrical trait, *Heredity*, **24**, 465–475.

JINKS, J. L. and PERKINS, J. M. (1970), A general method for the detection of additive, dominance and epistatic components of variation: III F_2 and backcross populations, *Heredity*, **25**, 419–429.

JINKS, J. L. and PERKINS, J. M. (1972), Predicting the range of inbred lines, *Heredity*, **28**, 399–403.

JINKS, J. L., PERKINS, J. M. and BREESE, E. L. (1969), A general method of detecting additive, dominance and epistatic variation for metrical traits: II Application to inbred lines, *Heredity*, **24**, 45–57.

JINKS, J. L., PERKINS, J. M. and GREGORY, S. R. (1972), The analysis and interpretation of differences between reciprocal crosses of *Nicotiana rustica* varieties, *Heredity*, **28**, 363–377.

JINKS, J. L. and POONI, H. S. (1976), Predicting the properties of recombinant inbred lines derived by single seed descent, *Heredity*, **36**, 253–266.

JINKS, J. L. and POONI, H. S. (1979), Non-linear genotype × environment interactions arising from response thresholds. I. Parents, F_1's and selections, *Heredity*, **43**, 57–70.

JINKS, J. L. and POONI, H. S. (1980), Comparing predictions of mean performance and environmental sensitivity of recombinant inbred lines based upon F_3 and triple test cross families, *Heredity*, **45**, 305–312.

JINKS, J. L. and POONI, H. S. (1981), Properties of pure-breeding lines produced by dihaploidy, single seed descent and pedigree breeding, *Heredity*, **46**, 391–395.

JINKS, J. L. and STEVENS, J. M. (1959), The components of variation among family means in diallel crosses, *Genetics*, **44**, 297–308.

JINKS, J. L. and TOWEY, P. M. (1976), Estimating the number of genes in a polygenic system by genotype assay, *Heredity*, **37**, 69–81.

JOHANNSEN, W. (1909), *Elemente der exakten Erblichkeitslehre*, Fischer, Jena.

JONES, D. F. (1945), Heterosis resulting from degenerative changes, *Genetics*, **30**, 527–542.

JONES, R. M. (1965), Analysis of variance of the half diallel table, *Heredity*, **20**, 117–121.

JONES, R. M. and MATHER, K. (1958), Interaction of genotype and environment in continuous variation: II Analysis, *Biometrics*, **14**, 489–498.

JUKES, T. H. (1966), *Molecules and Evolution*, Columbia University Press, New York and London.

KEARSEY, M. J. (1965), Biometrical analysis of a random mating population: A comparison of five experimental designs, *Heredity*, **20**, 205–235.

KEARSEY, M. J. (1970), Experimental sizes for detecting dominance variation, *Heredity*, **25**, 529–542.

KEARSEY, M. J. (1980), The efficiency of North Carolina Experiment III and the selfing, backcrossing series for estimating additive and dominance variation, *Heredity*, **45**, 73–82.

KEARSEY, M. J. and BARNES, B. W. (1970), Variation for metrical characters in *Drosophila* populations: II Natural selection, *Heredity*, **25**, 11–21.

KEARSEY, M. J. and JINKS, J. L. (1968), A general method of detecting additive, dominance and epistatic variation for metrical traits: I Theory, *Heredity*, **23**, 403–409.

KEARSEY, M. J. and KOJIMA, K. (1967), The genetic architecture of body weight and hatchability in *Drosophila melanogaster*, *Genetics*, **56**, 23–37.

KEMPTHORNE, O. (1957), *An Introduction to Genetical Statistics*, Wiley, New York.

KENDALL, M. G. (1957), *A Course in Multivariate Analysis*, Griffin, London.

KILLICK, R. (1971a), Sex-linkage and sex-limitation in quantitative inheritance. I. Random mating populations, *Heredity*, **27**, 175–188.

KILLICK, R. (1971b), The biometrical genetics of autotetraploids I. Generations derived from a cross between two pure lines, *Heredity*, **27**, 331–346.

LAW, C. N. (1967), The location of genetic factors controlling a number of quantitative characters in wheat, *Genetics*, **56**, 445–461.

LERNER, I. M. (1950), *Population Genetics and Animal Improvement*, University Press, Cambridge.

LERNER, I. M. (1954), *Genetic Homeostasis*, Oliver & Boyd, Edinburgh.

LERNER, I. M. and DEMPSTER, E. M. (1951), Alternation of genetic progress under continued selection in poultry, *Heredity*, **5**, 75–94.

LEWIS, D. (1954), Comparative incompatibility in Angiosperms and Fungi, *Advances in Genetics*, **6**, 235–285.

LOESCH, D. Z. (1979), Genetical studies of the palmar and sole patterns and some dermatoglyphic measurements in twins, *Ann. Hum. Genet.*, **43**, 37–53.

LUSH, J. L. (1943), *Animal Breeding Plans* (2nd edn), Iowa State College Press, Ames, Iowa.

LWOFF, A. (1962), *Biological Order*, MIT Press, Cambridge, Mass.

MARTIN, N. G., EAVES, L. J., KEARSEY, M. J. and DAVIES, P. (1978), The power of the classical twin study, *Heredity*, **40**, 97–116.

MARTIN, N. G., LOESCH, D. Z., JARDINE, R. and BERRY, H. S. (1981), Evidence for directional dominance in the genetics of finger ridge counts. In Press.

MATHER, K. (1936), Segregation and linkage in autotetraploids, *J. Genet.*, **32**, 287–314.

MATHER, K. (1941), Variation and selection of polygenic characters, *J. Genet.*, **41**, 159–193.

MATHER, K. (1942), The balance of polygenic combinations, *J. Genet.*, **43**, 309–336.

MATHER, K. (1943), Polygenic inheritance and natural selection, *Biol. Revs.*, **18**, 32–64.

MATHER, K. (1944), The genetical activity of heterochromatin, *Proc. Roy. Soc. B*, **132**, 308–332.

MATHER, K. (1946a), *Statistical Analysis in Biology* (2nd edn), Methuen, London.

MATHER, K. (1946b), Dominance and heterosis, *Amer. Nat.*, **80**, 91–96.

MATHER, K. (1946c), The genetical requirements of bio-assays with higher organisms, *Analyst*, **71**, 407–411.

MATHER, K. (1946d), Genes, *Sci. Jour. Roy. Coll. Sci.*, **16**, 63–71.

MATHER, K. (1948), Nucleus and cytoplasm in differentiation, *Symp. Soc. Exp. Biol.* **2**, 196–216.

MATHER, K. (1949a), *Biometrical Genetics* (1st edn), Methuen, London.

MATHER, K. (1949b), The genetical theory of continuous variation, *Proc. 8th Int. Cong. Genetics., Hereditas*, Suppl. Vol. pp. 376–401.

MATHER, K. (1953a), The genetical structure of populations, *Symp. Soc. Exp. Biol.* **7**, 66–95.

MATHER, K. (1953b), Genetical control of stability in development, *Heredity*, **7**, 297–336.

MATHER, K. (1954), The genetical units of continuous variation. *Proc. XIth Int. Cong. Genetics, Caryologia*, Suppl. Vol. pp. 106–123.

MATHER, K. (1956), Polygenic mutation and variation in populations, *Proc. Roy. Soc. B*, **145**, 293–297.

MATHER, K. (1960a), The balance sheet of variability, *Biometrical Genetics* (ed. O. Kempthorne), pp. 10–11, Pergamon Press, London.

MATHER, K. (1960b), Evolution in polygenic systems, *Int. Colloquium on Evolution and Genetics*, pp. 131–152, Acad. Naz. dei Lincei, Rome.

MATHER, K. (1961), Nuclear materials and nuclear change in differentiation, *Nature*, **190**, 404–406.

MATHER, K. (1966), Variability and selection, *Proc. Roy. Soc. B*, **164**, 328–340.

MATHER, K. (1967a), *The Elements of Biometry*, Methuen, London.

MATHER, K. (1967b), Complementary and duplicate gene interactions in biometrical genetics, *Heredity*, **22**, 97–103.

MATHER, K. (1971), On Biometrical Genetics, *Heredity*, **26**, 349–364.

MATHER, K. (1973), *The Genetical Structure of Populations*, Chapman and Hall, London.

MATHER, K. (1974), Non-allelic interaction in continuous variation of randomly breeding populations, *Heredity*, **32**, 414–419.

MATHER, K. (1975), Genotype × environment interactions. II. Some genetical considerations, *Heredity*, **35**, 31–53.

MATHER, K. (1979), Historical overview: quantitative variation and polygenic system, in *Quantitative Genetic Variation* (eds. J. N. Thompson Jr and J. M. Thoday), pp. 5–34, Academic Press, New York.

MATHER, K. and CALIGARI, P. D. S. (1974), Genotype × environment interactions. I. Regression of interaction on over-all effect of the environment, *Heredity*, **33**, 43–59.

MATHER, K. and CALIGARI, P. D. S. (1976), Genotype × environment interactions. IV. The effect of the background genotype, *Heredity*. **36**, 41–48.

MATHER, K. and EDWARDS, P. M. J. (1943), Specific differences in *Petunia*: III Flower colour and genetic isolation, *J. Genet.*, **45**, 243–260.

MATHER, K. and HANKS, M. J. (1978), Genetics of coxal chaetae in *Drosophila melanogaster*. I. Variation in gene action, *Heredity*, **40**, 71–96.

MATHER, K. and HARRISON, B. J. (1949), The manifold effect of selection, *Heredity*, **3**, 1–52 and 131–162.

MATHER, K. and JINKS, J. L. (1958), Cytoplasm in sexual reproduction, *Nature*, **182**, 1188–1190.

MATHER, K. and JINKS, J. L. (1963), Correlation between relatives arising from sex-linked genes, *Nature*, **198**, 314–315.

MATHER, K. and JINKS, J. L. (1977), *Introduction to Biometrical Genetics*, Chapman and Hall, London.

MATHER, K. and JONES, R. M. (1958), Interaction of genotype and environment in continuous variation: I Description, *Biometrics*, **14**, 343–359.

MATHER, K. and VINES, A. (1952), The inheritance of height and flowering time in a cross of *Nicotiana rustica, Quantitative Inheritance* (eds. E. C. R. Reeve and C. N. Waddington), pp. 49–80, H.M.S.O., London.

MATHER, K. and WIGAN, L. G. (1942), The selection of invisible mutations, *Proc. Roy. Soc. B*, **131**, 50–64.

McMILLAN, J. and ROBERTSON, A. (1974), The power of methods of the detection of major genes affecting quantitative characters, *Heredity*, **32**, 349–356.

MUKAI, T. (1979), Polygenic mutations, in *Quantitative Genetic Variation* (eds. J. N. Thompson Jr and J. M. Thoday), pp. 177–196, Academic Press, New York.

MUKAI, T. and COCKERHAM, C. C. (1977), Spontaneous mutation rates of enzyme loci in *Drosophila melanogaster, Proc. Nat. Acad. Sci. Wash.* **74**, 2514–2517.

MULLER, H. J. (1935), On the incomplete dominance of the normal alleles of white in *Drosophila, J. Genet.*, **30**, 407–414.

NANCE, W. E. and COREY, L. A. (1976). Genetic models for the analysis of data from the families of identical twins, *Genetics* **83**, 811–826.

NELDER, J. A. (1953), Statistical models in biometrical genetics, *Heredity*, **7**, 111–119.

NELDER, J. A. (1960), The estimation of variance components in certain types of experiment on quantitative genetics. *Biometrical Genetics* (ed. O. Kempthorne), pp. 139–158, Pergamon Press, London.

NILSSON-EHLE, H. (1909), *Kreuzunguntersuchungen an Hafer und Weizen*, Lund.

OPSAHL, B. (1956), The discrimination of interaction and linkage in continuous variation, *Biometrics*, **12**, 415–432.

PANSE, V. G. (1940a), The application of genetics to plant breeding: II The inheritance of quantitative characters and plant breeding, *J. Genet.*, **40**, 283–302.

PANSE, V. G. (1940b), A statistical study of quantitative inheritance, *Ann. Eugenics*, **10**, 76–105.

PAXMAN, G. J. (1957), A study of spontaneous mutation in *Drosophila melanogaster, Genetica*, **29**, 39–57.

PENROSE, L. S. (1951), Measurement of pleiotropic effects in phenylketonuria, *Ann. Eugenics*, **16**, 134–141.

PERKINS, J. M. and JINKS, J. L. (1968a), Environmental and genotype–environmental components of variability: III Multiple lines and crosses, *Heredity*, **23**, 339–356.

PERKINS, J. M. and JINKS, J. L. (1968b), Environmental and genotype–environmental components of variability: IV Non-linear interactions for multiple inbred lines, *Heredity*, **23**, 525–535.

PERKINS, J. M. and JINKS, J. L. (1970), Detection and estimation of genotype–environmental, linkage and epistatic components of variation for a metrical trait, *Heredity*, **25**, 157–177.

PERKINS, J. M. and JINKS, J. L. (1971), Analysis of genotype × environment interaction in triple test cross data, *Heredity*, **26**, 203–209.

PERKINS, J. M. and JINKS, J. L. (1973), The assessment and specificity of environmental and genotype–environmental components of variability, *Heredity*, **30**, 111–126.

PERKINS, J. M., EGLINGTON, E. and JINKS, J. L. (1971), The nature of the inheritance of permanently induced changes in *Nicotiana rustica, Heredity*, **27**, 441–457.

POONI, H. S. and JINKS, J. L. (1976), The efficiency and optimal size of triple test cross designs for detecting epistatic variation, *Heredity*, **36**, 215–227.

POONI, H. S. and JINKS, J. L. (1978), Predicting the properties of recombinant inbred lines derived by single seed descent for two or more characters simultaneously, *Heredity*, **40**, 349–361.

POONI, H. S. and JINKS, J. L. (1979), Sources and biases of the predictors of the properties of recombinant inbreds produced by single seed descent, *Heredity*, **42**, 41–48.

POONI, H. S. and JINKS, J. L. (1980), Non-linear genotype × environment interactions. II. Statistical models and genetical control, *Heredity*, **45**, 389–400.

POONI, H. S. and JINKS, J. L. (1981), The true nature of the non-allelic interactions in *Nicotiana rustica* revealed by association crosses. *Heredity*, **47**, 253–258.

POWERS, L. (1941), Inheritance of quantitative characters in crosses involving two species of *Lycopersicon, J. Agr. Res.*, **63**, 149–174.

POWERS, L. (1942), The nature of the series of environmental variances and the estimation of the genetic variances and the geometric means in crosses involving species of *Lycopersicon, Genetics*, **27**, 561–575.

POWERS, L. (1951), Gene analysis by the partitioning method when interaction of genes are involved, *Bot. Gaz.*, **113**, 1–23.

QUISENBERRY, K. S. (1926), Correlated inheritance of quantitative and qualitative characters in oats, *Bull. Agric. Exp. Sta., West Virginia*, 202.

RAO, C. R. (1952), *Advanced Statistical Methods in Biometric Research*, Wiley, New York.

RASMUSSON, J. M. (1935), Studies on the inheritance of quantitative characters in *Pisum*: I Preliminary note on the genetics of flowering, *Hereditas*, **20**, 161–180.

REES, H. and JONES, R. N. (1977), *Chromosome Genetics*, Edward Arnold, London.

REES, H. and THOMPSON, J. B. (1956), Genotypic control of chromosome behaviour in rye: III Chiasma frequency in homozygotes and heterozygotes, *Heredity*, **10**, 409–424.

RENDEL, J. M. (1979), Canalisation and selection, in *Quantitative Genetic Variation* (eds. J. N. Thompson Jr and J. M. Thoday), pp. 139–156, Academic Press, New York.

SAX, K. (1923), The association of size differences with seed-coat pattern and pigmentation in *Phaseolus vulgaris, Genetics*, **8**, 552–560.

SEARLE, S. R. (1966), *Matrix Algebra for Biologists*, Wiley, New York.

SEREBROVSKY, A. S. (1928), An analysis of the inheritance of quantitative transgressive characters, *Z.I.A.V.*, **48**, 229–243.

SHIELDS, J. (1962), *Monozygotic Twins*, Oxford University Press.

SIMCHEN, G. (1966), Monokaryotic variation and haploid selection in *Schizophyllum commune, Heredity*, **21**, 241–263.

SIMCHEN, G. and JINKS, J. L. (1964), The determination of dikaryotic growth-rate in the Basidiomycete, *Schizophyllum commune*: a biometrical analysis, *Heredity*, **19**, 629–649.

SMITH, H. F. (1936), A discriminant function for plant selection, *Ann. Eugenics*, **7**, 240–250.

SMITH, H. H. (1937), The relations between genes affecting size and colour in certain species of *Nicotiana, Genetics*, **22**, 361–375.

SNEDECOR, C. W. (1956), *Statistical Methods*, Iowa State College Press, Ames, Iowa.

SPICKETT, S. G. (1963), Genetic and developmental studies of a quantitative character, *Nature*, **199**, 870–873.

SPICKETT, S. G. and THODAY, J. M. (1966), Regular responses to selection: 3 Interaction between located polygenes, *Genetic Res.*, **7**, 96–121.

STRAUSS, F. S. and GOWEN, J. (1943), Heterosis: its mechanism in terms of chromosome units in egg production of *Drosophila melanogaster, Genetics*, **28**, 93.

STUDENT (1934), A calculation of the minimum number of genes in Winter's selection experiment, *Ann. Eugenics*, **6**, 77–82.

SVÄRDSON, G. (1944), Polygenic inheritance in *Lebistes, Arkiv. f. Zool.*, **36A**, 1–9.

THODAY, J. M. (1960), Effects of disruptive selection: III Coupling and repulsion. *Heredity*, **14**, 35–49.

THODAY, J. M. (1961), Location of polygenes, *Nature*, **191**, 368–370.

THODAY, J. M. (1973), The origins of genes found in selected lines, *Atti. del. Accad. del. Sci. Bologna. Mem. Ser.* III. No. **1**, 15–25.

THODAY, J. M., and BOAM, T. B. (1961), Regular responses to selection. I. Description of responses, *Genet. Res.*, **2**, 161–176.

TIMOFÉEFF-RESSOVSKY, N. W. (1934), Auslösung von Vitalitätsmutationen durch Röntgenbestrahlung bei *Drosophila melanogaster, Strahlentherapie*, **51**, 658–663.

TIMOFÉEFF-RESSOVSKY, N. W. (1935), Auslösung von Vitalitätsmutationen durch Röntgenbestrahlung bei *Drosophila melanogaster, Nach. Ges. Wiss. Göttingen Biologie N.F.*, **1**, 163–180.

TOWEY, P. M. and JINKS, J. L. (1977), Alternative ways of estimating the number of genes in a polygenic system by genotype assay, *Heredity*, **39**, 399–410.

VAN DER VEEN, J. H. (1959), Tests of non-allelic interaction and linkage for quantitative characters in generations derived from two diploid pure lines, *Genetica*, **30**, 201–232.

WADDINGTON, C. H. (1939), *An Introduction to Modern Genetics*, Allen & Unwin, London.

WALTERS, D. E. and GALE, J. S. (1977), A note on the Hayman analysis of variance for a full diallel table, *Heredity*, **38**, 401–407.

WARREN, D. C. (1924), Inheritance of egg size in *Drosophila melanogaster, Genetics*, **9**, 41–69.

WEARDEN, S. (1964), Alternative analyses of the diallel cross, *Heredity*, **19**, 669–680.

WHITE, M. J. D. (1954), *Animal Cytology and Evolution* (2nd edn), University Press, Cambridge.

WIGAN, L. G. (1944), Balance and potence in natural populations, *J. Genet.*, **46**, 150–160.

WIGAN, L. G. (1949), The distribution of polygenic activity on the X chromosome of *Drosophila melanogaster, Heredity*, **3**, 53–66.

WIGAN, L. G. and MATHER, K. (1942), Correlated response to the selection of polygenic characters, *Ann. Eugenics*, **11**, 354–364.

WOLSTENHOLME, D. R. and THODAY, J. M. (1963), Effects of disruptive selection: VII A third chromosome polymorphism, *Heredity*, **18**, 413–431.

WRIGHT, S. (1934a), An analysis of variability in number of digits in an inbred strain of guinea pigs, *Genetics*, **19**, 506–536.

WRIGHT, S. (1934b), The results of crosses between inbred strains of guinea pigs, differing in number of digits, *Genetics*, **19**, 537–551.

YATES, F. (1947), The analysis of data from all possible reciprocal crosses between a set of parental lines, *Heredity*, **1**, 287–301.

YATES, F. and COCHRAN, W. G. (1938), The analysis of groups of experiments, *J. Agric. Sci.*, **28**, 556–580.

Index

ADH, 35
Age, 243
Allard, 271
Amino-acids, 40
Analysis
  inclusive and exclusive —, 197, 199,
    200
  unweighted —, 154 *et seq.*, 169, 171,
    173, 198
  weighted —, 155, 162 *et seq.*, 170 *et
    seq.*, 196 *et seq.*
Analysis of variance, 9, 119, 147–53,
    154, 160, 196, 200, 202, 204, 235 *et
    seq.*, 244 *et seq.*, 251, 255, 260 *et
    seq.*, 270, 309, 358
Antigens, 30, 34, 38, 40, 57
Applied geneticists, 41, 48, 50
Aspergillus, 56, 309, 311
Association and dispersion (of genes),
    66, 85, 87, 91, 123, 126, 127, 287 *et
    seq.*, 308, 311, 316, 330, 342
Assortative mating, 230, 231
Autosomes, 8, 296, 297, 299
Avena (*see* Oates)

Backcrosses, 69 *et seq.*, 85, 88, 89, 93,
    96, 100, 111, 113, 135 *et seq.*, 138
    et seq., 145, 147, 179, 185, 205,
    208, 213, 279, 280, 293 *et seq.*, 302,
    308, 311, 313, 346, 355, 358, 360
Barigozzi, 32
Barley, 38, 49, 50, 198 *et seq.*, 202, 213,
    335, 336
Barnes, 302, 366
Bartlett, 122
Bateson, 29, 53, 364

Beadle, 44
Beale, 38
Beans, 2, 5, 6, 7
Beddows, 55
Benzer, 369
Biometricians and Mendelians, 2, 3
Biparental matings, 85, 97, 101, 102,
    141, 145, 146, 170 *et seq.*, 174, 185,
    188, 192 *et seq.*, 208, 213, 235 *et
    seq.*, 248
Birley, 35
Boam, 31
Breeders, plant and animal —, 36, 42
Breeding test, 2, 37, 57
Breese, 22, 23, 26, 27, 33, 55, 56, 175,
    268, 278, 353, 375
Brink, 58
Britten, 34, 43, 44
Broadhurst, 243, 353, 354
Brumby, 55
Bucio Alanis, 104, 105, 107, 108, 110,
    112, 113, 132, 148, 152, 353
Butcher, 309

Caligari, 12, 115, 124, 125, 274, 277,
    278, 279
Caten, 30, 309, 311
Cavalli, 72, 154
Chaetae
  abdominal —, 8, 22, 24, 26, 38, 39,
    365, 373, 375, 376
  coxal —, 34 *et seq.*
  sternopleural —, 17, 22, 26, 32, 39,
    153, 154, 192, 366, 373
Characters, 37
  — and genes, 43

389

genetical architecture of —, 38, 41
juvenile —, 54
sub —, 43, 44, 45 *et seq.*
super —, 45 *et seq.*
Charles, 316
Chiasma(ta), 59, 332, 333, 334, 370, 371
Chovnik, 34
Chromosome(s), 4, 17 *et seq.*, 23, 26, 28, 332, 333, 367, 368, 370, 371, 372, 374, 375, 376
— assays, 8 *et seq.*, 13, 14, 26, 297
B —, 31, 32
recombinant —, 7, 16, 23
special —, 42
Clarke, 38
Clayton, 43, 372
Clone, 55, 56, 57, 214, 309, 353
Cochran, 116
Cockerham, 34
Cocks, 30, 39
Collybia, 321, 334
Complementary (interaction), 84, 92, 127, 130, 180, 284 *et seq.*, 289, 364, 368
Computer
need for —, 170
— simulation, 357
Comstock, 147, 153, 251
Conditioning, 58
Connelly, 312
Cooke, 27, 153 *et seq.*, 163 *et seq.*, 192 *et seq.*, 196, 198, 297, 325 *et seq.*, 359
Cooper, 24
Corey, 234
Correlated response, 24 *et seq.*, 376
— statistics, 163, 170
Correlation(s), 43 *et seq.*, 215, 224, 230, 231 *et seq.*, 300, 349
— genotype-environment, 230
Cotton, 38, 339
Coughtrey, 280, 287
Coupling and repulsion (in linkage), 182 *et seq.*, 333 *et seq.*, 342, 347, 373, 374
Covariances, 139, 155, 166, 170, 178, 188, 192, 199, 219, 220, 221, 224,

236, 239, 244, 248, 266 *et seq.*, 281 *et seq.*, 288 *et seq.*, 350
Crick, 40
Croft, 321, 322
Cultural transmission, 232
Cytoplasm, 54 *et seq.*, 301

Darlington, 31, 55, 332, 333
Darwin, 2
Davidson, 34, 43
Davies, D.R., 54, 55
Davies, R.W., 23, 27
Dawson, 242, 246
Dempster, 24
de Vries, 2
Diallel(s), 214, 235, 250, 255 *et seq.*, 268 *et seq.*, 271 *et seq.*, 280 *et seq.*, 299, 307, 354, 358, 375
— table, 256, 260 *et seq.*, 281
— V_r, 266 *et seq.*, 283 *et seq.*, 288 *et seq.*, 375
— W_r, 266 *et seq.*, 283 *et seq.*, 288 *et seq.*, 375
— W_r^1, 267, 375
Dickinson, 253
Dikaryon, 312
Dioecious (organisms), 213
Discriminant functions, 49, 50, 198
Dispersion (of genes) (*see* Association)
DNA, 57
— nucleotides, 40
Dobzhansky, 29, 39
Dominance, 3, 4, 11, 12, 23, 52 *et seq.*, 55, 64, 65 *et seq.*, 125 *et seq.*, 136, 201, 257, 263, 268 *et seq.*, 279, 283, 312 *et seq.*, 330, 338, 355, 360, 363 *et seq.*, 374
phenotype of maximum —, 278
— ratio, 138, 150, 275, 343 *et seq.*
super (or over) —, 126, 201, 348, 374
Drosophila, 8 *et seq.*, 13, 15 *et seq.*, 20, 23 *et seq.*, 26 *et seq.*, 32, 33 *et seq.*, 38, 39, 43, 60, 61, 62, 115, 124, 153 *et seq.*, 164, 166, 170 *et seq.*, 192 *et seq.*, 268, 274, 277, 294 *et seq.*, 302, 325, 327 *et seq.*, 352, 355, 359, 365, 367, 369, 373, 375
Duplicate (interaction), 84, 91, 130,